SPRINGER
LAB MANUAL

Springer

Berlin
Heidelberg
New York
Barcelona
Hong Kong
London
Milan
Paris
Singapore
Tokyo

Stefan Surzycki

Basic Techniques in Molecular Biology

With 59 Figures

Springer

PROFESSOR DR. STEFAN SURZYCKI

Department of Biology
Indiana University
Bloomington IN 47405
USA
e-mail: surzycki@sunflower.bio.indiana.edu

ISBN 3-540-66678-8 Springer-Verlag Berlin Heidelberg New York

Library of Congress Cataloging-in-Publication Data
Surzycki, Stefan, 1936 – , Basic techniques in molecular biology / Stefan Surzycki.
p. cm. – (Springer lab manual) Includes bibliographical references and index.
   ISBN 3-540-66678-8 (alk. paper)
1. Molecular biology – Laboratory manuals. 2. Molecular genetics – Laboratory manuals.
I. Title. II. Series.   OH506.S89 2000   572.8 – dc21

Springer-Verlag Berlin Heidelberg New York
a part of Springer Science & Business Media

© Springer-Verlag Berlin Heidelberg 2000
Printed in Germany

Production: PRO EDIT GmbH, D-69126 Heidelberg
Cover design: design & production GmbH, D-69121 Heidelberg
Typesetting: Mitterweger & Partner, D-68723 Plankstadt
Printed on acid free paper 39/3111 5 4 3 2

Preface

Many laboratory manuals describing molecular biology techniques have already been published. What I have tried to do, to justify yet another laboratory manual, is to go beyond cookbook recipes for each technique. Thus, the description of each technique includes an overview of its general importance, historical background and theoretical basis for each step. This is done in the hope that users will acquire enough of an understanding of the theoretical mechanisms that they will be able to go on to design their own modifications and methods. Examples of some of the questions answered in this book are: How does 70% ethanol cause DNA and RNA to precipitate and why is it usually necessary to add salt to insure recovery of the nucleic acid? Which salt is best for this and why? What is the difference between the removal of protein from a nucleic acid preparation with chloroform:iso-amyl alcohol (CIA) or with phenol? Why is isoamyl alcohol added to the chloroform anyway? These are a few of the questions that one may not even think to ask when using these methods or adapting them to a research project but are very important to understand when experiments do not work or methods are being adapted to different systems.

This laboratory manual is directed at both students who are being introduced to molecular biology techniques for the first time, and experienced researchers who are very familiar with the methods but may want to learn more about the theoretical mechanisms involved in each procedure. It is also designed to help teachers develop laboratories that involve concepts of molecular biology. What began as an attempt to adapt for publication the manual I use to teach molecular biology to undergraduates has lead to two years of extensive research into the latest modifications and innovations of techniques for plasmid isolation, genomic DNA isolation, RNA isolation, Northern and Southern transfers, sequencing and PCR.

All of the procedures in this book have either been used extensively in the teaching of undergraduate laboratories and adult workshops or tested in my laboratory. Some methods were developed in my laboratory and are presented here for the first time. For example my impatient son, Stefan Surzycki, as an undergraduate at Indiana University developed a modification of the alkaline procedure for very rapid plasmid preparation in my laboratory. Along with the procedures, I also evaluate the various kits commercially available from different suppliers and make suggestions and comparisons based on my own experience. Emphasis is placed on the time and expense involved in the application of each technique. The descriptions of each step in protocols are very specific and detailed as to how to carry them out. These instructions may appear to be overly detailed, but they have been developed as a result of years of teaching undergraduates and trying to assure that the experiments work in inexperienced hands the first time they are performed.

If one were to read this manual from beginning to end, it would perhaps seem that there is a lot of repetition of the techniques described. After careful consideration I decided that the repetition is necessary in order to make the manual easy to use. This is because many methods presented in the book are similar but vary in different steps. A decision was made to present each procedure in its entirety to avoid the annoyance of being sent to other parts of the book to complete steps that were presented elsewhere.

Since this book is also intended to be a teaching manual for new inexperienced workers I am including my e-mail address so that the reader can contact me directly in case of difficulties with techniques and procedures described (surzycki@sunflower.bio.indiana.edu).

Finally I wish to thank my wife, Judy A. Surzycki, without whose help and encouragement this book would never have been written.

Bloomington, Winter 1999   STEFAN SURZYCKI

Contents

General Aspects of DNA Isolation and Purification

STEFAN SURZYCKI

1.1 Introduction

DNA isolation is an essential technique in molecular biology. Isolation of high-molecular weight DNA has become very important with the increasing demand for DNA fingerprinting, restriction fragment length polymorphism (RFLP), construction of genomic or sequencing libraries and PCR analysis in research laboratories and industry. Also, DNA isolation is the first step in the study of specific DNA sequences within a complex DNA population, and in the analysis of genome structure and gene expression. The quantity, quality and integrity of DNA will directly affect these results. DNA constitutes a small percentage of the cell material and is usually localized in a defined part of the cell. In prokaryotic cells, DNA is localized in the nucleoid that is not separated from the rest of the cell sap by a membrane. In eukaryotic cells, the bulk of DNA is localized in the nucleus, an organelle that is separated from the cytoplasm by a membrane. The nucleus contains about 90% of the total cellular DNA, the remaining DNA is in other organelles like mitochondria, chloroplasts or kinetochores. In viruses and bacteriophages, the DNA is encapsulated by a protein coat, and constitutes between 30 and 50 percent of the total mass of the virion. In prokaryotic and eukaryotic cells, DNA constitutes only about 1% of the total mass of the cell. The approximate composition of rapidly dividing *E. coli* cells and human cells (HeLa) is presented in Table 1.

The purpose of DNA isolation is to separate DNA from all the components of the cell listed in Table 1 resulting in a homogeneous DNA preparation that represents the entire genetic information contained within a cell. There is no difficulty in separating DNA from small molecules since the molecular weight of DNA is very large. Consequently, the main cellular components that have to be removed during DNA purification are protein and RNA. General requirements of effective DNA isolation are:

- The method should yield DNA without major contaminants, protein and RNA.

Table 1. Composition of typical prokaryotic and eukaryotic cells

Cell component	Percent of total cell weight	
	E. coli cells	HeLa cells
Water	70.0	70.0
Inorganic ions	1.0	1.0
Amino acids	0.4	0.4
Nucleotides	0.4	0.4
Lipids	2.2	2.8
Protein	15.0	22.3
RNA	6.0	1.7
DNA	1.0	0.85

- The method should be efficient; most of the cellular DNA should be isolated and purified. It also should be nonselective; all species of DNA in the cells should be purified with equal efficiency.

- The method should not physically or chemically alter DNA molecules.

- The DNA obtained should be of high molecular weight and with few single-stranded breaks.

- The method should be relatively fast and simple enough that it will not take a long time or much effort to prepare DNA. This is important since preparation of DNA is just the beginning of an experiment and not an end in itself.

There are several methods for isolation of DNA that, in general, fulfill most of the requirements listed above. All methods involve four essential steps:

- Cell breakage.

- Removal of protein and RNA.

- Concentration of DNA.

- Determination of the purity and quantity of DNA.

Two of the most common obstacles in obtaining a high yield of high molecular weight DNA are hydrodynamic shearing and DNA degradation by nonspecific DNases. To avoid these problems, some general precautions should be taken. All solutions should contain DNase inhibitors and all glassware, plastic pipettes tips, centrifuge tubes, and buffers should be sterilized. The method of cell breakage used should avoid strong forces that shear the DNA. DNA, in solution, should always be pipetted slowly with wide-bore pipettes (about 3 to 4 mm orifice diameter). The tip of the pipette should always be immersed in the liquid when pipetting DNA. The DNA solution should never be allowed to run down the side of a tube nor should it be vigorously shaken or vortexed. The use of molecular biology grade or ultra-pure chemical reagents is strongly recommended.

Each DNA purification task should begin with careful planning of the amount of DNA needed, the purity required and an estimation of the size of DNA molecules needed. Preparation of a genomic library requires 100 to 300 μg of large molecular weight DNA (more than 100 000 bp long), essentially devoid of protein contamination. Southern blot analysis of a single copy gene requires 5 to 10 μg of eukaryotic, genomic DNA per single gel lane. The purity and the size of this DNA is not as critical as it is for genomic library preparation. DNA of 50 000 to 100 000 bp is sufficient for most applications with purity achieved by standard DNA isolation methods. A single PCR reaction requires only a very small amount of DNA (50 to 500 ng) and can tolerate a considerable amount of contaminating proteins. Indeed, it is possible to achieve PCR amplification of DNA from a single lysed cell. However, care must be taken to remove excess RNA from the DNA preparation, because a large amount of RNA can severely inhibit the PCR reaction.

Table 2. Expected DNA yields from large scale DNA purification procedures of different cell types

Cell type	Cell number[a]	DNA present[b] (μg)	DNA obtained[c] (μg)	Yield (%)	Sample weight[d] (mg)	Purification yield (μg DNA/ mg sample)
Bacteria	2.5×10^{10}	120-200	160-180	80-90	30-40	5-6
Animal cells	5.0×10^{7}	250-400	200-300	50-80	50-100	3-7
Animal tissue	5.0×10^{8}	2500-4000	1000-2000	25-50	500-1000	1-4
Plant tissue	5.0×10^{7}	300-3200	60-800	20-30	600-4000	0.1-1.0

[a]-Approximate total cell number used in purification procedures.

[b]-Approximate total amount of DNA present in the indicated number of the cells.

[c]-Approximate total amount of DNA possible to isolate from the cell sample.

[d]-Approximate weight of the cell sample to use for DNA purification.

Table 2 presents data on the average yield of DNA usually achieved when DNA is isolated from animal, plant and bacterial cells. These data should serve only as a general guide in designing a DNA isolation procedure from a new, previously undescribed source. A list of nuclear DNA content from more than 100 species has been published and can serve as a guide to design purification procedure (Arumuganathan and Earle, 1991).

1.2 DNA Isolation Solutions

Material used for isolation of DNA (cells of fragmented tissue) should be suspended in isolation buffer (lysis buffer). An isolation buffer should:

- maintain the structure of DNA during breakage and purification steps.
- facilitate isolation of DNA, that is to make it easy to remove protein and RNA.
- inhibit DNA degradating enzymes present in the cells.
- prevent chemical modification of DNA molecules by cell sap components released by cell breakage.

The appropriate buffer concentration, pH, ionic strength and the addition of DNase inhibitors and detergents fulfill all of these requirements.

Breaking buffer pH, concentration and ionic strength

Tris buffer is most commonly used for DNA isolation due to its low cost and excellent buffering capacity. The concentration and pH of the buffer used should be sufficient to maintain the pH above pH 5 and below pH 12 after cell breakage. Low pH of the solution will result in DNA depurination and will cause the DNA to distribute into the phenol phase during deproteinization. The high end of the pH scale, above pH 12.0, is also detrimental, resulting in DNA strand separation. When Tris is used, the pH of the breaking buffer should be maintained between pH 7.6 and pH 9.0, the most effective buffering range for this compound. Use of Tris buffer for maintaining pH below 7.5 is not recommended because of its very low buffering capacity in this range. The concentration of the buffer depends on the type of cells used for the DNA preparation. Breakage of animal cells and tissue does not drastically change the pH of the solution, permitting use of low buffer concentration. Thus, Tris buffer used for breakage of animal cells, should be between 10 mM and 50 mM with the pH between 8 and 8.5.

Preparation of DNA from plant and bacteria requires much higher buffer concentration due to a large quantity of secondary metabolites present in bacterial cells and accumulated in plant cell vacuoles. The buffer used for breakage of these cells should be in the order of 100 to 200 mM at a pH above pH 8.

The ionic strength of the breaking solution is usually maintained by the addition of NaCl. The minimum concentration of sodium chloride should be at least 0.12 M. This salt concentration is sufficient to preserve the structure of DNA molecules and decrease the amount of water present in the phenol phase, facilitating deproteinization. The use of very high salt concentrations in breaking buffer is not recommended if phenol will be used for removing proteins. An excessive amount of sodium chloride will result in the inversion of organic and water phases during phenol extraction.

On the other hand, high salt concentration in the breaking solution is required when the CTAB (hexadecyltrimethylammonium bromide) detergent procedure is used for cell lysis and protein removal. This is because DNA forms stable insoluble complexes with CTAB at a salt concentration at and above 0.5 M. When the concentration of NaCl is maintained above 1.0 M, no complex formation occurs, and DNA remains soluble.

DNase inhibitors and detergents

All buffers used in DNA isolation should include DNase inhibitors. Two kinds of DNase inhibitors are in use, ethylenediaminetetraacetic acid (EDTA), and detergents.

The EDTA is a Mg^{++} ion chelator and a powerful inhibitor of DNases since most cellular DNases requires Mg^{++} ion as a cofactor for their activity. In addition, the presence of EDTA in extraction buffers inhibits Mg^{++} ion-induced aggregation of nucleic acids, to each other and to proteins. Different concentrations of EDTA are used in breaking buffers depending on the amount of DNase and Mg^{++} ions present in the cells. Isolation of DNA from bacterial, plant and some insect cells requires a very high concentration of EDTA, ranging between 0.1 to 0.5 M, because of the large amount of DNA degrading enzymes present in these cells. A concentration of 50 to 100 mM EDTA is usually sufficient to inhibit DNases present in animal cells. In fact a high concentration of EDTA (above 100 mM) in buffers used for mammalian cells is not recommended because it leads to a substantial decrease in yields.

When a high concentration of EDTA is used, it is important to remember that EDTA is precipitated with ethanol. Ethanol precipitation, commonly used to concentrate DNA, can result in a high concentration of EDTA in

DNA preparations. The often mentioned "impurities" in DNA preparations that inhibit DNA digestion with restriction enzymes or interfere with the action of other DNA modifying enzymes can be traced to nothing more than EDTA carry over from ethanol precipitation. Therefore, it is very important to precipitate DNA from solutions diluted enough so as not to precipitate EDTA, or to use a precipitation procedure that does not require centrifugation.

Along with EDTA, breaking buffers should contain detergents. Detergents are used for two purposes: as an inhibitor of DNases and, as compounds aiding in cell lysis and protein denaturation. Detergents used in DNA purification are: sodium or lithium deodecyl sulfate (SDS or LDS), sodium deodecyl sarcosinate (Sarcosyl), sodium 4 aminosalicylate, sodium tri-isopropylnaphthalene sulfonate and hexadecyltrimethyl ammonium bromide (CTAB).

Sodium and lithium deodecyl sulfate (SDS or LDS) are inexpensive and potent anionic detergents used at concentrations from 0.5 to 2.0%. Disadvantages of their use are: precipitation at low temperature; poor solubility in 70% ethanol, and poor solubility in solutions with a high salt concentration. SDS is especially poorly soluble in cesium salts frequently used in DNA purification by density gradient centrifugation. Poor solubility of SDS in ethanol can also result in an accumulation of this detergent in ethanol-precipitated DNA, leading to inhibition of enzymes used in subsequent work. Using lithium salt of the deodecyl sulfate alleviates some of these problems because lithium salt is more soluble in ethanol and high salt than sodium salt.

Buffers containing these detergents, as well as their stock solutions should never be sterilized by autoclaving. This results in the shortening of aliphatic chains, consequently decreasing their detergent activity.

Sarcosyl (N-laurylsarcosine, sodium salt) was introduced to DNA isolation procedures to overcome drawbacks of SDS. Sarcosyl is easily soluble at high salt concentrations (in particular cesium salts) and is frequently used when cesium density centrifugation is required for DNA purification. Moreover, this detergent does not precipitate at low temperature in the presence of ethanol and does not carry over into DNA samples. A disadvantage of its use is its somewhat weaker detergent activity. Consequently, Sarcosyl should be used at twice the effective SDS concentration. Moreover, some cells (for example, some plant cells) cannot be completely lysed by Sarcosyl alone, requiring the incorporation of a small amount of SDS, or other detergents, in the breaking buffers.

Sodium 4-aminosalicylate and sodium tri-isopropylnaphthalene sulfonate are detergents that do not precipitate at low temperatures even at very high concentration, and are fully soluble in 70% ethanol. They are very good inhibitors of nucleases and are frequently used in conjunction with SDS or Sarcosyl in the isolation of plant DNA. Sodium 4-aminosalicylate is used as a substitute for Sarcosyl when DNA preparations are precipitated with n-butanol because, unlike Sarcosyl, it is fully soluble in this alcohol. The drawback of these detergents is their rather weak detergent activity necessitating the use of highly concentrated solutions (5-8 %). Both of these detergents can be sterilized by autoclaving.

Cetyltrimethyl ammonium bromide (CTAB), commonly used as an antiseptic, is a powerful cationic detergent used for cell lysis. It is also a very good inhibitor of DNase because it forms CTAB:protein complexes that are enzymatically inactive. CTAB is used at concentrations between 1% to 10% in DNA purification from plants and bacteria. This detergent is soluble in 70% ethanol but precipitates at a temperature below 15 °C. Another drawback of its use is its poor solubility in water (10% w/v at room temperature) and high viscosity.

Cetyldimethylethyl ammonium bromide (CDA). This cationic detergent commonly used as a topical antiseptic, is a very good substitution for CTAB. The detergent is fully soluble in water, does not precipitate at low temperature and is not viscose in solution. All other characteristics of CDA are identical to those of CTAB. The major drawback of its use is its toxicity in powder form.

1.3 Breakage of Cells

Cell breakage is one of the most important steps in the isolation of DNA. The primary ways to break cells are chemical, mechanical and enzymatic. The mechanical means of cell breaking, such as sonication, grinding, blending or high pressure cannot be used for DNA preparations. These procedures apply strong forces to open cells that shear the DNA into small fragments. The best procedure to open the cells and obtain intact DNA is through application of chemical (detergents) and/or enzymatic procedures. Detergents can solubilize lipids in cell membranes resulting in gentle cell lysis. In addition, detergents have an inhibitory effect on all cellular DNAases and can denature proteins, aiding the removal of proteins from the solution. The lysis of animal cells is usually performed using anionic detergents such as SDS or Sarcosyl. These detergents are useful for isolating DNA from cells grown in cell cultures. To apply detergent lysis to tissue-derived cells, tissue is usually frozen in liquid nitrogen and gently crushed into small pieces that are accessible to detergent treatment.

Often plant and bacterial cells cannot be broken with detergent alone. To lyse these cells, they are first treated with enzymes that make the cell membrane accessible to detergents. Before the application of detergents, bacterial cells are treated with lysozyme. Plant cell walls can be removed by treatment with enzymes that partially or totally remove the cellulose-based cell wall. Since enzymatic treatment of plant cells is expensive and time consuming, it can be substituted by gentle grinding in liquid nitrogen. Treatment does not mechanically disrupt plant cells but forms small cracks in the cell wall, permitting detergent access to the cell interior and subsequent lysis of cell membranes.

1.4 Removal of Protein and RNA

The second purification step is the removal of proteins from the cell lysate. This procedure is called **deproteinization**. Removal of proteins from the DNA solution depends on differences in physical properties between nucleic acids and proteins. These differences are: differences in solubility, differences in partial specific volume and differences in sensitivity to digestive enzymes. Methods that exploit solubility differences are: solubility differences in organic solvents, solubility differences of salts or detergent complexes and differences in absorption to charged surfaces.

Deproteinization methods using organic solvents

Nucleic acids are predominantly hydrophilic molecules and are easily soluble in water. Proteins, on the other hand, contain many hydrophobic residues making them partially soluble in organic solvents. There are several methods of deproteinization based on this difference, and they differ largely by the choice of the organic solvent. The organic solvents most commonly used are phenol or chloroform containing 4% isoamyl alcohol. The method that uses phenol as the deproteinizing agent was first used by Kirby (1957) and is commonly referred to as the **Kirby method**. Use of chloroform isoamyl alcohol mixtures was introduced by Marmur (1961) and is usually referred to as the **Marmur method**. These methods have undergone many modifications since their first publication.

The application of phenol in the **Kirby method** is based on the following principle. Phenol is crystalline at room temperature, but in the presence of 20% water, it forms an aqueous solution containing phenol micelles surrounded by water molecules. Protein molecules generally contain many hydrophobic residues, which are concentrated in the center of the molecule. When an aqueous protein solution is mixed with an equal volume of phenol, some phenol molecules are dissolved in the aqueous phase (about 20% water and 80% phenol). Yet, the phenol molecules are extremely hydrophobic. Therefore they tend to be more soluble in the hydrophobic cores of the protein than in water. As a result, phenol molecules diffuse into the core of the protein, causing the protein to swell and eventually to unfold. The denatured protein, with its hydrophobic groups exposed and surrounded by micelles of phenol, are far more soluble in the phenol phase than in the aqueous phase. Therefore, most of the proteins are partitioned into the phenol phase or precipitated on the interphase with water. Nucleic acids do not have strong hydrophobic groups and are insoluble in phenol.

The phenol method does require mixing the phenol phase with the water phase. This introduces some shearing of DNA molecules. Since only relatively small amounts of protein can dissolve in a given volume of phenol, repeated extraction of the aqueous phase with phenol is required to remove all the protein. Because the phenol phase, at saturation, contains 20% water, every phenol extraction will remove 20% of the DNA into the phenol phase. Even more DNA is lost by entrapment in the interphase layer of precipitated proteins or when the pH of phenol drops below pH 8.0. Another drawback of the Kirby method is that oxidation products of phenol can chemically react with DNA (and RNA) molecules. In addition phenol is highly toxic and requires special disposal procedures.

To minimize these effects, several modifications have been introduced:

- The use of ionic detergents. These detergents, by unfolding the protein, help to expose hydrophobic regions of the polypeptide chains to phenol micelles, aiding partitioning of proteins into the phenol phase.

- Enzymatic removal of proteins before phenol extraction. This reduces the number of extractions needed, thus limiting the loss and shearing of DNA.

- Addition of 8-hydroxyquinoline to the phenol. This increases the solubility of phenol in water. In the presence of this compound, phenol remains liquefied at room temperature with only 5% of water. In addition, 8-hydroxyquinoline is easily oxidized and therefore it plays the role of antioxidant, protecting phenol against oxidation. Since the reduced form of 8-hydroxyquinoline is yellow and the oxidized form is colorless, the presence or absence of yellow color is an excellent visual indicator of the oxidation state of phenol.

- Removal of oxidation products from phenol and prevention of oxidation upon storage or during phenol extraction. Because water-saturated phenol undergoes oxidation rather easily, particularly in the presence of buffers such as Tris, phenol used for DNA purification is twice distilled, equilibrated with water and stored in the presence 0.1 % of 8-hydroxyquinoline.

- Adjusting pH of water-saturated phenol solution to above pH 8 by equilibration of the liquefied phenol with a strong buffer or sodium borate.

- Substituting phenol with similar but nontoxic reagents such as ProCipitate™ or StrataClean Resin® (CPG Inc; Stratagene Inc.).

DNA obtained by the modified Kirby method is usually of high molecular weight but contains about 0.5% protein impurities that could be removed by another method.

The application of chloroform:isoamyl (CIA) mixture in the **Marmur deproteinization method** is based on a different characteristic of this organic solvent. The chloroform is not miscible with water and therefore, even numerous extractions do not result in DNA loss in the organic phase. The deproteinization action of chloroform is based on the ability of denatured polypeptide chains to partially enter or be immobilized at the water-chloroform interphase. The resulting high concentration of protein at the interphase causes protein to precipitate. Since deproteinization action of chloroform occurs at the chloroform:water interphase, efficient deproteinization depends on the formation of a large interphase area. To accomplish this, one has to form an emulsion of water and chloroform. This can only be done by vigorous shaking since chloroform does not mix with water. An emulsifier, isoamyl alcohol, is added to chloroform to help form the emulsion and to increase the water-chloroform surface area at which proteins precipitate. The method is very efficient in the recovery of DNA during numerous extractions. However, due to its rather limited capacity for removing protein, it requires repeated, time-consuming extraction when large amounts of protein are present. Also chloroform extraction requires rather vigorous mixing that contributes to hydrodynamic shearing of large DNA molecules. Using the Marmur method, it is possible to obtain very pure DNA, but usually of limited size (20 000-50 000 bp). The method is useful for the preparation of DNA from viruses with small genomes or when DNA of high purity, but not high molecular weight, is required for experiments, such as PCR or probes for hybridization experiments that require shearing the DNA to 400 bp lengths.

A substantial improvement in the method can be accomplished by limiting the number of extractions. This saves time and limits DNA shearing.

This can be done by enzymatically removing most of the protein before chloroform isoamyl extraction. Another modification frequently used is combining phenol and chloroform extraction into one step.

In conclusion, the efficient and modern use of the Kirby and Marmur methods of deproteinization of DNA (and RNA) preparations, requires prior enzymatic digestion of most proteins in cell lysate. The method can be used without this preliminary step only when small amounts of protein contaminate the DNA solutions.

Deproteinization methods based on complex formation

The method is based on the property of some compounds or detergents to form insoluble complexes with proteins or DNA, as well as with other components of the cell. Compounds most frequently used are CTAB and potassium or sodium salt of ethyl xanthogenate (carbonodithioic acid, o-ethyl ester).

CTAB (**cetyltrimethyl ammonium bromide**) or CDA (**cetyldimethyl-ethyl ammonium bromide**)can form complexes with nucleic acids, proteins, and polysaccharides. At high NaCl concentrations of 1.0 M or above, nucleic acids do not form complexes with CTAB (CDA) and remain soluble in water, whereas complexes with proteins and polysaccharides are formed. These complexes are insoluble in water. The insoluble complexes are removed from solution by centrifugation or by a single CIA extraction, leaving DNA in the aqueous phase. DNA is collected from the aqueous phase by ethanol precipitation, and residual CTAB (CDA) is removed from precipitated DNA by washing with 70% ethanol.

There are three parameters upon which the successful application of CTAB (CDA) critically depends. First, the salt (NaCl) concentration of the solution. The lysate must be 1.0 M NaCl or above to prevent formation of insoluble CTAB:DNA complexes. Since the amount of water present in the cell pellet is difficult to estimate, the breaking solution containing CTAB (CDA) should be adjusted to at least 1.4 M NaCl.

Second, all the solutions and the cell extract containing CTAB should be kept at room temperature, because CTAB and CTAB:DNA complexes are insoluble below 15 °C. In particular, this is very important for centrifugation and ethanol washing steps. This is because, when other deproteinization procedures are used, centrifugation and washing are carried out at 4 °C and -20 °C, respectively. Application of these conditions to the CTAB procedure will result in total loss of the DNA. However, these precautions are not necessary when CDA is used instead of CTAB.

Third, not all commercially available preparations of CTAB work well in this procedure. This laboratory had most success with a CDA preparation called cetyldimethylethylammonium bromide. CTAB preparation called mixed alkyltrimethylammonium bromide also gives satisfactory results. Use of some other preparations, particularly highly purified ones, resulted in very low DNA yields frequently contaminated with polysaccharides.

The CTAB (CDA) procedure, because of its efficient removal of polysaccharides, is frequently used for purification of DNA from sources containing large amounts of polysaccharides. To this belong: all plant cells, and cells of a variety of gram-negative bacteria including those of the genera *Pseudomonas, Agrobacterium, Rhisobium* and *Bradyrhisobium.*

CTAB (CDA) is also used for the removal of a polysaccharides from DNA prepared by other methods. In this procedure, the DNA solution is adjusted to 1% CTAB (CDA) and 0.7 M or 1.0 M NaCl, extracted with chloroform: isoamyl mixture, precipitated and washed with ethanol. If necessary this procedure can be repeated several times, until all polysaccharides are removed. The presence of polysaccharides is indicated by the appearance of a white powdery precipitate at the water:chloroform interphase.

Potassium or sodium salt of ethyl xanthogenate (xanthogenates) are compounds that have been used by the paper industry for a long time because of their ability to solubilize polyhydric alcohols such as cellulose. These salts were recently introduced to DNA purification in a single step purification procedure of DNA from plant cells owing to their cellulose dissolving capability (Jhingan, 1992). Xanthogenates react with the hydroxyl groups of polysaccharides forming water soluble polysaccharide xanthates and thus rendering plant cells easily lysed by detergents. Xanthates can also react with amines of proteins (Bolth et al., 1997, Carr et al., 1975; Millauer and Edelmann, 1975) and single-stranded RNA, forming water-insoluble complexes, but do not react with double-stranded DNA. Insoluble protein or RNA xanthates are removed by centrifugation and DNA is separated from water-soluble cellulose xanthate by ethanol precipitation. DNA, isolated by this method, is devoid of contaminants such as polysaccharides, phenolic, compounds heavy metals, protein and single-stranded RNA and is suitable for restriction enzyme digestion and PCR analysis. The main remaining contaminant is double-stranded RNA and this can be removed by RNase treatment.

In our laboratory, we successfully used potassium ethyl xanthogenate with minor modifications, for the isolation of DNA from a variety of sources such as algae, animals, insects, and bacteria. Also this method was recently used successfully for isolation of DNA from various filamentous fungi (Ross, 1995). The xanthogenate salts proved to be especially useful for small scale preparations of DNA from numerous samples, in particular, DNA for PCR and small scale RFLP analysis. A disadvantage of this method is that DNA yields are rather low, usually 30% to 50% of the amount obtained by other methods, when used without mechanical disruption of the cells.

Deproteinization methods based on binding to silica

In the presence of a high concentration of salts, DNA will bind to a silica surface whereas proteins, due to the predominance of hydrophobic characteristics in high salt, will not. The method introduced by Vogelstain and Gillespie (1979), uses acid-washed glass beads as a source of silica. The bound DNA is washed to remove impurities and eluted from the glass beads by lowering the salt concentration. Since most proteins will not precipitate at high salt concentration in the presence of 50% ethanol, the washing media contains 50% ethanol and at least 0.1 M NaCl. The main advantage of the method is its simplicity because fewer manipulations are required compared to most other extraction methods.

However, there are numerous disadvantages of this method that limit its general usefulness. First, the method becomes cumbersome if used with large volumes of liquid, making it impractical for large scale DNA isolation. Second, the yield of DNA is somewhat lower, ranging from 50% to 75% of the starting material. Third, large molecular weight DNA is usually sheared to smaller fragments of less than 5 kb size due to binding to more than one silica particle.

Numerous improvements of this method have been introduced and are commercially available. Most of the improvements involve the modification of particles used for DNA binding. The binding surface of the particles has been activated to increase their capacity for DNA, and the particles have been made larger and more uniform in size to prevent DNA shearing. In addition, silica particle membranes are also used to facilitate quick removal of proteins by centrifugation of cell extracts through the membrane. These modifications make it possible to isolate DNA up to 50 kb with yields approaching 70% to 80% for large DNA fragments. These improvements, however, have not made it possible to increase the size of the sample that can be practically handled. Consequently, the method is used for fast preparations of genomic DNA from small samples when very large molecular weight molecules are not required, or for mini-isolation of plasmid DNA from bacterial cells.

Also this method has been adapted for isolation of DNA fragments from agarose gels. In these modifications chaotropic salts, such as sodium iodate or sodium perchloride are used to dissolve agarose and bind DNA to a silica surface.

Another use of silica for DNA purification is the modification of binding properties of silica by introduction of acid hydroxy groups on the silica. The hydroxylated silica, at neutral or near neutral pH, has a higher affinity for proteins than for nucleic acids. Stratagene Co. introduced the method in the form of a StrataClean Resin®, a slurry of hydroxylated silica particles. Mixing silica slurry with the aqueous phase containing DNA and proteins results in protein binding to the silica, leaving the DNA in the aqueous phase. Phases are separated by centrifugation in a similar way to phenol extraction. At least two or more extractions are required to completely remove proteins from the sample. However, unlike the phenol extraction, there does not appear to be any loss of DNA from the sample with each extraction. Another advantage of the StrataClean Resin® over phenol is that it is not toxic and therefore safe to use. The main limitations of this method are: the necessity for numerous extractions and its considerable cost, in particular, when applied to large volume samples. A useful modification of this method, is treatment with proteases prior to extraction with StrataClean Resin®. This modification makes the use of StrataClean Resin® somewhat less expensive and cumbersome, and increases the purity and quantity of the DNA isolated.

Deproteinization methods based on differences in partial specific volume

The partial specific volume, i.e., volume occupied by one gram of the substance, depends, in general, on the shape of the molecules. The partial specific volume of most proteins, predominantly globular molecules, is in the range of 0.70 to 0.73 cm^3/g. Whereas, the partial specific volume of DNA, rod-shaped molecules, is 0.55 cm^3/g, a much smaller value than that of proteins. The differences in partial specific volumes of these molecules can be exploited by the use of density gradient centrifugation. The density of a given substance is defined as the weight of 1 cm^3 of this substance (g/cm^3), i.e., it is the reciprocal value of its partial specific volume. The density of protein in water is, therefore, in the range of 1.36 g/cm^3 to 1.42 g/cm^3, whereas the density of DNA is about 1.818 g/cm^3.

Moreover, because the partial specific volume of RNA or ssDNA is different from that of dsDNA, these differences also serve to separate these molecules from DNA and proteins, as well as from each other. To create density gradients in which separation takes place, cesium salts (chloride or sulfate) are usually used to generate self-forming gradients during high speed centrifugation. The capacity of the method is limited by the requirement of high speed centrifugation and a considerable time required to create the gradient and to separate molecules. The method is usually used as the last step of a large scale DNA purification procedure to generate very pure DNA preparations. For most molecular biology applications it is not necessary to obtain DNA of such purity. Density gradient centrifugation is frequently used for single-step purification of RNA.

Deproteinization methods based on differences in sensitivity to enzymes

Proteins and nucleic acids are vastly different biological molecules that are recognized by different and highly specific digestive enzymes. Proteins can be removed from DNA or RNA preparations using a protease that can digest all proteins, i.e., general-purpose protease. Two such enzymes are in use, Proteinase K and Pronase. Both enzymes are very stable, general specificity proteases secreted by fungi. Commercial preparations of these enzymes are inexpensive and devoid of DNase contamination, making them safe to use in the purification of nucleic acids. These proteases are active in the presence of low concentrations of anionic detergent, high concentrations of salts, and EDTA and exhibit broad pH (6-10 pH) and temperature (50-67 °C) optima. They can digest intact (globular) and denatured (polypeptide chain) proteins and do not require any cofactors for their activities. Proteinase K and Pronase are usually used in DNA purification procedures at final concentrations of 0.1 to 0.8 mg/ml and stock solution can be stored in 50% glycerol at -20 °C for an indefinite time without substantial lost of activity. The difference between these two enzymes lies in their activities toward themselves; Pronase is a self-digesting enzyme, whereas Proteinase K is not. The fact that Proteinase K is not a self-digesting enzyme makes it a more convenient enzyme to use than Pronase, because it is unnecessary to continually add it during the prolonged course of the reaction. This accounts for greater popularity of Proteinase K rather than Pronase in published DNA purification procedures.

The major drawback in using these enzymes for deproteinization is that enzymatic treatment can only remove 80 to 90% of the proteins present. This is because protein digestion is an enzymatic reaction dependent on substrate and enzyme concentrations. In practice, the deproteinization rate depends only on the protein (substrate) concentration, because it is not practical to add a large amount of enzyme to accelerate the reaction at low substrate concentration. Therefore, as the reaction proceeds, the concentration of substrate decreases progressively, slowing the reaction rate. And, indeed, enzymatic reactions will reach completion only given infinite time. Conveniently, at high substrate concentrations and sufficient concentration of enzyme, the reaction proceeds at a maximal rate until 80 to 90% of the substrate has been removed. The reaction rate then becomes too slow to be practical for removal of remaining protein in a reasonable time.

In conclusion, enzymatic deproteinization should be used primarily when a large amount of protein is present, that is, right after cell lysis. The characteristics of enzymatic removal of proteins mentioned above make protease digestion an ideal and indispensable step in nucleic acid purification when applied before organic solvent extraction when the substrate (protein) concentration is very high.

Removal of RNA

The removal of RNA from DNA preparations is usually carried out using an enzymatic procedure. Consequently this procedure, does not remove all RNA, and therefore yields DNA preparations with a very small amount of RNA contamination. Two ribonucleases, that can be easily and cheaply prepared free of DNase contamination are used, ribonuclease A and ribonuclease T1.

Ribonuclease A (RNase A), isolated from bovine pancreas, is an endoribonuclease that cleaves RNA after C and U residues. The reaction generates 2':3' cyclic phosphate which is hydrolyzed to 3'-nucleoside phosphate producing oligonucleotides ending with 3' phosphorylated pyrimidine nucleotide. The enzyme is active under an extraordinarily wide range of experimental conditions, such as in the presence of detergent, at high temperatures, in the presence of phenol residues etc., and it is very difficult to inactivate. At NaCl concentrations below 0.3 M, the enzyme can cleave both single-stranded and double-stranded RNA, but at high salt concentrations (above 0.3 M) the enzyme can cleave only single-stranded RNA. To effectively remove all RNA, care must be taken to lower the NaCl concentration of DNA preparations to below 0.3 M before treatment with RNase. This is particularly important because some DNA purification procedures are initially carried out at high salt concentration. For this reason it is not recommended to use RNase during the first step of DNA purification. Complete removal of RNase from the preparation requires phenol and/or CIA extractions. However, the presence of a small amount RNase does not interfere with most DNA applications and consequently, its complete removal is not always necessary.

Ribonuclease T1 (RNase T1) isolated from *Aspergillus oryzea* is an endoribonuclease that is very similar to RNase A in its reaction conditions and stability. The enzyme cleaves double-stranded and single-stranded RNA after G residues, generating oligonucleotides ending in a 3' phosphorylated guanosine nucleotide. At high NaCl concentration, as with RNase A, the enzyme is active only on single-stranded RNA. Thus, the same precaution in its use should be applied as described for RNase A.

In conclusion, because of the RNA cleaving specificity of these enzymes, it is recommended that they be used **together** for complete RNA removal from DNA samples. The use of only one of these enzymes can result in contamination of DNA preparations with a large amount of oligonucleotides that will make the spectrophotometric measurement of DNA concentration practically impossible.

1.5 Concentrating the DNA

This step of DNA purification procedures serves two purposes. First, it concentrates high molecular weight DNA from deproteinization solutions and second, it removes nucleotides, amino acids and low-molecular-weight impurities remaining in the solution after cell breakage (see Table 1). This step can be implemented in two ways: by precipitating DNA with alcohols, or dialyzing and concentrating the DNA using compounds that actively absorb water.

Precipitation of DNA with alcohols

Two alcohols are used for DNA precipitation: ethanol and isopropanol. Alcohol precipitation is based on the phenomenon of decreasing the solubility of nucleic acids in water. Polar water molecules surround the DNA molecules in aqueous solutions. The positively charged dipoles of water interact strongly with the negative charges on the phosphodiester groups of DNA. This interaction promotes the solubility of DNA in water. Ethanol is completely miscible with water, yet it is far less polar than water. Ethanol molecules cannot interact with the polar groups of nucleic acids as strongly as water, making ethanol a very poor solvent for nucleic acids. Replacement of 95% of the water molecules in a DNA solution will cause the DNA to precipitate. Making a DNA solution of 95% ethanol concentration is not practical because it would require the addition of a large volume of 100% ethanol to the DNA solution. To precipitate DNA at a lower ethanol concentration, the activity of water molecules must be decreased. This can be accomplished by the addition of salts to DNA solutions. Moreover, the presence of salts will change the degree of charge neutralization of the DNA phosphates, eliciting extensive changes in the hydrodynamic properties of the DNA molecules (Eickbush and Moudrianakis, 1978). These changes concomitant with water elimination will cause the separation of the DNA phase, i.e., precipitation, at the moment of complete neutralization of DNA molecules.

DNA precipitation is customarily carried out with 70% ethanol (final concentration) in the presence of the appropriate concentration of sodium or ammonium salts. Table 3 presents a description of salts and the conditions used for DNA precipitation in 70% ethanol.

Table 3. Salts used in DNA precipitation

Salt	Stock Solution	Final Concentration
NaCl	3.0 M	0.3 M
Sodium acetate	3.0 M	0.3 M
Ammonium acetate	7.5 M	2.0 M
Lithium chloride	5.0 M	0.5 M

The use of each salt listed in Table 3 has its advantages and disadvantages. The major advantage of using sodium chloride, in addition to convenience and low cost, is that SDS remains soluble in ethanol in the presence of 0.2 M NaCl. The use of sodium chloride is therefore recommended if a high concentration of SDS has been used to lyse the cells. The disadvantage of NaCl is its limited solubility in 70% ethanol making it difficult to completely remove salt from the DNA samples. This is particularly true when the precipitated DNA is collected by centrifugation. High NaCl concentration in DNA preparations can interfere with the activity of many enzymes. When NaCl is used, the DNA should be spooled rather than centrifuged to collect precipitate, making NaCl particularly useful in large scale, high molecular weight DNA preparations.

Sodium acetate is more soluble in ethanol than NaCl and therefore is less likely to precipitate with the DNA sample. Its higher solubility in 70% ethanol makes it easier to remove from a DNA preparation by repeated 70% ethanol washes. Thus, sodium acetate is the most frequently used salt in DNA precipitation.

Ammonium acetate is highly soluble in ethanol and easy to remove from precipitated DNA due to the volatility of both, ammonium and acetate ions. The use of ammonium acetate, instead of sodium acetate, is also recommended to remove nucleotide triphosphates or small single or double-stranded oligonucleotides (less then 30 bp), since these molecules are less likely to precipitate in its presence at high ethanol concentrations. In addition, precipitation of DNA with ammonium acetate has proven to be more efficient for the removal of heavy metals, detergents and some unknown impurities that are potent inhibitors of restriction endonucleases and other enzymes used for DNA manipulation (Crouse and Amorese, 1987; Perbal, 1988). A disadvantage of ammonium acetate is the necessity to sterilize it by filtration rather than autoclaving due to the volatility of ammonium and acetate ions. The volatile nature of ammonium acetate requires also that a fresh stock solution be made every 1-2 months and stored refrigerated in tightly capped containers.

Lithium chloride, used in precipitation, has an advantage because of its high solubility in ethanol. Its use results in relatively salt-free DNA preparations. A disadvantage is the rather high cost of LiCl as compared to other salts used in DNA precipitation. High concentrations of LiCl are also used to separate mRNA and rRNA from double-stranded DNA and small RNAs in the absence of ethanol. At 4 M LiCl, large RNA molecules will precipitate out of solution leaving behind DNA and small molecular weight RNAs.

Customarily, ethanol precipitation is carried out at temperatures of -20 °C or lower. It is reasoned that low temperature and the presence of salts, further lower the activity of water molecules, facilitating more efficient DNA precipitation. However, a careful analysis of the efficiency of DNA precipitation at various temperatures and DNA concentrations, demonstrated that this step can be performed at room temperature without serious loss of DNA, even when the concentration of DNA in a sample is very low (Zeugin and Hartley, 1985; Crouse and Amorese, 1987). These authors carefully analyzed the recovery of precipitated DNA as a function of concentration, time and temperature when ammonium acetate is used (Crouse and Amarose, 1987). The best recoveries of DNA (DNA concentrations ranging 5 to 5000 ng/ml) were obtained when incubation was carried out at room or 4 °C temperatures, and the worst, when the solution was kept at -70 °C. The recovery of DNA at very low concentration (5 ng/ml) was not substantially different at the various temperatures and was largely dependent on time. Thus, in the procedures described in this book we do not recommend precipitating DNA at low temperatures.

Ethanol precipitation of small quantities of DNA (less than 1 µg) usually results in invisible pellets that are difficult to wash and dissolve. Recently a highly visible, inert carrier for precipitation of DNA and RNA was developed by Novagen, Inc. The reagent (Pellet Paint™) of vivid pink color helps to visualize and precipitate DNA at concentrations lower than 2 ng/ml. The presence of the reagent in the sample does not interfere with enzymes used in subsequent DNA manipulations or with electroporation and transformation (Novagen 69049-1 Application Bulletin, 1996). Moreover, when the DNA sample, containing Pellet Paint^{TM} reagent, is run in gel electrophoresis, the strongly fluorescing reagent moves toward positive electrode and does not hinder the visualization of DNA bands. Pellet Paint^{TM} can be used with both ethanol and isopropanol precipitation of DNA. The disadvantage of using this reagent is that it absorbs at 260 nm, making it difficult to determine the concentration of nucleic acid by A_{260} absorbance. This can be overcome by calculating the absorbance contribution of Pellet Paint to A_{260} in a given sample using the constant absorbance ratio of the reagent at 260 and 555 nm. The manufacturer for each sample of Pellet Paint^{TM} gives this ratio. However, applying the correction requires making an additional absorbance reading of the sample at 555 nm.

Precipitation of DNA with isopropanol has all the characteristics of precipitation with ethanol. The advantage of isopropanol over ethanol precipitation is that a much lower concentration of this alcohol is needed to precipitate DNA. DNA molecules will precipitate at an isopropanol concentration of 50% with salt concentrations identical to ethanol. Moreover, at this concentration of isopropanol, and at room temperature, small molecular weight DNA fragments will precipitate poorly, affording their removal from the sample. A disadvantage in the use of isopropanol is the difficulty in completely removing it from the sample because it is less volatile than ethanol. It is recommended, therefore, that isopropanol precipitation be followed by several 70% ethanol washes of the DNA pellet.

Concentrating DNA by dialysis

DNA can be concentrated, and small molecular weight molecules removed, by dialysis. This method should be used if isolation of very high molecular weight DNA (equal or larger than 200 kb) is desired. Alcohol precipitation can lead to DNA shearing, and indeed, it is difficult to obtain DNA larger than 150 kb using ethanol precipitation. Dialysis is carried out in pretreated dialysis tubing, against a large volume of buffer or water (dilution factor, 1000 or more) for a period of at least 24 hours. It is important to transfer the DNA solution to and from the dialysis tubing with wide-bore pipettes. To concentrate the DNA sample, dialysis bags are surrounded with a dry powder of a water-absorbing compound such as PEG or Aquacide II. The last compound, sold by Calbiochem Co., (cat. # 17851) is a better substance to use since, unlike PEG, it does not contain traces of alcohol that can accumulate in dialysis bags.

1.6 Single Step DNA Purification

These methods are especially useful when one needs to isolate DNA from a limited amount of sample, or from multiple samples. These procedures are usually modifications of the procedure first described by Chomczynski and Sacchi (1987). Chomczynski (1993) introduced a procedure that permits the single-step simultaneous isolation of RNA, DNA and proteins from animal cells and tissues. This method uses the TRI-Reagent™ (Macromolecular Resource Center, Inc) a monophasic solution of phenol and guanidine isothiocyanate in which the sample is homogenized. The aqueous phase is separated from the organic phase by the addition of chloroform. DNA remains in the aqueous phase and can be recovered by ethanol precipitation. Modification of this method has been published that greatly increases the quantity of DNA isolated (Tse and Etienne, 1995). Recently a new DNA single-step purification reagent, DNAzol®, was introduced (Chomczynski et al., 1997). This reagent contains quanidine thiocyanate and a detergent mixture that causes lysis of the cells and hydrolyses of RNA allowing selective precipitation of DNA from the cell lysate. This method permits isolation of genomic DNA from variety of cells and tissue resulting in the product that can be used in Southern analysis, dot blot, PCR and molecular cloning.

1.7 Purification of Plasmid DNA

Isolation of plasmid DNA from bacterial cells is an essential step for many molecular biology procedures. Many protocols for a large and small scale isolation of plasmid (minipreps) have been published. The plasmid purification procedures, unlike the procedures for purification of genomic DNA, should involve removal of not only protein but also another major impurity – bacterial chromosomal DNA. Thus, the task of plasmid purification substantially differs from that of the preparation of genomic DNA. Most plasmid DNA purification methods start from the preparation of a crude bacterial lysate and eventually employ the standard protein removal procedures already described above. To achieve separation of plasmid from chromosomal DNA these methods exploit the structural differences between plasmid and chromosomal DNA. Plasmids are circular supercoiled DNA molecules substantially smaller than bacterial chromosomal DNA. There are three basic methods of plasmid preparation: alkaline lysis introduced by Brinboim and Doly (Brinboim and Doly, 1979; Brinboim, 1983), lysis by boiling in the presence of detergent introduced by Holmes and Quigley (Holmes and Quigley, 1981) and application of affinity matrixes for plasmid or proteins introduced by Vogelstein and Gillespie (Vogelstein and Gillespie, 1979). Since their introduction, a myriad of modifications of the basic methods have been published to improve both yield and purity of plasmid DNA and decrease the time necessary for its purification.

In the alkaline lysis method, cells are lysed and DNA denatured by SDS and NaOH. Neutralization of the solution results in a fast reannealing of covalently closed plasmid DNA due to the interconnection of both single-stranded DNA circles. Much more complex bacterial chromosomal DNA cannot reaneal in this short time and forms a large, insoluble DNA network, largely due to interstrand reasociation at multiple sites along the long linear molecules. At the next step of the procedure, lowering the temperature results in precipitation of protein-SDS complexes. Subsequently both complexes, DNA and protein, are removed by centrifugation leaving plasmid molecules in the supernatant. If cleaner plasmid is desired, the remaining protein and RNA are removed by standard methods.

In the boiling method, high temperature and detergent lyse bacteria cells. Bacterial chromosomal DNA under these conditions, remains attached to the bacterial membrane. Subsequent centrifugation pellets chromosomal-DNA complexes while plasmid DNA remains in the supernatant. A recent modification of this procedure involves the lysis of bacterial cells using a microwave oven rather than a boiling water bath (Hultner and Cleaver, 1994; Wang et al., 1995). Further plasmid purification, if desired, can be carried out using standard deproteinization procedures and RNase treatment.

1.8 Determination of Concentration and Purity of DNA

The last step of any DNA isolation procedure is the evaluation of the results. For DNA this evaluation involves:

- Determination of DNA concentration.

- Evaluation of the purity of the DNA.

- Determination of DNA yield.

Ultraviolet (UV) spectrophotometry is most commonly used for the determination of DNA concentration. The resonanse structures of pyrimidine and purines bases are responsible for these absorptions. The DNA has a maximum and minimum absorbance at 260 nm and 234 nm, respectively. However these are strongly affected by the degree of base ionization and hence pH of the measuring medium (Beaven et al, 1955; Wilfinger et al, 1997). The relationship between DNA absorbance at 260 nm (A_{260}) and DNA concentration (N) is described by the following equation:

$$A_{260} = \varepsilon_{260} [N] \tag{1}$$

Thus giving:

$$N = \frac{A_{260}}{\varepsilon_{260}} \tag{2}$$

Where ε_{260} is the DNA extinction coefficient.

This coefficient for double-stranded DNA is usually taken as 0.02 μg^{-1} cm^{-1} when measured at neutral or slightly basic pH. Thus an absorbance of 1.0 at 260 nm gives DNA concentration 50 $\mu g/ml$ (1/0.02 = 50 $\mu g/ml$). The value of the absorption coefficient (ε_{260}) for double-stranded DNA varies slightly depending on the percent of GC, the concentration of DNA solutions having absorbance of 1 is not always 50 $\mu g/ml$. This slight variation is usually disregarded. The extinction coefficient of single-stranded DNA is 0.027 $\mu g^{-1} cm^{-1}$ giving a ssDNA concentration of 37 $\mu g/ml$ for an absorbance of 1 (1/0.027 = 37 $\mu g/ml$).

The linear relationship (equation 1) between absorbance at 260 nm and DNA concentration holds in a range between 0.1 to 2.0 absorbance units. Reliable measurements of DNA concentration can made for solutions of 0.5 to 100 $\mu g/ml$ using a standard UV spectrophotometer. Samples with an absorbance equal to or larger than 2 should be diluted before measuring. The measurement of DNA concentration at a lower range (A_{260} lower than 0.2) can be strongly affected by light scattering on dust particles present in the preparation. Measuring the absorbance at 320 nm (Schleif and Wensink 1981) should assess the degree of such contamination. At this wavelength, DNA does not absorb and any absorbance at 320 nm is due to light scattering. The absorbance at 320 nm should be less than 5% of the A_{260}.

Absorbance measurements at wavelengths other than 260 nm are used for determination of DNA purity. The relevant spectrum for this purpose lies between 320 nm and 220 nm. As discussed above, any absorbance at 320 nm indicates contamination of a particular nature. Proteins absorb maximally at 280 nm due to the presence of tyrosine, phenylalanine and tryptophan and absorption at this wavelength is used for detection of protein in DNA samples. This is usually done by determination of the A_{260}/A_{280} ratio. This ratio for pure double-stranded DNA was customarily taken to be between 1.8 to 1.9 (for example see Sambrook et al., 1989). However, recent revaluation of Warburg equations (Warburg and Christian, 1942), the basis for the use of this ratio, has shown that the purity evaluation using these values is at best misleading, if not erroneous, for estimation of protein contamination (Glasel, 1995; Manchester, 1995; Huberman, 1995; Held, 1997). The ratio A_{260}/A_{280} for pure double stranded DNA is 2. The ratio between 1.8 and 1.9 corresponds to 60% and 40% protein contamination, respectively (Glasel, 1995). Moreover, the A_{260}/A_{280} ratio is not a very sensitive indicator of protein contamination because the extinction coefficient for protein at 280 nm is 10 to 16 times lower than the extinction coefficient of DNA (see Table 4).

Table 4. Extinction coefficients for double stranded DNA and proteins

	Extinction Coefficient (10^{-3} μg^{-1} cm^{-1})		
	ε 280	ε 260	ε 234
DNA	10.0	20.0	9.0 - 10.0
Protein	0.63 - 1.0	0.57	5.62

One cannot use equation 1 to calculate DNA concentration precisely when protein is present in the sample since this equation is valid only when the A_{260}/A_{280} ratio is 2. However this calculation is possible at any A_{260}/A_{280} ratio using the following equation (derived from: Kalb and Bernlohr, 1977).

$$N = \frac{A_{260} - A_{280}/R}{\varepsilon_{260} - \varepsilon_{280}/R} \tag{3}$$

Where: R is ratio of extinction coefficient at 280 nm to 260 nm for protein ($p\varepsilon_{280}/p\varepsilon_{260}$); N is DNA concentration in μg; A_{260} and A_{280} are absorbance of DNA sample at 260 nm and 280 nm, respectively and, ε_{260} and ε_{280} are absorption coefficients for DNA at 260 nm and 280 nm, respectively.

Substituting values of absorption coefficients for DNA and protein from Table 4 gives;

$$N \ (\mu g \ ml^{-1}) = 70A_{260} - 40A_{280} \tag{4}$$

A better indicator of protein contamination in DNA samples is the ratio of A_{260}/A_{234}. DNA has an absorbance minimum at 234 nm, and protein absorbance is high due to the absorption maximum for peptide bonds at 205 nm (Scopes, 1974; Stoscheck, 1990). Since the ratio of DNA extinction coefficient at 234 nm (ε_{234}) to protein extinction coefficient at the same wavelength is between 1.5 to 1.8 (see Table 4), the A_{260}/A_{234} ratio is a very sensitive indicator of protein contamination. For pure nucleic acids, this ratio is between 1.8 and 2.0. The DNA concentration can be calculated from the absorbance at 260 nm and 234 nm using the following equation:

$$N = \frac{A_{260} - A_{234}/R}{\varepsilon_{260} - \varepsilon_{234}/R} \tag{5}$$

Where R is ratio of extinction coefficient at 234 nm to 260 nm for protein ($p\varepsilon_{234}/p\varepsilon_{260}$); N is DNA concentration in μg/ml; A_{260} and A_{234} is absorbance of DNA sample at 260 nm and 234 nm, respectively, and ε_{234} and ε_{260} is absorption coefficient for DNA at 234 nm and 260 nm, respectively.

Substituting values of extinction coefficients for DNA and protein from Table 4 gives;

$$N\ (\mu g\ ml^{-1}) = 52.6A_{260} - 5.24A_{234} \tag{6}$$

1.9 Storage of DNA Samples

DNA samples should be stored under the conditions that limit their degradation. Even under ideal storage conditions, one should expect about one phosphodiester bond break per 200 kb per year. For long-term storage, the pH of the buffer should be above 8.5 to minimize deamidation, and contain at least 0.15 M NaCl and 10 mM EDTA.

The following conditions will increase this rate and contribute to fast degradation of DNA during storage.

- DNase contamination.
  The most frequent source of this contamination is human skin. In spite of the low stability of most DNases, even short exposure to a very low concentration of these enzymes will result in substantial sample degradation. To avoid this contamination, it is necessary to avoid direct or indirect contact between sample and fingers by wearing gloves and using sterilized solutions and tubes. Since DNA is easily absorbed to glass surfaces only sterilized plastic tubes should be used for storage.

- Presence of heavy metals.
  Heavy metals promote breakage of phosphodiester bonds. Long term DNA storage buffer should contain 10 mM or more of EDTA, a heavy metal chelator. If EDTA is present, DNA can be stored as a precipitate in 70% ethanol. This storage condition is preferred if the sample is stored at 5 °C because it prevents bacterial contamination. For short-term storage of DNA, 1 mM to 2 mM EDTA concentration is sufficient and more convenient for every day work.

- Presence of ethidium bromide.

  The presence of ethidium bromide causes photooxidation of DNA with visible light in the presence of molecular oxygen. Since it is difficult to remove all ethidium bromide from DNA samples treated with this reagent, such samples should be always stored in the dark. Moreover, due to the ubiquitous presence of ethidium bromide in molecular biology laboratories, DNA samples can be easily contaminated with it. For this reason we recommend storage of all DNA samples in the dark.

- Temperature.

  The best temperature for short-term storage of high molecular weight DNA is between 4 °C and 6 °C. At this temperature, the DNA sample can be removed and returned to storage without cycles of freezing and thawing that cause DNA breakage. For very long-term storage (5 years or more) DNA should be stored at temperature -70 °C or below, providing the sample is not subjected to any freeze-thaw cycles. To avoid these cycles, the DNA should be aliquoted in small volumes in separate tubes permitting the withdrawal of a sample without repeating freezing and thawing of the entire DNA stock. Long or short term storage of high molecular weight DNA at -20 °C is not recommended. This temperature can cause extensive single and double strand breakage of DNA because, at this temperature, molecular bound water is not frozen.

References

Arumuganathan K, and Earle ED (1991) Nuclear DNA content of some important plant species. Plant Molecular Biology Reporter 9:208-218.

Beaven GH, Holiday ER, and Johnson EA (1955) Optical properties of nucleic acids and their components. p 493-553. In The nucleic acids Vol 1. Eds. Chargaff E and Davidson JN. Academic Press. New York.

Bolth FA, Crozier RD, and Strow LE (1975) Dialkylthiocarbamates. Chem Abstract, 83:27641K.

Brinboim HC (1983) A rapid alkaline extraction method for the isolation of plasmid DNA. In: Method Enzymol, vol 100. Academic Press, London, pp. 243-255.

Brinboim HC, and Doly J (1979) A rapid alkaline extraction procedure for screening recombinant plasmid DNA. Nucleic Acid Res 7:1513-1523.

Carr ME, Hofreiter BT and Russel CR (1975) Starch xanthate-polyethylenimine reaction mechanisms. J Polym Sci Polym Chem, 13:1441-1456.

Chomczynski P (1993) A reagent for the single-step simultaneous isolation samples of RNA, DNA and protein from cell and tissue samples. BioTechniques 15:532-537.

Chomczynski P, and Sacchi N (1987) Single-step method of RNA isolation by acid quanidinium thiocyanate-phenol-chlroform extraction. Anal Biochem 162:154-159

Chomczynski P, Mackey K, Drews R, and Wilfinger W (1997) DNAzol®: A reagent for the rapid Isolation of Genomic DNA. BioTechniques; 22:550-553.

CPG® Inc (1987) ProCipitate™ nucleic acid purification reagent. Instruction Manual. 20.2.

Crouse J and Amorese D (1987) Ethanol precipitation: Ammonium acetate as an alternative to sodium acetate. BRL Focus 9 (2):3-5.

Eickbush TH, and Moudrianakis EN (1978) The compaction of DNA helices into either continuous supercoils or folded-fiber rods and toroids. Cell 13:295-306.

Glasel JA (1995) Validity of Nucleic Acid Purities Monitored by 260nm/280nm Absorbance Ratios. BioTechniques. 18:62-63.

Held P (1997) The Importance of 240 nm Absorbance Measurement. The A_{260}/A_{280} ratio just isn't enough anymore. Biomedical Products 7:123.

Holmes DS and Quigley M (1981) A rapid boiling method for the preparation of bacterial plasmids. Anal Biochemistry 114:193-197.

Huberman JA (1995) Importance of measuring nucleic acid absorbance at 240 nm as well as at 260 and 280 nm. BioTechniques 18:636.

Hultner MI, and Cleaver JE (1994) A bacterial plasmid DNA miniprep using microwave lysis. BioTechniques 16:990-994.

Jhingan AK (1992) A novel technology for DNA isolation. Methods in Molecular and Cellular Biology 3:15-22.

Kalb VF, and Bernoohr RW (1977) A new spectrophotometric assay for protein in cell extracts. Anal Biochemistry 82:362-371.

Kirby K (1957) A new method for the isolation of deoxyribonucleic acids; evidence on the nature of bonds between deoxyribonucleic acid and proteins. Biochem J 66:495-504.

Manchester KL (1995) Value of A_{260}/A_{280} Ratios for Measurement of Purity of Nucleic Acids. BioTechniques. 19:208-210.

Marmur J (1961) A procedure for the isolation of deoxyribonucleic acid from microorganisms. J Mol Biol 3:208-218

Millauer H and Edelmann G (1975) O-alkyl thiocarbamates. Chem Abstract 83; 192612.

Perbal BV (1988) A Practical Guide to Molecular Cloning. 2nd edition. John Wiley & Sons.

Ross IK (1995) Non-Grinding Method of DNA Isolation from Human Pathogenic Filamentous Fungi Using Xanthogenates. BioTechniques 18:828-830.

Sambrook J, Fritsch E and Maniatis T (1989) Molecular Cloning: A Laboratory Manual. 2nd ed. Cold Spring Harbor Laboratory Press. Cold Spring Harbor. NY. Section E5.

Schleif RF and Wensink PC (1981) Practical method in molecular biology. New York. Springer-Verlag.

Scopes RK (1974) Measurement of protein by spectrophotometry at 205 nm. Anal Biochem 59:277-282

Stoscheck CM (1990) Quantitation of protein. Method in Ezymol. 182:50-68.

Tse CH and Etienne J (1995) Improved high-quality DNA extraction with tri-reagent: A reagent for single-step simultaneous isolation of RNA, DNA and protein from tissue samples. Method in Molecular and Cellular Biology 5:373-376.

Wang B, Merva M, Williams WV and Weiner DB (1995) Large-scale preparation of plasmid DNA by microwave lysis. BioTechniques 18:554-555.

Warburg O and Christian W (1942) Isolation and crystallization of enolase. Biochem Z 310:384-421.

Wilfinger WW, Mackey K and Chomczynski P (1997) Effect of pH and ionic strength of the spectrophotometric assessment of nucleic acid purity. BioTechniques 22:474-480.

Vogelstein B and Gillespie D (1979) Preparative and analitical purification of DNA from agarose. Proc Natl Acad Sci USA 76:615-619

Zeugin JA and Hartley JL (1985) Ethanol precipitation of DNA. BRL Focus 7(4):1-2.

Suppliers

Commercially available kits for genomic DNA isolation

Amersham Pharmacia Biotech. (www.apbiotech.com)
Two kits, RapidPrep Genomic DNA for cells and tissue and RapidPrep Genomic DNA for blood. These kits use anion-exchange resin in a spin column format to separate DNA from proteins. Some DNA shearing can be expected using these kits. Two other kits GenomicPreps Cells and Tissue DNA Isolation Kit and GenomicPrep Blood DNA Isolation Kit both use salt precipitation of proteins to purify DNA without the use of organic solvents. Some proteins remain in the preparation but they do not interfere with PCR, restriction enzyme digestion, Southern hybridization or RFLP analyses.

Amresco, Inc. (www.amresco-inc.com)
Amresco, Inc. offers a RapidGene genomic DNA purification kit for isolation of DNA from cells, tissues and gram-positive bacteria. This kit does not use organic solvent for separation of protein from DNA.

Bio 101 Co. (www.bio101.com)
The G Nome DNA isolation kits use proteases and salting out to remove proteins. DNA can be isolated from bacteria, yeast, animal or plant cells and tissue. Interesting DNA isolation kits for isolation of DNA from ancient DNA and from forensic samples is also offered. They use a silica matrix method for DNA purification.

Bio-Rad Laboratories (www.bio-rad.com)
Several kits for isolation of genomic DNA from different sources are offered. All kits use diatomaceous earth as the DNA binding matrix to separate proteins from nucleic acids in spin-column format. Shearing genomic DNA using these kits is not as extensive as when using silica particles as the DNA-binding matrix.

Clontech, Inc. (www.clontech.com)
Clontech provides a number of specialized kits for purification of DNA from cultured animal cells and tissue, plants, blood and viruses. These kits use silica membrane spin columns to bind DNA from cell extracts. Proteins are removed by repeated washing and DNA is released from the silica membrane by low salt elution. DNA larger than 50 kb is difficult to obtain using these kits.

CPG, Inc. (www.cpg-biotech.com)
Their DNA-Pure Genomic Kits use protein precipitation buffer to remove proteins from cell lysates. A high yield of genomic DNA is recovered but some protein contamination remains. This DNA is suitable for restriction digestion, Southern blot and PCR. A special kit is offered for purification of DNA from various species of yeast.

Dynal, Inc. (www.dynal.no)
Dynal is the company that introduced magnetic beads technology. Dynabeads, supermagnetic polysteren beads, are used to separate DNA from proteins. The use of magnetic beads eliminates the centrifugation step in DNA purification.

Epicentre Technologies (www.epicentre.com)
Two kits are offered, the MasterPure Genomic DNA Purification Kit and MasterAmp Buccal Swab DNA Extraction kit. The first kit can be used for isolation of DNA from bacteria, yeast cells, cultured cells and plant and animal tissues. The kit employs a proprietary reagent for cell lysis and a precipitation solution to recover DNA. The procedure can be carried out in microfuge tubes and completed in less than an hour. No organic solvents or DNA binding matrix is used. The second kit is used for extraction of human DNA from buccal swaps for PCR amplification. Yields range from 5 to 25 μg of DNA from a buccal sample.

Eppendorf 5' (www.eppendorf.com)
Five kits are offered. Rapid Micro-Genomic, Rapid Mini-Genomic and Large Scale Genomic isolation kits are designed to isolate DNA from mammalian cells, blood or tissue. They use standard, organic solvent procedures for removing proteins. Kits employ Phase Lock Gel tubes for fast separation of aqueous and organic phases by centrifugation. Two other kits for isolation of DNA from cells, bacteria and yeast separate DNA from proteins using DNA Capture Matrix in a spin filter format. Some shearing of DNA occurs due to binding DNA to silica particles.

Genosys Biotechnologies, Inc. (www.genosys.com)
DNA Isolator kit uses a proprietary solution of chaotropic agents and solvents for the isolation of high molecular weight DNA from samples of animal, plant, and bacterial origin. The reagents and procedure are essentially identical to the Single Step DNA Purification method of Chomczynski and Sacchi (1987). High yield of DNA can be isolated in less than 1 hour.

Invitrogen, Inc. (www.invitrogen.com)
Two kits are offered, Easy-DNA Kit and TurboGen Kit. Both kits can be used for purification of DNA from bacteria, cell suspension, yeast cells, plant and animal tissues. Easy-DNA kit uses proprietary lysis and extraction solutions together with chloroform extraction. TurboGen kit is based on application of CTAB (CDA) detergents and chloroform extraction to remove proteins.

Life Technologies BRL (www.lifetech.com)
This company offers a DNA purification kit DNAzol® developed by Chomczynski and Sacchi (1987).

LigoChem, Co. (www.ligochem.com)
This DNA purification kit uses a standard DNA purification protocol. However, phenol:CIA reagent is substituted by ProCipitate. ProCipitate is a polymer bridging network reagent which has proven to be an excellent substitute for phenol:CIA.

Macromolecular Resource Center, Inc. (www.mrcgene.com)
This company originally developed TRI-Reagent® and DNAzol® Reagent for genomic DNA isolation from animal cells and tissues. They also offer extra strength DNAzol®ES reagent for isolation of genomic DNA from plants and DNAzol®BD reagent, specifically designed for isolation of genomic DNA from whole blood.

Maxim Biotech, Inc. (www.maximbio.com)
BDtract and RDtract DNA isolation kits can be used for purification of high molecular weight DNA from whole blood, cultured cells, tissue, and bacteria. The protocol removes proteins by salt precipitation. The method precludes the need for phenol, chloroform or other organic extraction. Isolated DNA can be used directly for RFLP, restriction digests, cloning, Southern blotting and PCR amplification. Some proteins remain in the preparation, but they do not interfere with the above-mentioned DNA applications.

MBI Fermentas, Inc. (www.fermentas.com)
This company offers Genomic DNA Preparation Kit for isolation of genomic DNA from fresh or frozen human or animal blood, gram-positive and gram-negative bacteria, cell cultures, animal and plant tissues The procedure requires 20-25 minutes, involves extraction with chloroform alone and does not require either phenol extraction or proteinase digestion. DNA isolation is based on lysis of the cells with a subsequent selective DNA precipitation with detergent (CTAB). This DNA can be used for all molecular biology procedures.

Promega Co. (www.promega.com)
Promega offers several Wizard® Genomic DNA Purification Kits each designed for isolation of DNA from different sources (blood cells, culture cells and animal tissue, plant tissue, yeast, and gram positive and gram negative bacteria). These procedures use a four-step process: cell lysis, RNase digestion, and removal of proteins by salt precipitation and concentration of DNA by isopropanol precipitation. DNA purified by this system is suitable for a variety of applications, including amplification, digestion with restriction endonucleases and membrane hybridization. Some proteins remain in preparation but they do not interfere with these procedures.

Roche Molecular Biochemicals (biochem.roche.com)
Several kits are offered that are specially formulated for purification of DNA from various sources. Differential precipitation and filtration are used for removing proteins. Their kit for isolation of DNA from plants works exceptionally well with many different plant species.

Qiagen Co. (www.qiagen.com)
Company offers a series of genomic DNA isolation kits. A proprietary anion-exchange resin is used to separate proteins from DNA in spin-columns or gravity-flow columns. This allows the isolation of high yields of pure genomic DNA from a variety of sources in about 3 hours without using organic solvents. The isolated DNA is of sufficient purity for RFLP studies, Southern blot analysis and PCR.

Sigma Chemical Co. (www.sigma.com)
This company offers several kits for DNA purification. All kits use TRI-reagent(developed by Chomczynski and Sacchi (1987).

Stratagene, Inc. (www.stratagene.com)
This company offers several DNA extraction kits. These kits use a modification of a procedure based on separating contaminating protein from DNA by salt precipitation. The procedure involves digestion of cellular proteins and the subsequent removal of proteins by "salting out". A number of samples may be processed simultaneously. Isolated DNA contains some proteins but can be used directly for restriction digests, cloning, Southern blotting, PCR amplification, and other DNA analysis techniques.

US Biological, Inc. (www.usbio.net)
This company offers Genomic DNA Mini Prep Kit. The DNA can be extracted from isolated nuclei acids using a standard phenol:CIA method for deproteinization.

Worthington, Inc.(www.worthington-biochem.com)
This company offers DNA Isolation Kit formulated to isolate high molecular weight genomic DNA from blood, cell cultures or animal and plant tissues without the use of organic deproteinizing reagents such as phenol or chloroform. The method is based on the isolation of intact nuclei. Proteinase K is used to remove nuclear proteins.

Preparation of Genomic DNA from Animal Cells

STEFAN SURZYCKI

Introduction

This chapter describes three procedures for the isolation of animal DNA, one large scale procedure and two small scale procedures. All of these methods yield DNA suitable for restriction enzyme digestion, genomic library construction, Southern blot analysis and PCR. The large scale method will yield DNA in excess of 200 kb, whereas the small scale protocols will generate DNA of 50-150 kb.

The large scale procedure yields 200-300 µg of DNA from 50 to 100 mg of cells (c.a. 5×10^7 cells), grown in monolayer or suspension cultures, or from about 1-2 mg of solid tissue. The first procedure is a "hybrid" between the phenol and chloroform extraction methods, with the addition of Proteinase K digestion. The theoretical rationale for each procedure is described in detail in Chapter 1. Cells are washed in PBS and resuspended in cell lysis buffer containing Sarcosyl, EDTA and NaCl. Final concentrations of Sarcosyl, EDTA and NaCl are 1.2%, 0.1 M and 0.12 M, respectively. The majority of proteins are removed by Proteinase K treatment. The Proteinase K concentration used is very high (1000 µg/ml) to shorten the time of the treatment, and the volume is small (about 10 ml), to decrease the cost of treatment. RNA is removed with RNase A and RNase T1 treatment after lowering the salt concentration below 0.1 M, to optimize RNase activity (see Chapter 1 for a detailed explanation). Remaining proteins and RNases are removed by one phenol:CIA extraction. The DNA is concentrated and recovered by ethanol precipitation or dialysis.

The large scale method yields 50% to 80% high molecular weight DNA (150-200 Kb) that is relatively devoid of protein and RNA contamination. The procedure works with most animal cells and can be easily scaled up by a factor of ten. It can be also scaled down for isolating DNA from as little as 5 mg of cells in microfuge tubes. The final yield is about 30-40 µg of DNA, an amount sufficient for several Southern blots. A disadvantage of the procedure is the use of phenol and chloroform, which are health hazards and must be handled with caution. It is also costly to dispose of these solvents. The protocol involves one day's work and contains at least one stopping point, where the protocol can be interrupted for one or more days. This method is not suitable for isolation of DNA from multiple samples simultaneously, since this substantially increases the time necessary to perform the procedure.

The first small scale procedure quickly isolates DNA from a small amount of cells (about 1×10^6 to 1×10^7 cells) in a single step. This method uses xanthogenate salt (Jhingan, 1992) and can be accomplished in about 1 hour. Moreover, the actual time spent on the procedure is less than 15 minutes. Thus, this method is ideal for isolation of DNA from multiple samples for Southern blot analysis and PCR. The theoretical rationale for this procedure is described in Chapter 1.

The second small scale procedure uses recently introduced DNAzol® solution for single step DNA isolation (Chomczynski et al., 1997). The method can be finished in less than 30 minutes and results in isolation of 90% of DNA from cell or tissue material. The isolated DNA can contain partially degraded RNA that if necessary can be removed by RNase digestion. The reagent is commercially available from several sources. The DNA purified by this method from some sources was difficult to PCR without additional reprecipitation or purification.

Outline

A schematic outline of the procedure is shown in Figure 1.

Materials

Equipment
- Corex 25 ml centrifuge tubes Teflon-lined caps (Corex # 8446-25)
- 15 ml and 50 ml sterilized polypropylene conical centrifuge tubes (for example Corning # 25319-15; 25330-50)
- Glass hooks
- Glass hooks are made from Pasteur pipettes in the following way: First, place the end of pipette horizontally in to a Bunsen flame and seal it. Next, holding the pipette at a 45° angle, insert 0.5 cm of the tip into flame. The end of the pipette will slowly drop under gravity forming a hook.
- Proteinase K (Ambion # 2546)
- Ambion Inc. is a low cost source of Proteinase K. The enzyme is supplied in storage buffer containing 50% glycerol, at a concentration of

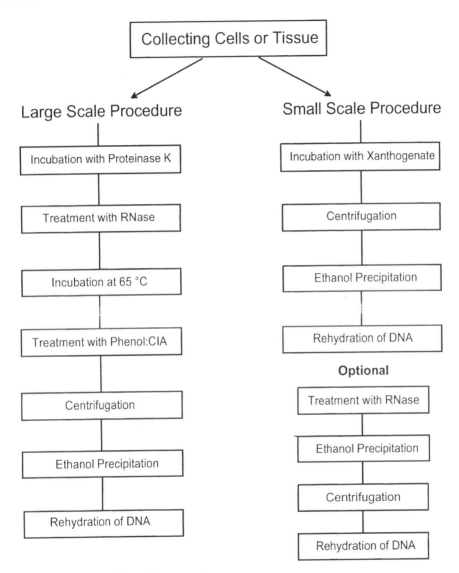

Fig. 1. Schematic outline of the procedure

20 mg/ml. Proteinase K solution should be stored at -20 °C. Proteinase K remains active for several years at - 20 °C.

- Ribonuclease A, DNase free (Ambion # 2272).
- Ambion is a source of inexpensive RNase supplied in storage buffer with 50% glycerol at a concentration of 1 mg/ml. The enzyme can be stored indefinitely in a - 20 °C freezer.

- Ribonuclease T1, DNase free (Ambion # 2280).
- Ambion is a source of inexpensive RNase T1. The enzyme is supplied in a storage buffer with 50% glycerol at concentration of 1000 units/µl and can be stored indefinitely in a - 20 °C freezer.
- Phenol, redistilled, water-saturated. (Ambion # 9712).
- Redistilled phenol is commercially available. There are a number of companies that supply high quality redistilled phenol. Water-saturated phenol is preferable to the crystalline form because it is easier and safer to prepare buffered phenol from it. Water-saturated phenol can be stored indefinitely in a tightly closed, dark bottle at -70 °C.
- Sarcosyl (N-lauroylsarcosine, sodium salt, (Sigma # L 5125)
- Cetyltrimethylammonium bromide (CTAB) (Sigma Co., # M7635)
- Cetyldimethylethyl ammonium bromide (DCA) (Sigma Co., # C0636)
- 8-hydroxyquinoline free base (Sigma Co., # H6878)

Note: Do not use hemisulfate salt of the 8-hydroxyquinoline.

- DNAzol® reagent (Molecular Research Center # DN 127)

Solutions   - **Phenol 8-HQ Solution**
Water-saturated, twice-distilled phenol is equilibrated with an equal volume of 0.1 M of sodium borate. Sodium borate should be used rather than the customary 0.1 M Tris solution because of its superior buffering capacity at pH 8.5, its low cost, its antioxidant properties and its ability to remove oxidation products during the equilibration procedure. Mix an equal volume of a water-saturated phenol with 0.1 M sodium borate in a separatory funnel. Shake until the solution turns milky. Wait for the phases to separate and then collect the bottom, phenol phase. Add 8-hydroxyquinoline to the phenol at a final concentration 0.1% (v/w). Phenol 8-HQ can be stored in a dark bottle at 4 °C for several weeks. For long term storage, store at -70 °C. At -70 °C the solution can be stored for several years.
Safety Note: Because of the relatively low vapor pressure of phenol, occupational systemic poisoning usually results from skin contact with phenol rather than from inhaling the vapors. Phenol is rapidly absorbed by and highly corrosive to the skin. It initially produces a white softened area, followed by severe burns. Because of the local anesthetic properties of the phenol, skin burns may not be felt until there has been serious damage. Gloves should be worn all the time when working with this chemical. Because some brands of gloves are soluble or permeable to phenol, they should be tested before use. If phenol is spilled on the skin, flush off immediately with a large amount of water and treat with a

70% aqueous solution of PEG 4000. (Hodge, 1994). **Do not use ethanol.**
Used phenol should be collected in a tightly closed glass receptacle and
stored in a chemical hood to await proper disposal.

- **Chloroform: Isoamyl Alcohol Solution (CIA)**
Mix 24 volumes of chloroform with 1 volume of isoamyl alcohol. Because chloroform is light sensitive and very volatile, the CIA solution
should be stored in a brown glass bottle, preferably in a fume hood.
Safety Note: Handle chloroform with care. Mixing chloroform with
other solvents can involve a serious hazard. Adding chloroform to a solution containing strong base or chlorinated hydrocarbons could result
in an explosion. Prepare CIA in a fume hood because isoamyl alcohol
vapors are poisonous. Used CIA can be collected in the same bottle as
phenol and discarded together.

- **3 M Sodium Acetate**
Sterilize the solution by autoclaving and store at 4° C

- **0.5 M EDTA Stock Solution, pH 8.5**
Weigh the appropriate amount of EDTA (disodium ethylendiamine-
tetraacetic acid, dihydrate) and add it to distilled water while stirring.
EDTA will not go into solution completely until the pH is greater than
7.0. Adjust the pH to 8.5 using concentrated NaOH, or by the addition of
a small amount of pellets, while mixing and monitoring the pH. Bring
the solution to the desired volume with water and sterilize by filtration.
The EDTA stock solution can be stored indefinitely at room temperature.

- **Sarcosyl Stock Solution, 20% (w/v)**
Dissolve 40 g of Sarcosyl in 100 ml of double distilled or deionized water.
Adjust to 200 ml with double distilled or deionized water. Sterilize by
filtration through a 0.22 µm filter. Store at room temperature.

- **PBS Solution, pH 7.4**
  - 137 mM NaCl
  - 2.7 mM KCl
  - 4.3 mM $Na_2HPO_4 \cdot 7H_2O$
  - 1.4 mM KH_2PO_4

Dissolve each salt in double distilled or deionized water. Be sure that one
salt has completely dissolved before adding the next. Adjust the pH to
7.4 with 1 N HCl. Autoclave for 20 minutes. Store at 4 °C. If desired PBS
can be made as 10 x concentrated stock solution.

- **Lysis Buffer**
  - 20 mM Tris-HCl, pH 8.5
  - 100 mM Na_2 EDTA

- 120 mM NaCl
- 1.2 % Sarcosyl

Add the appropriate amount of 1 M Tris HCl stock solution and 0.5 M EDTA stock to the water. Check the pH of the lysis buffer, and titrate it to pH 8.5 with concentrated NaOH, if necessary. Add NaCl and sterilize by autoclaving for 20 minutes. Store at 4 °C. Do not add Sarcosyl at this time. Sarcosyl will be added to cell suspension (step 3).

- **Dilution Buffer**
  - 20 mM Tris-HCl, pH 8.5
  - 100 mM Na$_2$ EDTA

  Prepare as described for lysis buffer.
- **TE Buffer**
  - 10 mM Tris HCl, pH 7.5 or 8.0
  - 1 mM Na$_2$EDTA, pH 8.5

  Sterilize by autoclaving and store at 4 °C.
- **Xanthogenate Buffer**
  - 12.5 mM potassium ethyl xanthogenate (Fluka Co. # 60045)
  - 100 mM Tris HCl, pH 7.5
  - 80 mM Na$_2$EDTA, pH 8.5
  - 700 mM NaCl

  Weigh the appropriate amount of NaCl and potassium ethyl xanthogenate and add to double distilled or deionized water. Add Tris-HCl and EDTA using stock solutions. Sterilize by filtration and store at room temperature.
- **CTAB/NaCl Solution**
  - 10% CTAB (or CDA)
  - 0.7 M NaCl

  Dissolve 4.1 g of NaCl in 80 ml of double distilled or deionized water and slowly add 10 g of CTAB. If necessary heat to 65 °C to fully dissolve CTAB. Adjust to a final volume of 100 ml. Store at room temperature.

Note: CDA can be substituted for CTAB. A 10% CDA solution is easier to use due to its low viscosity. See the description of other properties CDA as compared to CTAB in Chapter 1.

- **ATV Solution**
  - 0.14 M NaCl
  - 5 mM KCl
  - 7 mM NaHCO$_3$
  - 0.5 mM Na$_2$EDTA
  - 0.5% (w/v) Difco Trypsin (1:250)

- – 0.1% (w/v) Dextrose
- – Water to 1000 ml

Add all salts to distilled water and stir for 1 hour. Sterilize by filtration through 0.22 μm filter. Store at -20 °C.
- – **Loading Dye Solution (Stop Solution)**
  - – 15% Ficoll 400
  - – 5 M Urea
  - – 0.1 M Sodium EDTA, pH 8
  - – 0.01% Bromophenol blue
  - – 0.01% Xylene cyanol

Prepare at least 10 ml of the solution. Dissolve the appropriate amount of Ficoll powder in water by stirring at 40 - 50 °C. Add stock solution of EDTA, powdered urea and dyes. Aliquot about 100 μl into microfuge tubes and store in a -20 °C freezer.

Procedure

Collecting sample material

1. Grow cells as monolayers or in suspension as required. For monolayers, a confluent cell culture from one large bottle (75 cm² area) will yield approximately 50 to 80 mg of cells, that is, about 5×10^7 cells (see Table 2, Chapter 1). Suspension cultures, at saturation, will contain $1-2 \times 10^6$ cells/ml. For a large scale DNA preparation, use two to three 75 cm² TC bottles or 100 to 300 ml of suspension culture. Decant medium from the monolayer bottle. Add 10 ml of PBS to the bottle, tilt the bottle back and forth to rinse the cells, then discard the PBS. Washing with PBS removes serum left over from the spent medium. Serum components inhibit the activity of trypsin.

   Cell culture

2. Add 1.0 ml of ATV solution, swirl the trypsin around and incubate for about 3 minutes at room temperature. The amount of ATV needed is usually determined empirically. You may adopt the 3 minutes rule, that is, add just enough ATV to dislodge the cells from the surface of the bottle after 3 minutes incubation at room temperature. Too much ATV can damage the cells.

3. Once the cells begin to come off the surface, tap the flask against the palm of your hand or the bench top. Hold the flask up to the light to check that no patches of monolayer remain.

4. When the monolayer is completely dislodged, inactivate the ATV by adding 10 ml of medium containing serum. Aspirate the cells 3 to 4 times with a pipette to break up cell clumps. Transfer cells to a 50 ml polypropylene centrifuge tube and place the tube on ice. Collect cells from other bottles into the same tube.

5. Remove serum containing medium by centrifuging at 500 x g for 5 minutes at room temperature. Discard the supernatant.

6. Add 10 ml of cold PBS to the pellet and resuspend the cells by pipetting up and down. Collect the cells by centrifugation as described in step 5. Discard the supernatant and invert the tube on a paper towel to remove the remaining PBS. Proceed to step 1 of the large scale protocol.

Suspension cell culture

1. Transfer the appropriate amount of cells to a sterile 50 ml polypropylene tube and collect the cells by centrifuging for 5 minutes at 500 x g at room temperature. Discard the supernatant.

2. Add 10 ml of cold PBS to the pellet and resuspend the cells by pipetting up and down. Collect the cells by centrifugation as described in step 1. Discard the supernatant and invert the tube on a paper towel to remove the remaining PBS. Go to step 1 of the large scale protocol.

Solid tissues

1. Place a 250 ml plastic beaker into an ice bucket and pour liquid nitrogen into it. Harvest fresh tissue, using sterile scissors or a razor blade, in pieces less than 1 cm^3 size and immediately freeze in liquid nitrogen. Tissues harvested this way can be stored at -70 °C for several years.

2. Place a mortar into a second ice bucket and pour about 30 ml of liquid nitrogen into it. Cover the ice bucket with a lid and cool the mortar for 5 to 10 minutes.

3. Weigh about 500 to 1000 mg of frozen tissue as fast as possible, taking care that the tissue does not thaw in the process. Tare a plastic weighing boat. Pick frozen tissue fragments from the liquid nitrogen with forceps and place them in the weighing boat. Transfer the frozen tissue to the frozen mortar. Make sure that the mortar is always filled with liquid nitrogen.

4. Add a few chunks of dry ice to the mortar. The volume of dry ice used should be roughly 3 times the volume of tissue.

5. Grind the tissue with the mortar and pestle into a powder.

6. Make a "spatula" from a small plastic weighing boat by cutting the small boat in half along the diagonal. Use this spatula to transfer the powdered tissue to a 25 ml Corex tube. Allow the liquid nitrogen and dry ice to evaporate for a few minutes.

7. Add 10 ml of hot (65 °C) lysis buffer, and keeping the tube in a 65 °C water bath, mix the frozen tissue with the buffer as quickly as possible with a metal spatula. Continue from step 2 of the large scale protocol.

Human cheek cells

1. Pour 14 ml of PBS into a 15 ml conical centrifuge tube. Transfer the solution into a paper cup. Pour all the solution into your mouth, and swish vigorously for 20-40 seconds.

2. Expel the PBS back into the paper cup, then pour the solution from the paper cup back into a 50 ml conical centrifuge tube. Close the tube securely with the cap, and place the tube on ice.

3. Repeat step 1 one more time and collect cells from each mouth-wash in the same 50 ml tube.

4. Collect the cells by centrifugation at 500 x g for 10 minutes at room temperature.

5. Pour as much supernatant as possible back into the paper cup. Be careful not to disturb the cell pellet. Discard the supernatant from the tube into the sink. Invert the conical centrifuge tube with cells on a paper towel to remove the remaining PBS.

6. Continue the procedure from step 1 of large scale protocol.

Large scale procedure

Large scale protocol

1. Add 4 ml of lysis buffer, prewarmed to 65 °C, to the cells and mix gently.

2. Immediately add 0.25 ml of a stock solution of Proteinase K (20 mg/ml) and mix by inverting the tube several times.

3. Add 0.25 ml of 20% Sarcosyl and mix well by gently inverting the tube. Clearing of the "milky" cell solution and increased viscosity indicates lysis of the cells.

4. Incubate the mixture for 60 minutes in a 65 °C water bath.

5. Add 3 ml of dilution buffer and mix carefully. Hold the tube between your thumb and index finger and very quickly invert several times. Do not allow the lysate to run slowly down the side of the tube. This step lowers the NaCl concentration allowing single and double-stranded activities of both RNases (see Chapter 1 for full explanation).

6. Transfer the tube to a 37 °C water bath and add 50 µl of RNase A and 5 µl of RNase T1. Cap the tube, mix by quickly inverting several times. Incubate for 30 minutes at 37 °C.

7. Transfer the tube to a 65 °C water bath and continue the incubation for 40 minutes.

8. Add 13 ml of dilution buffer and mix by inverting as described in step 5. At this step the total volume of the preparation should be about 20 ml. Transfer 10 ml of the solution to two 25 ml Corex centrifuge tubes using a pipette with a wide-bore tip. When preparing DNA from less than 50 mg of cells add only 3 ml of dilution buffer to total volume of 10 ml and transfer the solution to one 25 ml Corex centrifuge tube.

Note: A wide-bore pipette can be prepared by cutting off the tip of a 10 ml plastic disposable pipette. Alternatively, insert pipette aid on the tip end of sterilized 10 ml glass pipette using the other end as the "intake" end.

9. Add 5 ml of phenol-8HQ solution and 5 ml of CIA solution to each tube. Close the tubes with Teflon-lined caps and mix using the procedure described in step 5. The solution should turn "milky" when properly mixed. This should not take more than 5-6 inversions.

10. Remove caps and place the tubes into a centrifuge. Centrifuge at 6000 rpm (4000 x g, SS-34 Sorvall rotor) for 5 minutes at 4 °C to separate the water and phenol phases. DNA will be in the top, aqueous layer.

11. Collect the aqueous phase using a wide-bore pipette. Avoid collecting the white **powdery looking** precipitate at the interphase. However, do collect as much of the viscose, bluish-white layer from the interphase as possible. This layer contains concentrated nucleic acids, not proteins. Record the volume of aqueous phase and transfer it into a 250 ml sterile Erlenmeyer flask. Place the wide-bore tip of the pipette at the bottom of the flask and slowly deliver the solution.

12. Add 0.1 volume (about 2 ml) of 3 M sodium acetate to the DNA solution and mix gently.

13. Add 2.5 volumes (55 - 60 ml) of 95% ethanol to the 250 ml Erlenmeyer flask containing DNA solution. Pour the appropriate volume of ethanol into a measuring cylinder and while keeping the Erlenmeyer flask at a 45° angle, carefully overlay the ethanol onto the viscose DNA solution. Since ethanol is less dense than the DNA solution, it will be the upper layer.

14. Gently mix ethanol with DNA solution by rolling the flask at a 45° angle in the palm of your hand. Continue until a small cotton-like clump of DNA precipitate forms at the interphase.

Note: If at this step, DNA does not form a clump and instead it forms several smaller fragments, do not try to collect them on a glass hook as described in step 15. Instead collect precipitated DNA by centrifugation as described in alternative procedure 1.

15. Insert the end of a glass hook into the precipitated DNA and swirl the hook in a circular motion to spool out the DNA. The DNA precipitate will adhere to the hook and in the beginning, will have a semi-transparent, gelatinous texture. Continue swirling the hook, allowing the end of hook to occasionally touch the bottom of the Erlenmeyer flask until all the ethanol is mixed with the aqueous phase. Note the appearance of the spooled DNA. As more DNA is collected on the hook, its initial gelatinous texture will become more compact and fibrous in appearance.

16. Transfer the hook with DNA into a 20 ml tube filled with 5-10 ml of 70% ethanol. Wash the DNA by gently swirling the glass hook. Pour out the 70% ethanol and repeat the wash two more times.

17. Drain ethanol from the precipitate by pressing it against the side of the tube. Transfer the hook to a sterile 5 ml plastic tube (Falcon 33) and add 500 μl of TE. Rehydrate the DNA slowly by moving the glass hook back and forth. More buffer may be necessary if the DNA does not dissolve in a few hours. To speed up the rehydratation of DNA, incubate the solution in a 65 °C water bath for 10 to 15 minutes moving the tube gently every 2 to 3 minutes.

18. Determine the concentration of DNA by measuring absorbance at 260 nm. Initially use a 1:10 dilution of the DNA. The absorbance reading should be in the range 0.1 to 1.5 OD_{260}. A solution of 50 μg/ml has an absorbance of 1 in a 1-cm path cuvette. Special care must be taken to dilute the viscose solution of DNA when micropipettors are used. Most micropipettors will not measure the volume of a very viscose solution correctly.

To prepare a 1:10 dilution of DNA, add 1 ml of PBS to a microfuge tube. Prepare a wide-bore, yellow tip by cutting off 5 to 6 millimeters from the end of the tip with a razor blade. Withdraw 100 μl of PBS from the tube and mark the level of the solution on the tip with a marking pen. Using the same tip withdraw DNA solution to the mark and add it to the tube containing PBS. Pipette up and down several times to remove the viscose DNA solution from the inside of the pipette tip.

Note: DNA concentration should never be measured in water or TE buffer. For discussion of determination of DNA concentration and its dependence on ionic strength and pH see Chapter 1.

19. Determine the purity of DNA by measuring the absorbance at 280 nm and 234 nm. Calculate 260/280 and 260/234 ratios. See Chapter 1 for a detailed discussion of the validity of this method and calculation of DNA concentration in a sample with a small amount of protein contamination.

20. Check the integrity of the DNA by running 2 to 5 μl of the DNA on a low percentage agarose gel (0.3% - 0.5%). Add 10 μl of TE buffer and 5 μl of stop solution to the microfuge tube, mix by pipetting up and down. Prepare a wide-bore tip as described in note to step 18. Withdraw 2 to 5 μl of TE buffer and mark the level of solution in the tip. Discard the TE solution and use the tip to add the DNA to the tube with TE and stop solution. Mix by stirring the solution with the yellow tip, **do not** mix by pipetting up and down. Load the DNA into the agarose gel using the same wide-bore tip. Run electrophoresis at 1 to 0.6 V/cm for 30 minutes. Use bacteriophage lambda DNA as the size marker. The DNA should appear as a high-molecular weight band, running with or slower than the DNA standard. Degraded DNA will appear as a smear running down the gel. Contaminating RNA will be visible as the diffuse band running slightly faster than bromophenol blue dye.

Alternative procedure 1

If DNA precipitate is not visible after the addition of ethanol in step 14, use following procedure to recover DNA.

1. Mix ethanol with the DNA solution thoroughly by gently swirling. Place the solution into two sterile 25 ml Corex centrifuge tubes. Collect the DNA by centrifugation at 8000 x g for 10 minutes at 4 °C. Use a swinging bucket rotor for this centrifugation (for example, 7000 rpm in a Sorvall HB-4 rotor or Beckman JA rotor).

2. Discard supernatant and fill the tubes with the remaining ethanol mixture. Centrifuge as described above and discard the supernatant. The DNA will appear at the bottom of the tubes as white precipitate.

3. Add 5 ml of cold 70% ethanol to the tube and wash the pellet by carefully rolling the tube at a 45° angle in the palm of your hand. Take care not to dislodge the DNA pellet from the bottom of the tube during this procedure. **Never vortex the tube**. Discard 70% ethanol and drain the tube well by inverting it over a paper towel for a few minutes. Repeat the wash one more time. Add 250 µl of TE to each tube and rehydrate the DNA pellet by gently swirling. Using a cut off P1000 tip, combine the resuspended DNA from both tubes. More buffer may be necessary if the DNA does not dissolve in a few hours. To speed up rehydration of the DNA, incubate the solution in a 65 °C water bath for 10 to 15 minutes. Follow the large scale protocol from step 18.

Small scale procedures

Two small scale protocols are described; the first uses xanthogenate salts and the second, commercially available reagent DNAzol®.

1. Use between 1×10^6 and 1×10^7 cells for a single preparation (see Table 2, Chapter 1 for details of cell weights and expected yields). If trypsinized or suspension cell culture is used, centrifuge the sample at room temperature for 5 minutes at 500 x g.

2. Discard the supernatant and centrifuge the tube for 5-6 seconds. Using a capillary micropipette tip, remove remaining medium from the tube.

3. Add 800 µl of xanthogenate buffer and gently resuspend the cells by pipetting up and down. Transfer the cells to a 1.5 ml microfuge tube and close the tube tightly.

4. Place the tube into a 65 °C water bath and incubate it for 40 minutes, mixing occasionally.

5. Centrifuge the tube for 5 minutes at room temperature. Using a cut-off 1000 µl pipette tip, transfer 400 µl to each of two microfuge tubes. Discard the pellet.

6. Add 1000 µl of 95% ethanol (2.5 volumes) to each tube. Mix well by inverting several times. At this point the preparation can be stored at -70 °C for an indefinite time.

Small scale
protocol 1

7. Place the tube in a centrifuge, orienting the attached end of the lid away from the center of rotation. Centrifuge the tube at maximum speed for 5 minutes at room temperature.

8. Remove tubes from the centrifuge. Pour off ethanol into an Erlenmeyer flask by holding the tube by the open lid and gently inverting the end, touch the tube edge to the rim of the flask and drain the ethanol. You do not need to remove all the ethanol from the tube. Return the tubes to the centrifuge in the same orientations as before.

Note: When pouring off ethanol, do not invert the tube more than once because this could disturb the pellet.

9. Wash the pellet with 700 μl of cold 70% ethanol. Holding the P1000 Pipetman vertically (see Figure 3) slowly deliver the ethanol to the side of the tube opposite the pellet. **Do not start the centrifuge,** in this step the centrifuge rotor is used as a "tube holder" that keeps the tube at an angle convenient for ethanol washing. Withdraw the tube from the centrifuge by holding the tube by the lid. Remove ethanol as before. Place the tube back into the centrifuge and wash with 70% ethanol one more time.

Note: This procedure makes it possible to quickly wash the pellet without centrifugation and vortexing. Vortexing and centrifuging the pellet is time consuming and frequently leads to substantial loss of material and shearing of DNA.

10. After the last ethanol wash, collect the ethanol remaining on the sides of the tube by centrifugation. Place the tubes back into the centrifuge with the side of the tube containing the pellet facing away from the center of rotation and centrifuge for 2-3 seconds. For this centrifugation you do not need to close the lids of the tubes. Remove collected ethanol from the bottom of the tube using a Pipetman (P200) equipped with capillary tip.

Fig. 2.

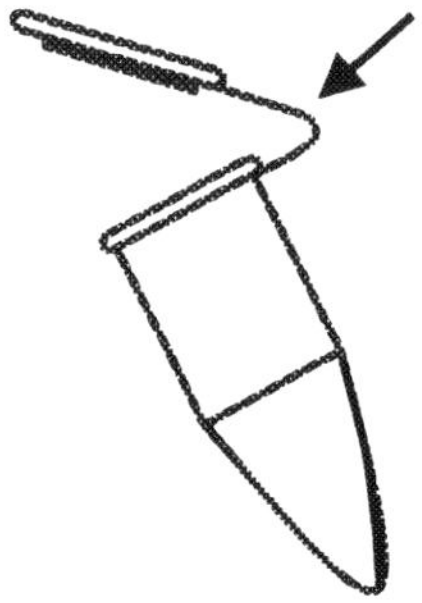

Fig. 3.

Note: Never dry the DNA pellet in a vacuum. This will make rehydration of the DNA very difficult if not impossible.

11. Add 150 µl of TE to each tube and resuspend the pelleted DNA. Use a yellow tip (P-200 Pipetman) with a cut off end for this procedure. Gently pipette the buffer up and down directing the stream of the buffer toward the pellet. If the pellet does not dissolve in several minutes, place the tube in a 60° - 65 °C water bath and incubate for 10 to 20 minutes mixing occasionally.

12. Combine the DNA from both tubes into one microfuge tube using a cut-off micropipette tip. Combined volume should be about 300 µl.

13. Add 10 µl of DNase-free RNase A and 1 µl of RNase T1 and mix by inverting the tube 2 to 3 times. Place the tube into a 37 °C water bath and incubate for 30 minutes.

14. Add 150 µl of 7.5 M ammonium acetate (half the volume).

15. Add 900 µl of 95% ethanol (2 x total volume). Mix well by inverting the tube 4 to 5 times and centrifuge for 10 minutes at room temperature. Pour off the supernatant using the technique described in step 8.

16. Wash the pellet two times with 700 µl of cold 70% using the technique described in step 9. Be sure to remove residual ethanol with a capillary tip after brief centrifugation. **Do not dry the DNA pellet in vacuum.**

17. Dissolve the DNA pellet in 25 µl of TE or water. Use a cut-off yellow tip for this procedure. Incubate the solution for 10 to 20 minutes in a 65 °C water bath to dissolve the pelleted DNA completely. Store the DNA sample at-70 °C.

Small scale protocol 2

1. Use between 1×10^6 and 1×10^7 cells for a single preparation (see Table 2, Chapter 1 for details of cell weights and expected yields). To collect cells grown in monolayer culture, add 1 ml of reagent per 10 cm^2 of culture plate area. Do not trypsinize the monolayer cells. If cells are grown in suspension culture use 1.5 ml of cells. Centrifuge the sample at room temperature for 5 minutes at 500 x g.

2. Add 900 µl of DNAzol® and gently resuspend the cells by inverting the tube several times. If the pellet is difficult to dissolve, pipette the solution up and down using a cut-off 1000 µl pipette tip.

3. Centrifuge the tube for 10 minutes at room temperature to remove cell debris. Using a cut-off 1000 µl pipette tip, transfer the supernatant into fresh tube. The centrifugation step is optional.

4. Add 470 µl of 95% ethanol (0.52 volume) to the tube and mix by inverting the tube 5 to 8 times.

5. Incubate at room temperature for 1 to 3 minutes.

6. Place the tube in a centrifuge, orienting the attached end of lid away from the center of rotation. Centrifuge the tube at maximum speed for 5 minutes at room temperature.

7. Remove tubes from the centrifuge. Pour off ethanol into an Erlenmeyer flask by holding the tube by the open lid and gently inverting the end, touch the tube edge to the rim of the flask and drain the ethanol. You do not need to remove all the ethanol from the tube. Return the tubes to the centrifuge in the same orientations as before.

Note: When pouring off ethanol do not invert the tube more than once because this could disturb the pellet.

8. Wash the pellet with 1000 µl of 95% ethanol. Holding the P1000 Pipetman vertically (see Figure 3) slowly deliver the ethanol to the side of the tube opposite the pellet. **Do not start the centrifuge**, in this step the centrifuge rotor is used as a "tube holder" that keeps the tube at an angle convenient for ethanol washing. Withdraw the tube from the centrifuge by holding the tube by the lid. Remove ethanol as before. Place the tube back into the centrifuge and wash with 95% ethanol one more time.

Note: This procedure makes it possible to quickly wash the pellet without centrifugation and vortexing. Vortexing and centrifuging the pellet is time consuming and frequently leads to substantial loss of material and shearing of DNA.

9. After the last ethanol wash, collect the ethanol remaining on the sides of the tube by centrifugation. Place the tubes back into the centrifuge with the side of the tube containing the pellet facing away from the center of rotation and centrifuge for 2-3 seconds For this centrifugation you do not need to close the lids of the tubes. Remove collected ethanol from the bottom of the tube using a Pipetman (P200) equipped with capillary tip.

Note: Never dry the DNA pellet in a vacuum. This will make rehydration of the DNA very difficult if not impossible.

10. Dissolve the DNA pellet in 25 µl of TE or water. Use a yellow tip with a cut-off end for this procedure. Incubate the solution for 10 to 20 minutes in a 65 °C water bath to dissolve the pelleted DNA completely. Store the DNA sample at -70 °C.

11. Determine the concentration of DNA by measuring absorbance at 260 nm. Initially use a 1:10 dilution of the DNA. The absorbance reading should be in the range 0.1 to 1.5 OD_{260}. A solution of 50 µg/ml has an absorbance of 1 in a 1-cm path cuvette. Special care must be taken to dilute the viscose solution of DNA when micropipettors are used. Most micropipettors will not measure the volume of a very viscose solution correctly.
To prepare a 1:10 dilution of DNA, add 1 ml of PBS to a microfuge tube. Prepare a wide-bore, yellow tip by cutting off 5 to 6 millimeters from the end of the tip with a razor blade. Withdraw 100 µl of PBS from the tube and mark the level of the solution on the tip with a marking pen. Using the same tip withdraw DNA solution to the mark and add it to the tube containing PBS. Pipette up and down several times to remove the viscose DNA solution from the inside of the pipette tip.

Note: DNA concentration should never be measured in water or TE buffer. For discussion of determination of DNA concentration and its dependence on ionic strength and pH see Chapter 1.

12. Determine the purity of DNA by measuring the absorbance at 280 nm and 234 nm. Calculate 260/280 and 260/234 ratios. See Chapter 1 for a detailed discussion of the validity of this method and calculation of DNA concentration in a sample with a small amount of protein contamination.

Results

- Large scale protocol.
  A typical spectrum of the DNA purified from human cheek cells is shown in Figure 4. The DNA concentration is 0.5 µg/µl. The total amount of DNA was 250 µg. Since one human cell contains approximately 660×10^{-14} g DNA and, in this experiment, 5×10^7 cells were used for DNA purification, the DNA yield from this isolation is 75.7% ($250/330 \times 100$). The DNA is high molecular weight and does not contain RNA. Figure 5 presents the results of *Bam H* I and *Bgl* II digestion of the isolated DNA.

- Small scale protocol.
  The amount of DNA isolated from human cheek cells (single mouth wash), using the small scale protocol 1, is between 4 and 10 µg. This DNA is of sufficient quality for restriction enzyme digestion and PCR analysis. The small amount of RNA remaining does not interfere. The 260/280 ratio is between 1.8 and 1.9, indicating very low contamination with proteins. Figure 6 presents the results of PCR using Alu TPA-25 specific primers and DNA isolated from five individuals.

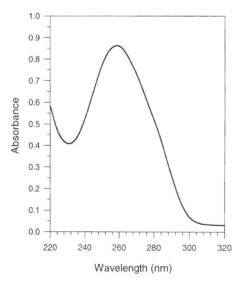

Fig. 4. Absorption spectrum of DNA purified from human cheek cells. DNA was diluted 20 x in PBS and scanned using a UV spectrophotometer. The 260/280 absorbance ratio was 1.7 and the 260/234 absorbance ratio was 1.8.

Fig. 5. Gel electrophoresis of human DNA isolated using the large scale procedure. Lane 1 has 1 µg, and lane 4 has 5 µg of undigested DNA, respectively. Lanes 2 and 3 show DNA (5 µg) digested with *Bam* HI and *Bgl* II restriction enzymes, respectively. Std lane is a molecular weight standard of 1 kb ladder from Life Technologies Gibco BRL.

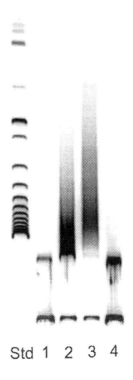

Troubleshooting

Problems encountered in large and small scale protocols can be classified into two groups: problems resulting in inadequate purity of the product and problems resulting in inadequate DNA yields.

Problems with DNA purity; large scale procedure

Impure DNA can manifest itself by one or all of the following features:

- low 260/280 or 260/234 ratio. Use recovery procedure 1 to remedy this problem.

- presence of a large amount of ethidium bromide positive material glowing in the sample well of the agarose gel. This is usually associated with a reduced amount, or total absence, of DNA on the gel in spite of a high 260 nm absorbance. Use recovery procedure 2.

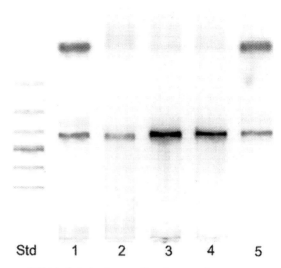

Fig. 6. PCR of human DNA isolated using small scale procedure. DNA was isolated from four individuals and PCR was done with Alu TPA-25 primers. Products were electrophoresed on a 2.5% MetaPhor gel for 1 hour. The expected fragment size with an Alu insertion is 400 bp. The expected fragment size without the insertion is 100 bp. Individuals in lanes 2, 3 and 4 are homozygous for the presence of Alu (400 bp fragment only). Individuals in lanes 1 and 5 are heterozygous and have both the 400 bp and 100 bp fragments. Standards are: 1000 bp, 700 bp, 525 bp, 500 bp 400 bp, 300 bp and 200 bp. The diffused bands ahead of the 100 bp band are primer dimers.

- partial digestion with restriction endonuclease enzyme. Use recovery procedure 3.

- large amount of RNA visible on the gel. Use recovery procedure 4.

- degradation of DNA visible as a smear running down the gel. Use recovery procedure 5.

Recovery procedure 1   A low 260/280 and/or 260/234 ratio indicates protein contamination and usually is caused by low activity of Proteinase K or inadequate mixing of phenol and aqueous phases. The low activity of Proteinase K is usually indicated by the presence of a large amount of "foamy" material at the interphase after the first phenol extraction (step 9). Whereas inadequately mixed phenol and aqueous phases do not have a uniformly "milky" appearance. Contaminating proteins can be removed using the following procedure:

1. Transfer the DNA sample to a 25 ml Corex tube with a wide-bore pipette tip, and dilute it to a final volume of 5 ml using lysis buffer.

2. Add 3 ml of phenol-8HQ and 3 ml of CIA, close the tube and mix by rapidly inverting the tube 5 or 6 times until solution turns "milky".

3. Follow purification steps 10 through 17 of the large scale protocol.

The presence of the ethidium bromide stainable material in the wells of the agarose gel indicates the presence of DNA-protein aggregates that are unable to enter the gel. This is frequently the result of too much EDTA in the lysis buffer. The DNA-protein complexes can be removed using the following recovery procedure.

Recovery procedure 2

1. Transfer the DNA sample to a 25 ml Corex tube (with wide-bore pipette tip) and dilute it to a final volume of 2 ml with lysis buffer prepared with half the amount of EDTA. Add NaCl to a final concentration of 0.3 M.

2. Add 0.15 ml of Proteinase K solution (20 mg/ml) and incubate the mixture for 60 minutes at 55 °C.

Note: Extracting the DNA with phenol:CIA will not remove DNA:protein complexes. Proteinase K treatment is required for successful removal of proteins from such complexes.

3. Add 3 ml of TE buffer and mix gently.

4. Add 3 ml of Phenol-8HQ and 3 ml of CIA, close the tube and mix by rapidly inverting the tube 5 to 6 times.

5. Follow purification steps 10 through 17 of the large scale protocol.

Difficulty with restriction endonuclease digestion of DNA is usually associated with the presence of polysaccharides, phenolic, or other related secondary metabolites in the preparation. If the amount of these impurities is small, they can be removed easily by precipitating DNA in the presence of 2 M ammonium acetate (Cruse and Amorese, 1987, Perbal, 1988). If this treatment does not overcome the problem, use the following procedure for the removal of contaminating macromolecules.

Recovery procedure 3

1. Transfer the DNA solution to a 25 ml Corex tube using a wide-bore pipette tip. Dilute the DNA preparation with TE to a final volume of 2 ml.

2. Adjust the NaCl concentration of the solution to 0.7 M by adding 5 M NaCl.

3. Add 0.1 volume of the CTAB/NaCl (CDA/NaCl) solution and mix by gently inverting the tube several times.

4. Incubate the tightly closed tube at 65 °C for 20 minutes.

5. Add an equal volume of CIA to the solution, mix thoroughly by rapidly inverting the tube 5 to 6 times.

6. Centrifuge the tube at 6000 rpm (4000 x g, SS-34 Sorvall rotor) for 10 minutes at 4 °C and transfer the top, aqueous phase into a new Corex tube with a wide-bore pipette. Repeat the CIA extraction two more times or until the white powdery interphase, containing contaminating macromolecules, is no longer visible.

7. Transfer the aqueous phase into a new, sterile 25 ml Corex tube using the wide-bore pipette. Record the sample volume.

8. Add 2.5 volumes (about 15 ml) of 95% ethanol and mix by inverting the tube two to three times.

9. Collect the precipitated DNA by centrifugation at 4 °C for 10 minutes at 10 000 rpm (16 000 x g). Use a swinging bucket rotor for this centrifugation. Discard the supernatant and remove remaining ethanol by inverting the tube over a paper towel.

10. Wash the DNA pellet with cold 70% ethanol. Add 5 ml cold 70% ethanol to the tube, being careful not to disturb the pellet. Rinse the tube sides and the pellet by gently swirling ethanol. Discard the ethanol and repeat the washing one more time. Remove all ethanol by inverting the tube over paper towel for 5 to 10 minutes.

11. Dissolve the pellet in 100 µl of TE. First, add first 50 µl of TE buffer using a yellow tip with a cut off end and resuspend the pellet by pipetting up and down. Transfer the DNA into a microfuge tube and repeat the procedure with second 50 µl of TE buffer. Store the DNA at -70 °C.

Recovery procedure 4

RNA contamination can occur when using material particularly rich in RNA. It can be remedied by repeating RNase treatment in step 6 of the main procedure.

1. Transfer the DNA solution to 25 ml Corex tube using a wide-bore pipette. Dilute the DNA preparation with TE to final volume of 2 ml.

2. Add 50 µl of RNase A and 5 µl of RNase T1. Cap the tube, mix gently and incubate at 37 °C for 30 minutes.

3. Add 8 ml TE buffer, 5 ml of phenol-8HQ and 5 ml of CIA, close the tube and very quickly invert the tube back and forth five times until the solution turns "milky".

4. Follow steps 10 through 17 of the large scale protocol.

Avoiding any manipulation that can shear DNA can usually eliminate the presence of small amounts of low molecular weight DNA in the preparation. DNA is mechanically sheared when pipettes with too narrow openings are used or by allowing the DNA solution to run down the side of the tube. Inverting tubes too slowly during organic extraction will also result in substantial shearing of DNA molecules. To remedy these problems the purification should be repeated and care taken to handle the DNA solution more carefully.

Recovery procedure 5

The presence of small molecular weight DNA in the preparation also can result from insufficient inhibition of DNase activity. This can be remedied by repeating the purification using lysis buffer with a higher concentration of EDTA. Try increasing EDTA concentration by increments of 50 mM in the lysis buffer.

Problems with DNA recovery; large scale procedures

In the large scale procedure, a low yield of DNA can result from:

- Inadequate lysis of the cells. This problem will be noticeable after the addition of Sarcosyl. Adequate lysis of the cells results in a drastic increase in viscosity of the solution.

- Too high concentration of the EDTA in the lysis buffer. In this case, a low yield of nucleic acid will be apparent during precipitation with ethanol (step 14). Lowering the concentration of the EDTA in the lysis buffer solves this problem. Try lowering EDTA concentration by 20 mM increments.

- Contamination of a stock solution with DNase. Preparation of fresh stock solutions is the best remedy.

Problems with DNA purity; small scale procedure

Poor quality of DNA isolated with xanthogenate generally manifests itself by the inhibition of restriction enzyme activity or inability to perform PCR. The problem is usually associated with residual amounts of xanthogenate salt in the DNA due to inadequate washing with 70% ethanol. Even a very small amount of this salt will inactivate proteins. Xanthogenate salt can be removed using the following procedure:

Recovery procedure

1. Add TE buffer to the DNA to bring the volume to 100 μl.

2. Add 50 μl of 7.5 M ammonium acetate (half the volume) and mix well by inverting the microtube several times.

3. Add 300 μl of 95% ethanol (2 x total volume) and mix well. Centrifuge for 15 minutes at room temperature.

4. Wash the pelleted DNA as described in steps 9 and 10 of the small scale protocol 1 and resuspend DNA in 20 μl of TE buffer using the procedure described in step 17.

Problems with DNA recovery; small scale procedures

If DNA yield is low when using small scale protocol 1, this can be remedied by one of the following:

- Freezing cell pellet in liquid nitrogen and grinding it to a fine powder using a glass homogenizer (Jhingan, 1992).

- Increasing concentration of the EDTA in xanthogenate buffer from 10 mM to 80 mM.

- Decreasing concentration of sodium xanthogenate to 6.25 mM in the xanthogenate buffer. DNA samples prepared by DNAzol® procedure frequently contain residual amounts of RNA. This can be removed by:
  - Centrifugation in step 3 of small scale protocol 2.
  - Removing RNA by RNase treatment as described in steps 11 to 15 of small scale protocol 1.

References

Chomczynski P, Mackey K, Drews R, Wilfinger W 1997 DNAzol®: A reagent for the rapid isolation of genomic DNA. BioTechniques 22:550-553.

Crouse J, Amorese D 1987 Ethanol precipitation: Ammonium acetate as an alternative to sodium acetate. BRL Focus 9 (2): 3-5.

Hodge, R 1994 In: Protocols for nucleic acids analysis by nonradioactive probes. Method in Molecular Biology Volume 28. Ed. Isaac PG. Humana Press. Totowa, New Jersey.

Jhingan AK 1992 A novel technology for DNA isolation. Methods in Molecular and Cellular Biology 3:1-22.

Perbal BV 1988 A Practical guide to molecular cloning. 2nd Edition. John Wiley & Sons.

Preparation of Genomic DNA from Plant Cells

STEFAN SURZYCKI

Introduction

The isolation of undegraded DNA from plants is notoriously difficult, presenting problems that are not, in general, encountered in purification of DNA from animal or bacterial cells. Plant cells have a very tough cell wall, the breakage of which requires the application of vigorous physical force that can shear large DNA molecules. Also, plant cells accumulate large amounts of secondary metabolites in their vacuoles that either co-purify with nucleic acids (e.g., polysaccharides) or interact with DNA (polyphenols, oxalic acid, etc.) rendering the final product unusable. Finally, plant cells contain large amounts of very active nucleic acid degrading enzymes, making the time between cell rupture and inactivation of these nucleases a critical factor.

This chapter describes DNA extraction methods that, to a large extent, overcome these difficulties and can be used for both frozen and fresh plant material. Two methods are described, both of which yield DNA suitable for restriction enzyme digestion, genomic library construction, Southern blot analysis and PCR. A large scale procedure uses between 0.6 to 1.0 g of plant material and yields 500 to 800 µ g of DNA with average size of 150 Kb. A small scale protocol yields 20 to 50 µg of DNA of comparable size from 100 to 500 mg of fresh plant tissue.

The advantage of the large scale method is that the DNA isolated is high molecular weight (150 Kb) and relatively devoid of protein, polysaccharides and RNA. The procedure gives very good results with most plant tissue and can be easily scaled up. Also the procedure can be scaled down, using as little as 10 to 15 mg of tissue. A disadvantage of the procedure is the use of CIA that is a health hazard and must be handled with caution. StrataClean Resin® (Stratagen) can be substituted for the CIA, but this increases the cost considerably. The procedure can be completed in one day, but the protocol has a convenient stopping point, permitting its interruption for one or more days.

The small scale procedure uses xanthogenate salts that simultaneously dissolves the cellulose wall and removes protein (Jhingan, 1992). This method is ideal for isolation of DNA from multiple plant samples and does not require mechanical breakage of the cell. The theoretical rationale for this procedure is described in Chapter 1. DNA isolated by both of these procedures can be stored indefinitely at -70 °C.

Outline

An outline of the procedure is shown in Figure 1.

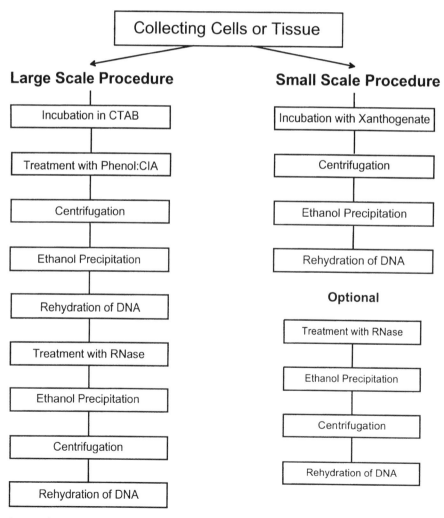

Fig. 1. Schematic outline of the procedures.

Materials

- Corex 25 ml centrifuge tubes with Teflon-lined caps (Corex # 8446-25)
- 15 ml and 50 ml sterilized polypropylene conical centrifuge tubes (for example Corning # 25319-15; 25330-50)
- Glass hooks
- Glass hooks are made from Pasteur pipettes as follows: Place the end of pipette horizontally in a Bunsen burner flame and seal it. Next, holding the pipette at a 45° angle, insert 0.5 cm of the tip into the flame. The end of the pipette will slowly drop under gravity forming a hook.
- Proteinase K (Ambion # 2546)
- Ambion is a source of inexpensive Proteinase K. The enzyme is supplied in storage buffer containing 50% glycerol at a concentration of 20 mg/ml (2000 U/µl). Proteinase K solution should be stored at -20 °C. Proteinase K remains active at -20 °C for several years.
- Ribonuclease A, DNase free (Ambion # 2272).
- Ambion is a source of inexpensive RNase supplied in storage buffer with 50% glycerol at concentration of 1 mg/ml. The enzyme can be stored indefinitely at - 20 °C.
- Ribonuclease T1, DNase free (Ambion # 2280).
- Ambion is a source of inexpensive RNase T1. The enzyme is supplied in a storage buffer with 50% glycerol at concentration of 1000 units/µl and can be stored indefinitely at - 20 °C.
- Phenol, twice-distilled, water-saturated. (Ambion # 9712).
  The author recommends using commercially available, twice distilled phenol. There are a number of companies that supply high quality, twice distilled phenol. It is better to buy water-saturated phenol rather than the phenol in crystalline form. This makes it easier and safer to prepare buffered phenol. Water-saturated phenol can be stored indefinitely in a tightly-closed, dark bottle at -70 °C.
- Cetyltrimethylammonium bromide (CTAB) (Sigma Co. # M 7635)
- Cetyldimethylethyl ammonium bromide (CDA) (Sigma Co. # C 0636)
- 8-hydroxyquinoline free base (Sigma Co. # H6878)

Note: Do not use the hemisulfate salt of the 8-hydroxyquinoline.

- Caylase M3 (Cayla Co. Toulouse France, or equivalent)
- Caylase M3 is a mixture of enzymes that hydrolyze polysaccharides and is commonly used for the preparation of plant protoplasts.

Solutions   –   **Phenol 8-HQ Solution**
Water-saturated, twice-distilled phenol is equilibrated with an equal vo-lume of 0.1 M of sodium borate. Sodium borate should be used rather than the customary 0.1 M Tris solution because of its superior buffering capacity at pH 8.5, its low cost, and its antioxidant properties. Sodium borate has the ability to remove oxidation products from phenol during the equilibration procedure. Mix water-saturated phenol with an equal volume of 0.1 M sodium borate in a separatory funnel. Shake until the solution turns milky. Wait for the phases to separate and then collect the bottom phenol phase. Add 8-hydroxyquinoline to the phenol to a final concentration of 0.1% w/v. Phenol 8-HQ can be stored in a dark bottle at 4 °C for several weeks. For long term storage, the solution can be stored at -70 °C for several years.
Safety Note: Because of the relatively low vapor pressure of phenol, oc-cupational systemic poisoning usually results from skin contact rather than from inhaling the vapors. Phenol is rapidly absorbed by and is highly corrosive to the skin. It initially produces a white softened area, followed by severe burns. Because of the local anesthetic properties of phenol, skin burns may not be felt until there has been serious da-mage. Gloves should be worn all the time when working with this che-mical. Because some brands of gloves are soluble or permeable to phe-nol, they should be tested before use. If phenol is spilled on the skin, flush off immediately with a large amount of water and treat with a 70% solution of PEG 300 in water (Hodge, 1994). **Do not use ethanol.** Used phenol should be collected into a tightly-closed, glass receptacle and stored in a chemical hood until it can be properly disposed of.

–   **Chloroform:Isoamyl Alcohol Solution (CIA)**
Mix 24 volumes of chloroform with 1 volume of isoamyl alcohol. Be-cause chloroform is light sensitive, the CIA solution should be stored in a brown glass bottle, preferably in a fume hood.
Safety Note: Handle chloroform with care. Mixing chloroform with other solvents can involve a serious hazard. Adding chloroform to a so-lution containing strong base or chlorinated hydrocarbons could result in an explosion. Prepare the CIA solution in a hood because isoamyl alcohol vapors are poisonous. Used CIA can be collected in the same bottle as the phenol solution and discarded together.

–   **3 M Sodium Acetate**
Sterilize the solution by autoclaving and store at 4 °C.

–   **0.5 M EDTA Stock Solution, pH 8.5**
Weigh the appropriate amount of EDTA (disodium ethylendiamine-tetraacetic acid, dihydrate) and add it to distilled water while stirring.

EDTA will not go into solution completely until the pH is greater than 7.0. Adjust the pH to 8.5 using a concentrated NaOH solution or by the addition of a small amount of pellets while mixing and monitoring the pH. Bring the solution to a final volume with water and sterilize by filtration. The EDTA stock solution can be stored indefinitely at room temperature.

- **Lysis Buffer**
- 100 mM Tris-HCl, pH 8.0
- 100 mM Na_2 EDTA
- 1.4 M NaCl
- 2% (w/v) CTAB or CDA
- 2% (v/v) β-mercaptoethanol

  Add CTAB (or CDA) and NaCl powders to 70 ml of double distilled or deionized water. Stir to dissolve and heat to 60 °C if necessary. Add the appropriate amounts of 1 M Tris-HCl and 0.5 M EDTA stock solutions. Check the pH of the lysis buffer and titrate it to pH 8.0 with concentrated NaOH, if necessary. Adjust the volume to 100 ml with double distilled or deionized water. Sterilize by filtration and store at room temperature. Add β-mercaptoethanol just before use.

Note: CDA is a very good substitution for CTAB, see a description of its properties in Chapter 1.

- **TE Buffer**
- 10 mM Tris-HCl, pH 7.5 or 8.0
- 1 mM Na_2EDTA, pH 8.5

  Sterilize by autoclaving and store at 4 °C.
- **Xanthogenate Buffer**
- 12.5 mM potassium ethyl xanthogenate (Fluka Co. # 60045)
- 100 mM Tris-HCl, pH 7.5
- 80 mM Na_2EDTA, pH 8.5
- 700 mM NaCl

  Weigh the appropriate amount of NaCl and potassium ethyl xanthogenate and add them to double distilled or deionized water. Add Tris-HCl and EDTA using stock solutions. Sterilize by filtration and store at room temperature. Solution can be stored for several months.
- **CTAB/NaCl Solution**
- 10% CTAB (or CDA)
- 0.7 M NaCl

  Dissolve 4.1 g of NaCl in 80 ml of double distilled or deionized water and slowly add 10 g of CTAB. If necessary, heat to 65 °C to fully dissolve CTAB. Adjust to a final volume of 100 ml. Store at room temperature.

CDA can be substituted for CTAB in this solution. CDA solutions are easier to prepare because a 10% CDA solution is less viscose (see comparison of CTAB and CDA reagents in Chapter 1).

- **Loading Dye Solution (Stop Solution)**
- 15% Ficoll 400
- 5 M Urea
- 0.1 M Sodium EDTA, pH 8
- 0.01% Bromophenol blue
- 0.01% Xylene cyanol

Prepare at least 10 ml of the solution. Dissolve an appropriate amount of Ficoll powder in water by stirring at 40-50 °C. Add stock solution of EDTA, powdered urea and dyes. Aliquot about 100 μl into microfuge tubes and store at -20 °C.

Procedure

Large scale procedure

This procedure is a combination of the CTAB procedure originally described by Murray and Thompson (1980) and modified by Saghai-Maroof et al., (1984). However, a Proteinase K digestion has been added to the procedure. Plant tissue is washed, harvested and quickly frozen in liquid nitrogen. Cells are partially opened by gentle grinding in a liquid nitrogen-dry ice mixture, and DNA is isolated in buffer containing 2% CTAB. Water-insoluble, protein:CTAB complexes are removed by one CIA extraction. The DNA is concentrated and recovered by ethanol precipitation or dialysis. If necessary RNA is removed with RNase A and RNase T1 treatment of the redisolved or dialyzed DNA. The theoretical rationale of this procedure is described in detail in Chapter 1

Large scale protocol

1. Harvest plant material, preferably young plant tissue, with a large number of meristematic cells. This is because the young plant cells contain fewer secondary metabolites. If necessary, rinse tissue with cold sterile water to remove adhering debris and blot dry with a paper towel.

2. Place a 250 ml plastic beaker into an ice bucket and pour the liquid nitrogen into it. Cut plant material into small pieces, less than 1 cm^3 in size, and immediately freeze it in liquid nitrogen. Tissues harvested this way can be stored at -70 °C for several years.

3. Place a mortar into a second ice bucket and pour about 30 ml of liquid nitrogen into it. Cover the bucket with a lid and cool the mortar for 5 to 10 minutes.

4. Weigh about 0.6 to 4.0 g of frozen plant material as fast as possible. Place a plastic weighing boat on the balance and tare it. Pick frozen tissue fragments from the liquid nitrogen with forceps and place them in the weighing boat. Transfer the frozen plant tissue to the frozen mortar. Make sure that the mortar is always filled with liquid nitrogen.

5. Add a few chunks of dry ice to the mortar. The volume of dry ice should be roughly 3 times the volume of tissue.

6. Grind the plant tissue into a white powder.

7. Make a "spatula" from a small plastic weighing boat by cutting the small boat in half along the diagonal. Use this spatula to transfer the powder to two 25 ml Corex tubes. Allow the liquid nitrogen and dry ice to evaporate for a few minutes.

8. Add 5 ml of hot (65 °C) lysis buffer to one tube, and keeping the tube in a 65 °C water bath, mix the powder with the buffer with a metal spatula as quickly as possible. Repeat this procedure with the second tube. Tightly cap the tubes.

9. Incubate the mixture for 60 minutes at 65 °C in a water bath with occasional gentle swirling.

10. Cool the tubes to room temperature and add 5 ml of phenol-8HQ solution and 5 ml of CIA solution to each tube. Cap the tubes with Teflon-lined caps and mix carefully. Hold the tube between your thumb and index finger and very quickly invert it several times. Do not allow the lysate to run down the side of the tube. The solution should turn "milky" when properly mixed. This should not take more than 5 to 6 quick inversions of the tube.

11. Centrifuge at 6000 rpm (4000 x g, SS-34 Sorvall rotor) for 5 minutes at 4 °C to separate the water and phenol phases. DNA will be in the top, aqueous layer.

12. Collect the aqueous phase using a wide-bore pipette. Avoid collecting the white **powdery looking** precipitate at the interphase. However, do collect as much of the viscose, bluish-white layer from the interphase as possible. This layer contains concentrated nucleic acids, not proteins. Record the volume of aqueous phase and transfer it to a sterile

250 ml Erlenmeyer flask. Place the wide-bore end of the pipette at the bottom of the flask and slowly deliver the solution.

Note: A wide-bore pipette can be prepared by cutting off the tip of a 10 ml plastic disposable pipette. Alternatively, insert the tip end of sterilized 10 ml glass pipette into your pipette aid and use the other end as the "intake" end.

13. Add 2.5 volumes (55 to 60 ml) of 95% ethanol to the DNA solution. Pour the appropriate volume of ethanol into a measuring cylinder, and keeping the Erlenmeyer flask at a 45° angle, carefully overlay the viscose DNA solution with the ethanol. Since ethanol is less dense than the DNA solution, it will be the upper layer.

14. Gently mix ethanol with DNA solution by rolling the flask at a 45° angle in the palm of your hand. Continue until a small clump of precipitated DNA forms at the interphase.

Note: If at this step, DNA does not form a clump and instead it forms several small fragments, do not try to spool them with a glass hook as described in the next step. Instead collect precipitated DNA by centrifugation as described in alternative procedure 1.

15. Insert the end of a glass hook into the precipitated DNA and swirl the hook in a circular motion to spool out the DNA. The DNA precipitate will adhere to the hook, and in the beginning, will have semi-transparent, gelatinous texture. Continue swirling until all of the ethanol is mixed with the aqueous phase, occasionally touching the bottom of the flask. Note the appearance of the spooled DNA. As more DNA is collected on the hook, its initial gelatinous texture will become more compact and fibrous in appearance.

16. Transfer the DNA to a 20 ml tube containing 5-10 ml of 70% ethanol. Wash the DNA by gently swirling the glass hook. Pour out the 70% ethanol and repeat the wash two more times.

17. Drain the ethanol from the DNA by pressing it against the side of the tube. Transfer the hook to a fresh 25 ml Corex tube and add 5 ml of TE buffer. Rehydrate the DNA slowly by moving the glass hook back and forth. More buffer may be necessary if the DNA does not dissolve within a day. To speed up the hydration, incubate the solution in a 65 °C water bath for 10 to 15 minutes moving the tube gently every 2 to 3 minutes.

Note: The procedure can be stopped at this step and the DNA stored for several days at 4 °C.

18. Add 50 µl of RNase A and 5 µl of RNase T1. Cap the tube, mix gently and incubate at 37 °C for 30 minutes.

19. Add 50 µl of Proteinase K and continue incubation for another hour at 37 °C.

20. Remove the tube from the water bath and add 5 ml of CIA. Mix by rapidly inverting the tube five times until the solution turns "milky".

21. Centrifuge the tube at 6000 rpm (4000 x g, Sorvall SS-34 rotor) for 10 minutes at 4 °C to separate phases.

22. Collect the upper aqueous phase using a wide-bore pipette and transfer it to a sterile 50 ml Erlenmeyer flask. Place the wide-bore end of the pipette at the bottom of the flask and slowly deliver the solution. Record the volume of aqueous phase.

23. Add 1/2 volume (2.5 ml) of 7.5 M ammonium acetate to the DNA solution. Mix gently.

24. Precipitate the DNA by the addition of 15 ml (2 x the total volume) of 95% ethanol. Layer the ethanol as described previously and spool the DNA onto a glass hook.

Note: If at this step the DNA does not precipitate in visible, cotton-like strings, collect the precipitated DNA by centrifugation as described in alternative procedure 1.

25. Transfer the hook with DNA into a tube containing 10 ml of 70% ethanol. Wash the DNA by gently moving the glass hook in the ethanol. Discard the ethanol and repeat the washing two more times. Drain the ethanol from DNA by pressing it against the side of the tube.

26. Transfer the hook to a 1.5 ml microfuge tube and add 200 µl of TE buffer. Rehydrate the DNA slowly by moving the glass hook back and forth. More buffer may be necessary if the DNA will not dissolve within one day. To speed up rehydratation of the DNA, incubate the solution in a 65 °C water bath for 10 to 15 minutes moving the tube gently every 2 to 3 minutes. Transfer DNA to a plastic storage tube using a P1000 Pipetman fitted with a cut off tip.

27. Determine the concentration of DNA by measuring absorbance at 260 nm. Initially use a 1:10 dilution of the DNA. The absorbance reading should be in the range of 0.1 to 1.5 OD_{260}. A solution of 50 µg/ml has an absorbance of 1 in a 1-cm path cuvette.

Special care must be taken to dilute the viscose DNA solution when micropipettors are used. Most micropipettors will not measure the volume of a very viscose solution correctly. To prepare a 1:10 dilution of the DNA, add 1 ml of PBS to a plastic microfuge tube. Prepare a wide-bore, yellow tip by cutting off 5 to 6 millimeters from the end with a razor blade. Withdraw 100 µl of PBS from the tube and mark the level of the solution on the tip with a marking pen. Discard the solution. Using the same tip withdraw DNA solution to the mark and add it back to the tube containing PBS. Pipette in and out several times to remove the viscose DNA solution from the inside of the pipette tip.

Note: DNA concentration should never be measured in water or TE buffer. For discussion of determination of DNA concentration and its dependence on ionic strength and pH see Chapter 1.

28. Determine the purity of DNA by measuring the absorbance at 280 nm and 234 nm. Calculate 260/280 and 260/234 ratios. See Chapter 1 for a detailed discussion of the validity of this method and calculation of DNA concentration in a sample with a small amount of protein contamination.

29. Check the integrity of the DNA by running 2.0 to 5 µl of the DNA on a low percentage agarose gel. (0.3% - 0.5%). Add 10 µl of TE buffer and 5 µl of loading dye (stop solution) to a microfuge tube, mix by pipetting up and down. Prepare a wide-bore tip as described in step 28. Withdraw 2 to 5 µl of TE buffer and mark its level on the yellow tip. Discard the solution and use the tip to add the DNA to the tube with TE and stop solution. Mix by swirling the solution with yellow tip, **do not** mix by pipetting up and down. Load the DNA into the agarose gel using the same wide-bore tip. Run electrophoresis at 0.6 to 1 V/cm for 30 minutes in TAE buffer. Use bacteriophage lambda DNA as the size marker. The DNA should appear as a high-molecular weight band running with or slower than the DNA standard. Degraded DNA will appear as a smear running down the gel. RNA contamination will be visible as a diffuse band running slightly faster than bromophenol blue dye.

Alternative procedure 1

1. Mix ethanol thoroughly with the DNA solution by gently swirling. Place the solution into a 25 ml Corex centrifuge tube. Collect the DNA by centrifugation at 10 000 rpm (16 000 x g) for 10 minutes at 4 °C. Use a swinging bucket rotor for this centrifugation.

2. Discard the supernatant. The DNA will appear at the bottom of the tubes as a white precipitate.

3. Add 5 ml of cold 70% ethanol to the tube and wash the pellet by carefully rolling the tube at a 45° angle in the palm of your hand. Be careful not to dislodge the DNA pellet from the bottom of the tube during this procedure. **Never vortex the tube.** Discard the 70% ethanol and drain the tube by inverting it over a paper towel for a few minutes. Repeat the wash one more time. Rehydrate the DNA pellet in 5 ml of TE buffer. From this point, follow steps 18 through 27 of the large scale procedure.

Small scale procedure

This procedure permits an efficient and fast isolation of DNA from a small amount of material in a single step. The method uses xanthogenate salt for DNA purification, essentially as described by Jhingan (1992).

1. Harvest 100 to 500 mg of plant material and transfer it to a 15 ml Corex tube. Add 2 to 5 ml of liquid nitrogen to the tube and freeze the sample until the tissue is very brittle. Crush the material to a coarse powder using a glass rod.

Small scale protocol

2. Add 2 ml of xanthogenate buffer and gently resuspend the powdered material.

3. Cap the tube and place the tube in a 65 °C water bath. Incubate the lysate for minimum 40 minutes, mixing occasionally.

4. Centrifuge the tube for 10 minutes at room temperature at 10 000 rpm (16 000 x g) in a swinging bucket rotor (e.g., Sorvall HP-4 rotor). Collect the supernatant in a new 15 ml Corex tube. Use a cut off 1000 µl pipette tip for this transfer. Discard the pellet.

5. Add 5 ml of 95% ethanol (2.5 volumes) to the tube. Mix well by inverting several times. At this point the preparation can be stored at -70 °C for an indefinite time.

6. Collect precipitated DNA by centrifugation for 5 minutes at 10 000 rpm (16 000 x g) in a swinging bucket rotor at room temperature. Discard the ethanol. Centrifuge the tube once more for 1 minute and remove remaining ethanol using capillary tip.

7. Add 300 µl of TE buffer and resuspend the pelleted DNA by pipetting it up and down. Use a yellow tip with a cut off end for this procedure. Transfer the DNA to a 1.5 ml microfuge tube.

8. Add 10 μl of DNase-free RNase A and 1 μl of RNase T1 and mix by inverting the tube 2-3 times. Place the tube into a 37 °C water bath and incubate for 30 minutes.

9. Add 150 μl of 7.5 M ammonium acetate (half the volume).

10. Add 900 μl of 95% ethanol (2 x the total volume). Mix well by inverting the tube 4-5 times.

11. Place the tube in a microfuge, orienting the attached end of the microfuge tube lid pointing away from the center of rotation. Centrifuge at maximum speed for 10 minutes at room temperature. (See Figure 2, Chapter 2)

12. Remove tubes from the centrifuge. Pour off ethanol into an Erlenmeyer flask by holding the tube by the open lid and gently flipping the end, touch the tube edge to the rim of the flask. Allow ethanol to drain. You do not need to remove all of the ethanol from the tube. Place the tubes back into the centrifuge in the same orientation as before.

Note: When pouring off ethanol do not invert the tube more than once because this could dislodge the pellet.

13. Wash the pellet with 700 μl of cold 70% ethanol in the following way. Holding the P1000 Pipetman vertically (See Figure 3, Chapter 2) slowly deliver the ethanol to the side of the tube opposite the pellet. **Do not start the centrifuge,** in this step the centrifuge rotor is used as a "tube holder" which keeps the tube at an angle convenient for ethanol washing. Withdraw the tube from the centrifuge by holding the tube by the lid. Remove ethanol as before. Place the tube back into the centrifuge and wash with 70% ethanol one more time.

Note: This procedure quickly washes the pellet without centrifugation and vortexing. Vortexing and centrifuging the pellet are time consuming and frequently lead to substantial loss of material and to DNA shearing.

14. Collect the ethanol remaining on the sides of the tube by centrifugation for 2 to 3 seconds. Place the tubes back into the centrifuge with the side of the tube containing the pellet facing away from the center of rotation and start the centrifuge. You do not need to close the lid of the tubes for this centrifugation. Remove ethanol from the bottom of the tube using a Pipetman (P200) equipped with a capillary tip.

Note: Never dry the DNA pellet in a vacuum. This will make the rehydration of the DNA very difficult if not impossible.

15. Resuspend the pelleted DNA in 50 µl of sterile distilled or deionized water. Use a cut off yellow tip for this procedure. Gently pipette the water up and down directing the stream of the water toward the pellet. If the pellet does not dissolve in several minutes, place the tube into a 65 °C water bath and incubate for 10-20 minutes mixing occasionally. Store the DNA sample at -70 °C.

Results

- Large scale protocol.
  A typical spectrum of DNA purified from 2.0 g of young corn leaves is shown in Figure 2. Figure 3 presents the results of agarose gel electrophoresis. The concentration of isolated DNA was 1.3 µg/µl. The total amount of DNA isolated was 150 µg. The DNA yield is 0.07 µg/mg tissue used. Since 1 gram of a young corn tissue has approximately 1.5×10^7 cells, and one corn cell contains 6.7×10^{-12} grams of DNA, the expected amount of the DNA to be isolated from 2 grams of tissue is approximately 200 µg. Thus, the efficiency of this purification is 75% (150/ 200 x 100). The DNA is high molecular weight and does not contain RNA. Figure 3 presents the results of *BamH* I and *Bgl* II digestion of this DNA. The patterns show a diffuse smear starting at 25 kb and lacks smears at the low molecular weight end of the gel. This attests to the high molecular weight of the DNA and its complete digestion by restriction endonucleases.

- Small scale protocol.
  DNA isolated from corn leaves, using the small scale protocol, is of sufficient quality for restriction enzyme and PCR analysis. The small amount of RNA present does not interfere with these analyses. The 260/280 ratio is between 1.8 and 1.9, indicating very little protein contamination. Figure 4 presents the results of gel electrophoresis of 2 µl of DNA isolated from two plants. Half of the sample was treated with RNase. Treated and not treated samples were electrophoresed on a minigel. Figure 5 presents the results of PCR using DNA isolated by the xanthogenate method and corn specific primers. The template used was the DNA sample shown in line 2 of Figure 4. Different amounts of 10 x diluted DNA template were used in amplification.

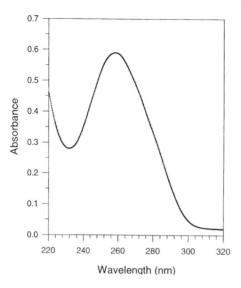

Fig. 2. Absorption spectrum of DNA purified from corn leaves. DNA was diluted 40 x in PBS and scanned using a UV spectrophotometer. The 260/280 absorbance ratio was 1.9 and the 260/234 absorbance ratio was 2.0.

Troubleshooting

Problems encountered in large and small scale protocols of DNA purification can be classified into two groups: problems resulting in inadequate DNA purity, and problems resulting in an inadequate yield of the DNA.

Problems of DNA purity in the large scale protocol

Insufficient purity of the DNA can manifest itself by one or all of the following features:

- Low 260/280 and/or 260/234 ratio. Use recovery procedure 1 to remedy this problem.

- Presence of a large amount of ethidium bromide stained material glowing in the sample well of the test gel. This is usually associated with a reduced amount or total absence of DNA on the gel in spite of a high 260 nm absorbance. Use recovery procedure 2 to remedy this problem.

- Partial digestion or total resistance to digestion with restriction endonuclease enzymes. Use recovery procedure 3 or recovery procedure 4 to correct this problem.

- Large amount of RNA visible after gel electrophoresis. Use recovery procedure 5 to correct this problem.

Fig. 3. Gel electrophoresis of corn DNA isolated using the large scale procedure. Lanes 1 has 1 µg and lane 2 has 5 µg of undigested DNA, respectively. Lanes 3 and 4 show DNA (5 µg) digested with *Bam* Hi and *Bgl* II restriction enzymes, respectively. Standards are a 1 kb DNA ladder from Life Technologies Gibco BRL.

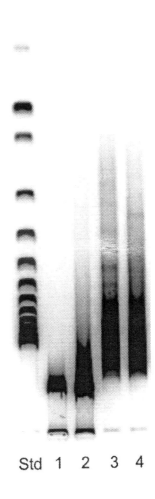

Std 1 2 3 4

- Degraded DNA visible as a smear running down the gel. Use recovery procedure 6 to remedy this problem.

A low 260/280 and 260/234 ratio indicates protein contamination and usually results from low activity of Proteinase K or inadequate mixing of phenol and aqueous phases. Low activity of Proteinase K is usually indicated by the presence of a large amount of "foamy" material at the interphase after the first phenol extraction. Whereas inadequately mixed phenol and aqueous phases do not have a uniformly "milky" appearance.

Recovery procedure 1

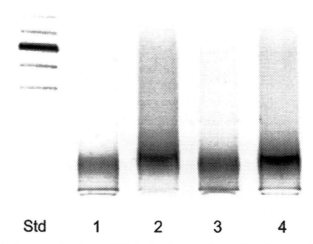

Fig. 4. Gel electrophoresis of corn DNA isolated using the small scale procedure. DNA was isolated from two plants and half of each sample was treated with RNase. Lanes 1 and 3 have 1 µl of RNase treated samples from plant 1 and plant 2, respectively. Lanes 2 and 4 show DNA samples from the same plants that were not treated with RNase. Standards are a 5 kb DNA ladder from Life Technologies Gibco BRL. The most intensely stained band is 10 kb DNA.

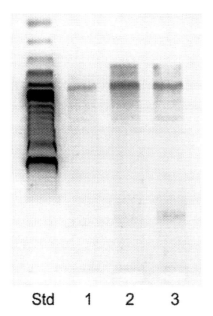

Fig. 5. PCR of corn DNA isolated using the small scale procedure. DNA template is DNA from the sample shown in lane 2 of Figure 3. Products were electrophoresed on a 2.5% Meta-Phor gel for 1 hour. The expected fragment size is 520 bp. Lane 1 is the product of a reaction run with 1 µl of 10 x diluted DNA. Lane 2 has 2 µl and lane 3 has 3 µl of the same DNA, respectively. Standards are a 100 bp DNA ladder from Life Technologies Gibco BRL. The most intense band is 600 bp. The high molecular weight band in lane 3 is template DNA.

Contaminating proteins can be removed using the following procedure:

1. Transfer the DNA sample to a 25 ml Corex tube (using a wide-bore pipette tip) and dilute it to a final volume of 5 ml using lysis buffer.

2. Add 3 ml of phenol-8HQ and 3 ml of CIA, close the tube and mix by quickly inverting the tube 5 to 6 times until the solution turns "milky". Centrifuge for 10 minutes at 10 000 rpm (12 000 x g; Sorvall SS-34 rotor) at 4 °C to separate water and phenol phases.

3. Collect the aqueous phase, precipitate DNA with 2.5 volumes of 95% ethanol, wash the precipitate in 70% ethanol and resuspend DNA in TE buffer as described previously.

The presence of ethidium bromide stained material in the wells of the agarose gel indicates the presence of DNA-protein aggregates that are unable to enter the gel. This is frequently the result of too much EDTA in the lysis buffer. The DNA-protein complexes can be removed using the following procedure:

Recovery procedure 2

1. Transfer the DNA sample to a 25 ml Corex tube (using wide-bore pipette tip) and dilute it to a final volume of 2 ml with lysis buffer containing half the amount of EDTA. Add NaCl to a final concentration of 0.3 M.

2. Add 150 µl of Proteinase K solution and incubate the mixture for 60 minutes at 60 °C.

Note: Organic extraction of DNA with phenol:CIA will not remove DNA: protein complexes. Proteinase K treatment is required for the successful removal of proteins from such complexes.

3. Add 3 ml of TE buffer and mix gently.

4. Add 3 ml of phenol 8-HQ solution and 3 ml of CIA. Close the tube and mix using the technique described previously. Centrifuge for 5 minutes at 10 000 rpm (12 000 x g) using a fixed angle rotor (e.g., Sorvall SS-34) at 4 °C to separate the aqueous and phenol phases.

5. Collect the aqueous phase, precipitate DNA with 2.5 volumes of 95% ethanol, wash the precipitate with 70% ethanol and resuspend the DNA in TE buffer as described previously.

Difficulties with restriction endonuclease digestion of DNA are usually associated with the presence of polysaccharides, phenolic compounds or other secondary metabolites in the preparation. This usually is apparent

Recovery procedure 3

during ethanol precipitation, resulting in a viscous, brown, "jelly-like" pellet. If the amount of these impurities is small, they can be removed easily by precipitation of the DNA in the presence of 2 M ammonium acetate (Cruse and Amorese, 1987, Perbal, 1988). If this treatment does not help, use the following procedure to remove contaminating macromolecules from the DNA sample.

1. Transfer the DNA solution to a 25 ml Corex tube with a wide-bore pipette tip. Dilute the DNA preparation with TE to a final volume of 2 ml.

2. Adjust the NaCl concentration of the DNA solution to 0.7 M with 5 M NaCl.

3. Add 0.1 volume of CTAB/NaCl (or CDA/NaCl) solution and gently mix together by inverting the tube several times.

4. Close the tube tightly and incubate it at 65 °C for 20 minutes.

5. Add an equal volume of CIA and mix thoroughly using the procedure described in step 2 of recovery procedure 1.

6. Centrifuge the tube at 6000 rpm (4000 x g) for 10 minutes at 4 °C and collect the top, aqueous phase into a new Corex tube with a wide-bore pipette. Repeat the CIA extraction procedure two more times or until the white powdery interphase, containing contaminating macromolecules, is removed.

7. Transfer the aqueous phase into a new sterile 25 ml Corex tube with a wide-bore pipette. Record the sample volume.

8. Add 2.5 volumes (about 15 ml) of 95% ethanol and mix by inverting the tube two to three times.

9. Collect the precipitated DNA by centrifugation at 4 °C for 10 minutes at 10 000 rpm (16 000 x g). Use a swinging bucket rotor for this centrifugation. Discard the supernatant and remove ethanol by inverting the tube over paper towel.

10. Wash the DNA pellet with cold 70% ethanol. Add 5 ml of cold 70% ethanol being careful not to disturb the pellet. Rinse the tube sides and the pellet by gently swirling the ethanol. Discard the ethanol and repeat the ethanol wash one more time. Remove all ethanol by inverting the tube over a paper towel for 5 to 10 minutes.

11. Dissolve the pellet in 100 µl of TE as follows: Add 50 µl of TE buffer using a yellow tip with a cut off end and resuspend the pellet by pipetting up and down. Transfer the DNA sample to a microfuge tube. Add another 50 µl of TE buffer to the tube that contained the pellet and wash the sides of the tube. Join both resuspended DNA samples and store at -70 °C.

Interfering polysaccharides can be removed from the DNA solution by enzymatic degradation of these molecules using glycoside hydrolases. These enzymes will degrade polysaccharides without affecting the DNA molecules (Rether et al., 1993). The enzyme commonly used is caylase M3, an enzyme used for the preparation of fungal and plant protoplasts.

Recovery procedure 4

1. Transfer the DNA solution to a 25 ml Corex tube using a wide-bore pipette tip. Dilute the DNA preparation to a final volume of 2 ml with 0.05 M potassium acetate buffer, pH 5.5.

2. Add caylase M3 to a final concentration of 200 mg/ml and incubate at 37 °C for at least 3 hours

3. Add 3 ml of 50 mM Tris buffer, pH 8.0 and mix gently by inverting the tube several times.

4. Add an equal volume of CIA to the solution, mix thoroughly using the procedure described in step 2 of recovery procedure 1.

5. Centrifuge the tube at 6000 rpm (4000 x g) for 10 minutes at 4 °C and transfer the top, aqueous phase with a wide-bore pipette to a clean Corex tube. Repeat the CIA extraction procedure one more time.

6. Transfer the aqueous phase into a sterile 25 ml Corex tube using a wide-bore pipette. Record sample volume.

7. Add 2.5 ml (half the total volume) of 7.5 M ammonium acetate and mix gently.

8. Add 2 volumes (about 15 ml) of 95% ethanol and mix by inverting the tube two to three times.

9. Collect the precipitated DNA by centrifugation at 4 °C for 10 minutes at 10 000 rpm (16 000 x g). Use a swinging bucket rotor for this centrifugation. Discard the supernatant and remove ethanol by inverting the tube over a paper towel.

10. Wash the DNA pellet with cold 70% ethanol. Add 5 ml of cold 70% ethanol to the tube, being careful not to disturb the pellet. Rinse the sides of the tube and the pellet by gently swirling the ethanol. Discard the ethanol and repeat the washing one more time. Remove all ethanol by inverting the tube over a paper towel for 5 to 10 minutes.

11. Dissolve the pellet in 100 μl of TE as follows: Add 50 μl of TE buffer using a yellow tip with a cut-off end and resuspend the pellet by pipetting up and down. Transfer the DNA sample to a microfuge tube. Repeat the procedure with an additional 50 μl of TE buffer. Join both resuspended samples and store the DNA at -70 °C.

Recovery procedure 5

RNA contamination can occur because of inadequate RNase treatment in step 6. It can be remedied by repeating RNase digestion as follows:

1. Transfer the DNA solution to a 25 ml Corex tube using a wide-bore pipette. Dilute the DNA preparation with TE to final volume of 2 ml.

2. Add 50 μl of RNase A and 5 μl of RNase T1. Cap the tube, mix gently and incubate at 37 °C for 30 minutes.

3. Add 8 ml TE buffer, 5 ml of phenol-8HQ and 5 ml of CIA, close the tube and very quickly invert the tube back and forth 5 times until the solution turns "milky".

4. Follow the large scale protocol from step 11.

Recovery procedure 6

Avoiding any manipulation that can shear DNA can usually eliminate the presence of small amounts of low molecular weight DNA in the preparation. DNA is mechanically sheared when pipettes with too narrow openings are used or by allowing the DNA solution to run down the side of the tube. Inverting tubes too slowly during organic extraction will also result in substantial shearing of DNA molecules. To remedy these problems, the purification should be repeated and care taken to handle the DNA solution more carefully.

The presence of small molecular weight DNA in the preparation also can result from insufficient inhibition of DNase activity. This can be remedied by repeating the purification using lysis buffer with a higher concentration of EDTA. Try increasing EDTA concentration by increments of 50 mM in the lysis buffer.

Problems of DNA yield in the large scale protocol

In the large scale protocol a low DNA yield can result from:

- Inadequate lysis of the cells. Increasing the time or intensity of grinding will help this problem.

- Too high concentration of the EDTA in the lysis buffer. In this case, a low yield of nucleic acid will be apparent during ethanol precipitation. Lowering the concentration of EDTA in the lysis buffer will remedy this problem. Try lowering EDTA concentration by 20 mM increments.

- Contamination of one of the stock solutions with DNase. Preparation of fresh stock solutions is the best remedy.

Problems of DNA purity in the small scale protocol

Poor quality DNA, isolated with xanthogenate, generally manifests itself by inhibition of restriction enzyme activity or the inability to perform PCR. The problem is usually associated with residual amounts of xanthogenate salt in the DNA preparation due to inadequate washing with 70% ethanol. Even a very small amount of this salt will inactivate proteins. The remaining xanthogenate salt can be removed using the following procedure:

1. Add TE buffer to the DNA to bring the volume to 100 μl.

2. Add 50 μl of 7.5 M ammonium acetate (half the volume) and mix well by inverting the microtube several times.

3. Add 300 μl of 95% ethanol (2 x the total volume) and mix well. Centrifuge for 15 minutes at room temperature.

4. Wash the pelleted DNA as described in steps 12 through 14 of the small scale protocol and resuspend DNA in 30 μl of TE buffer.

Recovery procedure 7

Problems of DNA yield in the small scale protocol

If DNA yield is low, it can be increased by one of the following:

- Increasing the concentration of EDTA in the xanthogenate buffer from 80 to 100 mM.
- Increasing the incubation time at 65 °C to 2 hours.
- Decreasing the concentration of sodium xanthogenate to 6.25 mM in the xanthogenate buffer.

References

Crouse J and Amorese D 1987 Ethanol precipitation: Ammonium acetate as an alternative to sodium acetate. BRL Focus 9 (2): 3-5.

Hodge R 1994 In: Protocols for nucleic acids analysis by nonradioactive probes. Method in Molecular Biology Volume 28. Ed. Isaac PG. Humana Press. Totowa, New Jersey.

Jhingan AK 1992 A novel technology for DNA isolation. Methods in Molecular and Cellular Biology 3:1-22.

Murray MG and Thompson WF 1980. Rapid isolation of high molecular weight DNA. Nucleic Acids Res. 8:43221-4325

Perbal BV 1988 A Practical guides to molecular cloning. 2nd edition. John Wiley & Sons Inc.

Rether B, Delmas G and Laouedj A 1993 Isolation of polysaccharide free DNA from plants. Plant Molecular Biology Reporter 11 (4) 333-337.

Saghai-Maroof MF, Soliman KM, Jorgensen RA and Allard RW 1984 Ribosomal DNA spacer-length polymorphism in barley: Mendelian inheritance, chromosomal location, and population dynamics. Proc Natl Acad Sci USA 81: 8014-8018.

Preparation of Genomic DNA from Bacteria

STEFAN SURZYCKI

Introduction

This chapter describes two procedures for the isolation of chromosomal DNA from *E. coli*. These procedures can be used for most gram-negative and gram-positive bacteria or modified to isolate DNA from organisms other than bacteria.

A large scale procedure yields 200-300 μg of DNA from 2-4 x 10^{10} cells of freshly grown bacteria or from 20 mg to 30 mg of frozen cells. The first procedure is a "hybrid" between the phenol and chloroform extraction methods. The theoretical rationale of each procedure is described in detail in Chapter 1. The cells are resuspended in a cell collection buffer (Buffer I) and treated with lysozyme. The pH of the buffer is adjusted to 8.5 to maintain the pH necessary for maximal lysozyme activity. The buffer also includes 0.9% glucose to prevent cell lysis during lysozyme treatment and 10 mM sodium EDTA to inhibit DNases. Cells are lysed by the addition of lysis buffer containing 0.1% SDS and 0.08 M EDTA. The majority of proteins are removed by Proteinase K treatment. The Proteinase K concentration used is high (600 μg/ml), to shorten time of the treatment and the volume of the cell lysis buffer used is small (5 ml), to lower the cost. Remaining proteins are removed by a single phenol:CIA extraction, DNA is recovered by ethanol precipitation and dissolved in low salt buffer to optimize RNA removal with RNase A and RNase T1. Finally, RNases are removed by phenol extraction and DNA is concentrated by ethanol precipitation.

This method gives high molecular weight DNA (150-200 Kb) that is relatively devoid of protein and RNA contamination with very high yields (80% to 90%). The procedure is fast and its cost is low. Moreover, the procedure gives very good results with most bacterial species and can be easily scaled up. A disadvantage of the procedure is the use of phenol and chloroform that are health hazards and must be handled with caution and properly disposed of. StrataClean Resin® can be used instead of phenol and chloroform but this considerably increases the cost.

The protocol can be completed in one day and contains at least one stopping point where the procedure can be interrupted for one or more days. The method is not suitable for isolation of DNA from multiple samples simultaneously since this substantially increases the time necessary to perform it.

For efficient isolation of pure DNA from a small amount of bacterial cells in a single step, a method using xanthogenate salt is presented. The procedure is a modification of the procedure described by Jhingan (Jhingan. 1992). This procedure can be accomplished in one hour, however the actual time spent on the procedure is less than 15 minutes. Thus, this method is ideal for isolation of DNA from multiple small samples or from a single bacterial colony. The theoretical rationale for this procedure is described in Chapter 1. DNA isolated by both of these procedures can be stored indefinitely at -70 °C (see storing DNA samples Chapter 1).

Outline

An outline of the procedure is shown in Figure 1.

Materials

– 25 ml Corex centrifuge tubes with Teflon-lined caps (Corex # 8446-25)
– 50 ml Oak Ridge polypropylene centrifuge tubes with caps. (For example Nalgene® # 21009) Only polypropylene plastic tubes should be used because they are resistant to phenol and chloroform.
– Glass hooks
  Glass hooks are made from Pasteur pipettes. Place the end of the pipette horizontally into flame and seal it. Holding the pipette at a 45° angle, insert 0.5 cm of the tip into the flame. The end of the pipette will slowly drop under gravity, forming a hook.
– 8-hydroxyquinoline free base (Sigma Co. # H6878)
– Proteinase K (Ambion # 2546)
  Ambion Inc. is a source of inexpensive Proteinase K. The enzyme is supplied in storage buffer containing 50% glycerol, at a concentration of 20 mg/ml. Proteinase K remains active for years when stored at -20°C.
– Ribonuclease A, DNase free (Ambion # 2272).
  Ambion is a source of inexpensive RNase supplied in storage buffer with 50% glycerol at concentration of 1 mg/ml. The enzyme can be stored indefinitely at - 20°C.

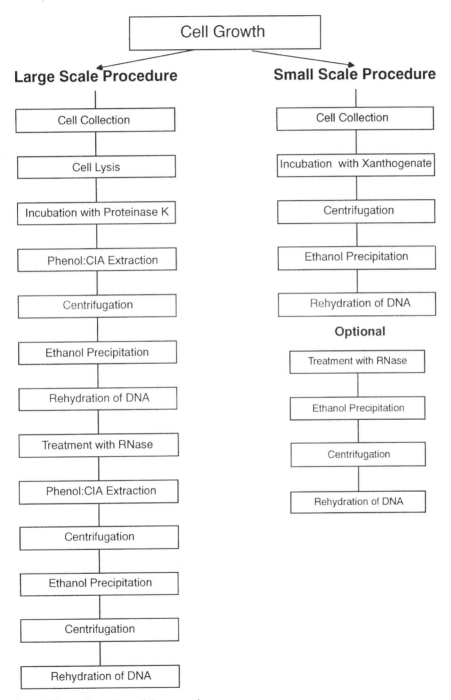

Fig. 1. Schematic outline of the procedure

- Ribonuclease T1, DNase-free (Ambion # 2280).
  Ambion is a source of inexpensive RNase T1. The enzyme is supplied in a storage buffer with 50% glycerol at a concentration of 1000 units/µl and can be stored indefinitely at - 20°C.
- Lysozyme (Sigma Co. # L 6876)
  Lysozyme powder should be stored at - 20°C.
- Phenol, redistilled, water-saturated. (Ambion # 9712).
  Using commercially available redistilled phenol is recommended. There are a number of companies that supply high-quality redistilled phenol. It is better to buy water-saturated phenol rather than the phenol in crystalline form. This makes it easier and safer to prepare buffered phenol. Water-saturated phenol can be stored indefinitely in a tightly closed, dark bottle at -70°C.

Media
- **Terrific Broth Medium (TB)**

Bacto-tryptone	12 gm
Yeast Extract	24 gm
Glycerol	4 ml
Water to	900 ml
Phosphate Stock Solution	100 ml

Mix first three ingredients in water and autoclave for 20 minutes, cool to room temperature and add the Phosphate Stock Solution to final volume of 1000 ml. Store at 4 °C.

- **Phosphate Stock Solution**
  - 0.72 M Potassium phosphate dibasic (K_2HPO_4)
  - 0.17 M Potassium phosphate monobasic (KH_2PO_4)
    Dissolve in distilled water and autoclave for 20 minutes. Store at 4 °C.

Solutions
- **Phenol 8-HQ Solution**
  Water-saturated phenol is equilibrated with an equal volume of 0.1 M of sodium borate. Sodium borate should be used rather than the customary 1 M Tris solution because of its superior buffering capacity at pH 8.5, its low cost, its antioxidant properties, and its ability to remove phenol oxidation products during the equilibration procedure. Mix an equal volume of a water-saturated phenol with 0.1 M sodium borate in a separatory funnel. Shake well until the solution turns milky. Wait until the phases separate and collect the bottom, phenol phase. Add the sodium salt of 8-hydroxyquinoline to the phenol to final concentration of 0.1%

(w/v). **Do not use the hemisulfate salt of 8-hydroxyquinoline.** Phenol 8-HQ can be stored in a dark bottle at 4 °C for several weeks. For long term storage keep it at -70 °C. Phenol can be stored at -70 °C for several years. **Safety Note:** Because of the relatively low vapor pressure of phenol, occupational systemic poisoning usually results from skin contact with phenol rather than from inhaling it. Phenol is rapidly absorbed by the skin and is highly corrosive. It initially produces a white softened area, followed by severe burns. Because of the local anesthetic properties of the phenol, skin burns may not be felt until there has been serious damage. Gloves should be worn when working with this chemical. Because some brands of gloves are permeable to phenol, they should be tested before use. If phenol is spilled on the skin, flush it off immediately with a large amount of water and treat with a 70% aqueous solution of PEG 4000 (Hodge, 1994). **Do not use ethanol.** Used phenol should be collected into tightly closed, glass receptacles and stored in a chemical hood to await disposal as hazardous waste.

- **Chloroform:Isoamyl Alcohol Solution (CIA)**
  Mix 24 volumes of chloroform with 1 volume of isoamyl alcohol. Because chloroform is light-sensitive and very volatile, the CIA solution should be stored in a brown glass bottle, preferably in a fume hood.
  Safety Note: Handle chloroform with care. Mixing chloroform with other solvents can involve a serious hazard. Adding chloroform to a solution containing strong base or chlorinated hydrocarbons could result in an explosion. Isoamyl alcohol vapors are poisonous. Handle the CIA solution in a fume hood. Used CIA can be collected in the same bottle as phenol and discarded together.

- **3 M Sodium Acetate**
  Sterilize the solution by autoclaving and store at 4 °C

- **0.5 M EDTA Stock Solution**
  Weigh an appropriate amount of EDTA (disodium ethylendiamine-tetraacetic acid, dihydrate) and stir into distilled water. EDTA will not go into solution completely until the pH is greater than 7.0. Adjust the pH to 8.5 using concentrated NaOH solution or by adding a small amount of pellets while monitoring the pH. Bring the solution to the final volume with water and sterilize by filtration. EDTA stock solution can be stored indefinitely at room temperature.

- **SDS Stock Solution, 10% (w/v)**
  Add 10 g of powder to 70 ml of distilled water and dissolve by stirring slowly. Add water to a final volume of 100 ml and sterilize by filtration through a 0.22 μm filter. Store at room temperature.

Safety Note: SDS powder is a nasal and lung irritant. Weigh the powder carefully and wear a face mask.

– **Lysozyme Solution**

Lysozyme	20 mg
Buffer I	1.0 ml

Place lysozyme powder into a 1.5 ml microfuge tube and add 1 ml of cold buffer I. Vortex to dissolve the lysozyme. Place the tube into an ice bucket. Lysozyme must be prepared fresh before use.

– **Loading Dye Solution (Stop Solution)**
  – 20% Ficoll 400
  – 5 M Urea
  – 0.1 M Sodium EDTA, pH 8
  – 0.01% Bromophenol blue
  – 0.01% Xylene cyanol

Prepare at least 10 ml of the solution. Dissolve the appropriate amount of Ficoll powder in water by stirring at 40 to 50 °C. Add the stock solution of EDTA, powdered urea and dyes. Aliquot about 100 µl into microfuge tubes and store at -20 °C.

Buffers – **Buffer I**
  – 25 mM Tris HCl, pH 8.5
  – 10 mM Na_2EDTA, pH 8.0
  – 0.9% glucose

Add the appropriate amounts of 1 M Tris HCl, 0.5 M EDTA and glucose powder to the water. Sterilize by filtration, do not autoclave. Store at 4 °C.

– **Lysis Buffer (5 x)**
  – 50 mM Tris-HCl, pH 8.5
  – 0.4 M Na_2 EDTA
  – 0.5% SDS

Add the appropriate amount of 1 M Tris HCl and 0.5 M EDTA to the water. Check the pH and titrate it to pH 8.5 with concentrated NaOH if necessary. Sterilize by autoclaving for 20 minutes. After sterilization, add the appropriate amount of SDS using a 10% stock solution. Store at 4 °C.

Note: Before use, the lysis buffer should be warmed to dissolve precipitated SDS.

- **TE Buffer**
  - 10 mM Tris HCl, pH 7.5 or 8.0
  - 1 mM Na_2EDTA

    Sterilize by autoclaving and store at 4°C.
- **Xanthogenate Buffer**
  - 12.5 mM Potassium ethyl xanthogenate (Fluka Co. # 60045)
  - 100 mM Tris HCl, pH 7.5
  - 80 mM Na_2EDTA, pH 8.5
  - 700 mM NaCl

    Weigh the appropriate amount of NaCl and potassium ethyl xanthogenate and add them to the water. Add Tris HCl and EDTA using stock solutions. Sterilize by filtration and store at room temperature. Solution can be stored for several months.
- **CTAB/NaCl Solution**
  - 10% CTAB (or CDA)
  - 0.7 M NaCl

    Dissolve 4.1 g of NaCl in 80 ml of double distilled or deionized water and slowly add 10 g of CTAB. If necessary, heat to 65 °C to fully dissolve CTAB. Adjust to a final volume of 100 ml. Store at room temperature. CDA can be substituted for CTAB in this solution. CDA solutions are easier to prepare because a 10% CDA solution is less viscose (see comparison of CTAB and CDA reagents in Chapter 1).

Procedure

Large scale procedure

1. Prepare a 50 ml Erlenmeyer flask containing 10 ml of TB (Terrific Broth) medium and inoculate it with 0.1 ml from an overnight culture of bacteria.

   Large scale protocol

Note: If a medium other than the TB is used, a larger volume of bacteria should be grown. Concentration of cells per ml at stationary phase in TB medium is approximately 3 to 4 times greater than the concentration obtained in less rich media.

Note: To provide a large surface-to-volume ratio for aeration, culture cells in an Erlenmeyer flask with volume of medium **one-quarter** the total flask volume. Shaker speed should be no less than 300 rpm.

2. Grow the cells at 37 °C overnight (at least 17 hours) on an orbital shaker.

3. Place the Erlenmeyer flask containing cells on ice for a few minutes. Pour the cold cells into chilled 50 ml polypropylene centrifuge tubes.

4. Collect the cells by centrifugation at 5000 rpm (3000 x g) for 5 minutes at 4°C.

5. Stop the centrifuge and discard the supernatant. Be very careful not to disturb the cell pellet.

6. Suspend the cell pellet in 3 ml of cold buffer I and place the cells on ice.

7. Add 1 ml of freshly prepared lysozyme solution to the cell suspension (final concentration 4 mg/ml). Incubate the cells on ice for 20 minutes.

Note: To lyse gram-negative bacteria 4 mg/ml lysozyme is sufficient but time and temperature can vary from 20 minutes at room temperature (e.g., for *E. coli*) to 4 h at 37°C depending on the organism. To monitor progresses of lysozyme treatment withdraw 100 µl sample periodically and mix with 100 µl of lysis buffer. If the cells are lysed, go to step 8.

Note: To lyse gram-positive bacteria, 8 mg/ml of lysozyme should be used, and incubation should be carried out at 37 °C. Check for lysis at 20 minute intervals as described in Note 1.

8. Add l ml of lysis buffer, pre-warmed to 55 °C, to the cells and mix gently. Immediately add 0.15 ml of a stock solution of Proteinase K (20 mg/ml).

9. Incubate the mixture for 60 minutes at 55 °C.

10. Add 35 ml of TE buffer to the suspension and mix gently. Hold the tube between thumb and index finger and very quickly invert back and forth. This prevents lysate from running down the side of the tube. Using a wide-bore pipette, transfer the solution to two 25 ml Corex centrifuge tubes, 10 ml to each tube.

Note: A wide-bore pipette can be prepared by cutting off a tip of a 10 ml plastic disposable pipette. Alternatively, insert a pipette aid onto the tip end of a sterilized 10 ml glass pipette and use the other end as the "intake" end.

Note: From this point, the procedure describes the purification of DNA from one half of the volume of initially grown bacteria, i.e., approximately 2-4 x 10^{10} cells. If twice as much DNA is desired, purification can be carried out on the other half of the cells.

11. Add 6 ml of phenol-8HQ solution and 6 ml of CIA solution to each tube. Close the tubes with Teflon-lined caps and mix by quickly inverting several times. The solution should turn "milky" when properly mixed. This should not take more than 5-6 inversions.

12. Remove caps from the tubes and place them into a centrifuge. Centrifuge at 6000 rpm (4000 x g) for 5 minutes at 4 °C to separate the aqueous and phenol phases. DNA will be in the top, aqueous layer.

13. Collect the aqueous phase using a wide-bore pipette. Avoid collecting the white **powdery** precipitate at the interphase. However, do collect as much of the viscose, bluish-white layer from the interphase as possible. This layer contains concentrated nucleic acids, not proteins. Record the volume of aqueous phase and transfer it to a 250 ml sterile Erlenmeyer flask. Place the wide-bore end of the pipette at the bottom of the flask and slowly release the solution.

14. Add 0.1 volume (i.e., about 2 ml) of 3 M sodium acetate to the solution and mix gently.

15. Add 2.5 volumes (60 ml) of 95% ethanol to the 250 ml Erlenmeyer flask. Pour the appropriate volume of ethanol into a measuring cylinder and, keeping the Erlenmeyer flask at a 45° angle, carefully overlay the viscose DNA solution with the ethanol. Since ethanol is less dense than the DNA solution, it will be the upper layer.

16. Gently mix ethanol with the DNA solution by rolling the flask at a 45° angle in the palm of your hand. Continue until a small clump of precipitated DNA forms at the interphase.

17. With a glass hook, swirl in circular motions to spool out the DNA. The precipitated DNA will adhere to the hook and will have a semi-transparent gelatinous appearance. Continue swirling until all DNA is collected and the ethanol is completely mixed with the water. Note the appearance of the spooled DNA. As more DNA is collected on the hook, its initial gelatinous texture will become more compact and fibrous in appearance.

18. Transfer the hook with spooled DNA into a 20 ml tube filled with 5 ml of 70% ethanol. Wash the DNA by gently swirling the glass hook with the DNA. Pour out the 70% ethanol and repeat the washing two more times.

19. Drain all ethanol from the precipitate by pressing it against the side of the tube. Transfer the hook with DNA to a clean 25 ml Corex tube.

20. Add 5 ml of TE buffer to the tube. Submerge the coated end of the hook into the buffer. Dissolve the DNA by gently moving the hook in TE buffer. Allow the DNA to rehydrate at room temperature for at least 20 to 60 minutes. Do not break or remove the glass hook before the entire DNA has dissolved.

Note: At this step, the procedure can be stopped and the DNA stored for several days at 4 °C. If storing is desired cover the tube with Parafilm **with the glass hook still inside.** Storing the DNA will help to dissolve it. High molecular weight DNA can take several days to completely rehydrate.

21. Remove the glass hook from the DNA solution and add 50 μl of RNase A and 5 μl of RNase T1. Cap the tube, mix gently and incubate the DNA solution at 37 °C for 30 minutes.

22. Remove the tube from the water bath and add 5 ml of TE to it. Mix gently and add 5 ml of phenol 8-HQ and 5 ml CIA. Mix by inverting until the solution turns "milky".

23. Centrifuge at 6000 rpm (4000 x g) for 10 minutes at 4 °C to separate phenol and aqueous phases.

24. Collect the top aqueous phase using a wide-bore pipette and transfer it to a new 25 ml Corex tube. Add 10 ml of CIA and mix by inverting the tube several times until the water phase is mixed with the CIA.

25. Centrifuge at 6000 rpm (4000 x g) for 5 minutes at 4 °C.

26. Collect the upper aqueous phase using a wide-bore pipette and transfer it to a sterile 50 ml Erlenmeyer flask. Place the wide-bore end of the pipette at the bottom of the flask and slowly release the solution. Record the volume of aqueous phase.

27. Add 1/10 volume (1 ml) of 3 M sodium acetate to the DNA solution. Mix gently.

28. Precipitate the DNA by the addition of 30 ml (2.5 volume) of 95% ethanol. Layer the ethanol as described in steps 15 and 16 and spool the DNA onto a hook as described in step 17.

29. Transfer the hook with DNA into a tube containing 10 ml of 70% ethanol and move gently to wash the DNA. Discard ethanol and repeat this wash two more times. Drain the ethanol completely by pressing the DNA against the side of the tube.

30. Transfer the hook with DNA to a sterile 5 ml plastic tube (Falcon 33) and add 500 μl of TE. Rehydrate the DNA slowly by moving the glass hook back and forth. More buffer may be necessary if DNA will not dissolve within one day. For faster rehydration, incubate the DNA solution in a 65 °C water bath for 10 to 15 minutes moving the tube gently every 2 to 3 minutes.

31. Determine the concentration of DNA by measuring absorbance at 260 nm. Initially use a 1:10 dilution of the DNA. For correct calculation of DNA concentration, the absorbance reading should be in the range of 0.1 to 1.5 OD_{260}. A solution of 50 μg/ml has an optical density of 1 in a 1-cm path cuvette.

    Special care must be taken to dilute the viscose DNA solution precisely when micropipettors are used. Most micropipettors will not correctly measure the volume of very viscous solutions. To prepare a 1:10 dilution of the DNA, add 1 ml of PBS to a microfuge tube. Prepare a wide-bore yellow tip by cutting off 5 to 6 millimeters from the end with a razor blade. Withdraw 100 μl of PBS from the tube and mark the level of the solution on the tip with a marking pen. Discard the PBS solution. Using the same tip, withdraw DNA solution to the mark and add it to the tube with PBS. Pipette up and down several times to wash the viscose DNA solution from inside the pipette tip.

Note: DNA concentration should never be measured in water or TE buffer. For discussion of determination of DNA concentration and its dependence on ionic strength and pH see Chapter 1.

32. Determine the purity of DNA by measuring the absorbance at 280 nm and 234 nm. Calculate 260/280 and 260/234 ratios. See Chapter 1 for a detailed discussion of the validity of this method and calculation of DNA concentration in a sample with a small amount of protein contamination.

33. Check the integrity of the DNA by running 2 to 5 μl of the DNA on a low percentage agarose gel. (0.4% - 0.5%). Add 10 μl of TE buffer and 5 μl of loading dye (stop solution) to a microfuge tube, mix by pipetting up and down. Prepare a wide-bore tip as described in step 30. Withdraw the appropriate amount (2 to 5 μl) of TE buffer and mark its level on the tip. Draw DNA solution to the mark on the cut-off tip and add it to the tube with TE and stop solution. Mix by pipetting up and down and load the DNA into the agarose gel using the same wide-bore tip. Run electrophoresis at 0.6 to 1.0 V/cm for 30 minutes. Use bacteriophage lambda DNA as a size standard. The DNA should appear as a

high-molecular weight band running with or slower than the DNA standard. Degraded DNA appears as a smear running down the gel. Contaminating RNA will be visible as a diffuse band running slightly faster than the bromophenol blue dye.

Small scale procedure

Small scale protocol

DNA can be isolated using overnight culture or cells from single colony.

1. Place from 0.2 to 1.0 ml of bacterial cells from an overnight culture into a 1.5 ml microfuge tube. Centrifuge at full speed for 5 minutes at room temperature.
   To isolate DNA from a bacterial colony grown on agar collect half of a colony of 2-3 millimeters in size on a toothpick and suspend it directly in xanthogenate buffer, as described in step 3 of this protocol. Follow the protocol from this point.

2. Discard the supernatant and re-centrifuge the empty tube for 5-6 seconds. Using a capillary tip, remove the remaining medium from the tube.

3. Add 300 µl of xanthogenate buffer to the tube and gently suspend the cells by pipetting up and down. Close the tube tightly.

4. Place the tube into a 65 °C water bath and incubate it for 40 minutes with occasional gentle shaking.

5. Centrifuge the tube at room temperature for 5 minutes. Transfer the supernatant to a new microfuge tube. Use a cut off yellow tip for this transfer. Discard the pellet.

6. Add 750 µl of 95% ethanol (2.5 volumes) to the tube. Mix well by inverting several times. At this point the preparation can be stored at -70 °C for an indefinite time.

7. Place the tube into the centrifuge, orienting attached end of the tube lid away from the center of rotation. Centrifuge at maximum speed for 5 minutes at room temperature.

8. Remove the tube from centrifuge. Pour off the ethanol by holding the tube by the open lid and gently lifting the end, touch the tube to the rim of an Erlenmeyer flask. Hold the tube to drain the ethanol. You do not need to remove all of the ethanol from the tube. Place the tubes back into the centrifuge in the same orientation as before (See Figure 2, Chapter 2).

Note: When pouring off ethanol do not invert the tube more than once because this could dislodge the pellet.

9. Wash the pellet with 700 µl of cold 70% ethanol. Holding a P1000 Pipetman vertically, slowly deliver ethanol to side of the tube opposite the pellet. Hold the Pipetman as shown in Figure 3, Chapter 2. **Do not start the centrifuge,** in this step the centrifuge rotor is used as a "tube holder" which keeps the tube at an angle convenient for ethanol washing. Withdraw the tube from the centrifuge by holding the tube by the lid. Pour off the ethanol by holding the tube by the open lid and gently lifting the end, touch the tube to the rim of an Erlenmeyer flask. Hold it to drain the ethanol. You do not need to remove all of the ethanol from the tube. Place the tube back into centrifuge and repeat the 70% ethanol wash one more time.

Note: This procedure makes it possible to quickly wash the pellet without centrifugation and vortexing. Vortexing and centrifuging the pellet is time consuming and frequently leads to substantial loss of material.

10. Place the tube back into a microfuge, making sure that the side containing the pellet faces away from the center of rotation. Without closing the tube lid start centrifuge until it reaches 500 rpm (1-2 seconds) to collect the remaining ethanol at the bottom of the tube. Remove all ethanol using a Pipetman (P200) equipped with a capillary tip.

Note: Never **dry the DNA pellet** in a vacuum. This will make dissolving the DNA very difficult, if not impossible.

11. Add 20-30 µl of TE or water to the tube to dissolve DNA. Use a cut-off yellow tip for this procedure. Gently pipette the buffer up and down, directing the stream of the buffer toward the pellet. If the pellet does not dissolve in several minutes, place the tube at 60-65 °C and incubate for 10 to 20 minutes with occasional mixing.

12. Determine the concentration of DNA by measuring absorbance at 260 nm. Initially use a 1:10 dilution of the DNA. For correct calculation of DNA concentration, the absorbance reading should be in the range of 0.1 to 1.5 OD_{260}. A solution of 50 µg/ml has an optical density of 1 in a 1-cm path cuvette.
    Special care must be taken to dilute the viscose DNA solution precisely when micropipettors are used. Most micropipettors will not correctly measure the volume of very viscous solutions. To prepare a 1:10 dilution of the DNA, add 1 ml of PBS to a microfuge tube. Prepare a wide-bore yellow tip by cutting off 5 to 6 millimeters from the end with a razor

blade. Withdraw 100 µl of PBS and mark the level of the solution on the tip with a marking pen. Discard the PBS solution. Using the same tip, withdraw DNA solution to the mark and add it to the tube with PBS. Pipette up and down several times to wash the viscose DNA solution from inside the pipette tip.

Note: DNA concentration should never be measured in water or TE buffer. For discussion of determination of DNA concentration and its dependence on ionic strength and pH see Chapter 1.

13. Determine the purity of DNA by measuring the absorbance at 280 nm and 234 nm. Calculate 260/280 and 260/234 ratios. See Chapter 1 for a detailed discussion of the validity of this method and calculation of DNA concentration in a sample with a small amount of protein contamination.

Removing RNA   If the DNA preparation is contaminated with a large quantity of RNA, remove it using the following procedure.

1. Dilute the DNA to 100 µl with TE and place the tube into a 37 °C water bath.

2. Add 10 µl of DNase-free RNase A and 1 µl of RNase T1 to the tube and incubate for 30 minutes at 37 °C.

3. Add 50 µl of 7.5 M ammonium acetate (half the volume). Mix well by inverting 3-4 times.

4. Add 300 µl of 95% ethanol (2 x the total volume) to the tube. Mix well by inverting 4-5 times and centrifuge for 10 minutes at room temperature. Pour off supernatant using the technique described in step 8.

5. Wash the pellet 2 times with 700 µl of cold 70% ethanol using the procedure described in steps 8 and 9.

6. Dissolve the DNA pellet in 20-30 µl of TE or water. Use a cut off yellow tip for this procedure. Incubate the solution for 10 - 20 minutes in a 60 °C to 65 °C water bath until the DNA is dissolved completely. Store the DNA at -70 °C.

Results

- Large scale protocol.
  A typical spectrum of DNA purified from 3×10^{10} cells of *E. coli* is shown in Figure 2. The concentration of the DNA was 0.5 µg/µl. The total amount of DNA isolated was 250 µg (500 µl x 0.5 µg/µl). Since one cell of *E. coli*, at stationary phase, contains 5.15×10^{-15}g of DNA (or approximately 100 µg/10^{10} cells) the DNA yield is 83% (250/300 x 100). The DNA is high molecular weight and does not contain RNA. Figure 3 also presents the results of *Bam* HI digestion of DNA isolated from four bacterial strains. The patterns show multiple diffused bands starting at 25 kb and little or no smearing at the low molecular end of the gel. This attests to the high molecular weight of the DNA and its complete digestion by restriction endonucleases.

- Small scale protocol.
  DNA isolated from *E. coli* cells, using the small scale protocol, is of sufficient quality for restriction enzyme and PCR analyses. The small

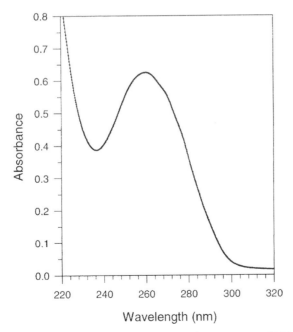

Fig. 2. Absorption spectrum of DNA purified from *E. coli* strain CSH66. DNA was diluted 20 x in PBS and scanned using a UV spectrophotometer. The 260/280 absorbance ratio was 1.95 and the 260/234 absorbance ratio was 1.87

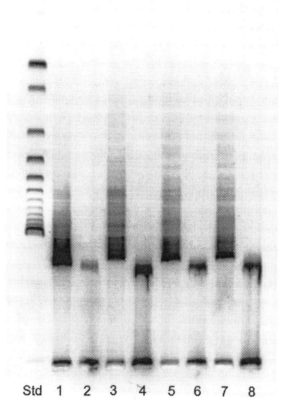

Fig. 3. Gel electrophoresis of *E. coli* DNA isolated from four different strains using the large scale procedure. Lanes 1, 4, 6 and 8 have 1 μg of undigested DNA. Lanes 1, 3, 5, and 7 have 5 μg of DNA digested with *Bam* HI restriction enzyme. Standards are molecular weight standards of 1 kb DNA ladder from Life Technologies Gibco BRL.

amounts of RNA present do not interfere with these analyses. The 260/ 280 ratio is between 1.8 and 1.9 indicating very low protein contamination. Figure 4 presents the results of gel electrophoresis of 2 μl of DNA purified from single colony of four bacterial strains.

Troubleshooting

Problems encountered in the application of both a large and small scale protocols can be classified into two groups: problems resulting in inadequate purity of the product or the problems resulting in inadequate yield of the DNA.

Fig. 4. Gel electrophoresis of *E. coli* DNA isolated from a single colony using the small scale procedure. DNA was isolated from four strains of bacteria. Each lane was loaded with 2 µl of DNA and electrophoresed for 40 minutes. Arrows indicate the small amount of RNA present in some samples. Standards are a 1 kb DNA ladder from Life Technologies Gibco BRL.

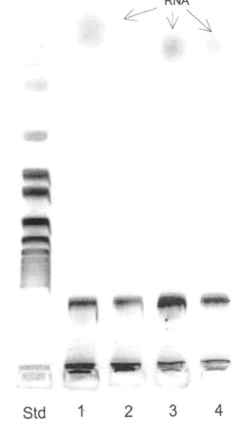

Problems with DNA purity; large scale procedure

Impurity of the DNA can manifest itself by one or all of the following features:

- Low 260/280 (260/234) ratio. Use recovery procedure 1 to remedy this problem.

- Presence of a large amount of ethidium bromide stained material glowing in the sample well of the agarose gel. This is usually associated with a reduced amount or total absence of DNA on the gel in spite of a high reading of 260 nm absorbance. Use recovery procedure 2 to remedy this problem.

- Partial digestion or total resistance to digestion with restriction endonuclease enzymes. Use recovery procedure 3 to remedy this problem.

- Large amount of RNA visible on the gel. Use recovery procedure 4 to remedy this problem.

- Degradation of DNA, visible as a smear running the length of the gel. Use recovery procedure 5 to remedy this problem.

Recovery procedure 1

A low 260/280 ratio usually indicates contamination with proteins and results from either inadequate Proteinase K digestion or insufficient mixing of phenol and water phases at step 11. Low Proteinase K activity is usually indicated by the presence of a large amount of "foamy" material at the interphase after the first phenol extraction (step 11). Inadequately mixed phenol and aqueous phases do not have a uniformly "milky" appearance. Protein contaminants can be removed using the following procedure:

1. With a wide-bore pipette, transfer the DNA sample to a 25 ml Corex tube and dilute it to a final volume of 10 ml with lysis buffer.

2. Add 5 ml of phenol-8HQ and 5 ml of CIA. Close the tube and invert to mix until the solution turns "milky".

3. Follow purification steps 12 through 32 of the large scale protocol.

Recovery procedure 2

Ethidium bromide-stained material in the wells of the agarose gel are DNA-protein aggregates that are unable to enter the gel. This is frequently caused by too much EDTA in the lysis buffer. The DNA-protein complexes can be removed using the following procedure:

1. Transfer the DNA sample to a 25 ml Corex tube (with a wide-bore pipette tip) and dilute it to a final volume of 2 ml with lysis buffer prepared with half the amount of EDTA. Add NaCl to a final concentration of 0.3 M.

2. Add 0.15 ml of Proteinase K solution and incubate the mixture for 60 minutes at 55 °C to 60 °C.

Note: Re-extracting the DNA with phenol:CIA mixture will not remove DNA:protein complexes. Proteinase K treatment is required for successful removal of proteins from such complexes.

3. Add 8 ml of TE buffer and mix gently.

4. Add 5 ml of phenol-8HQ and 5 ml of CIA, close the tube, and mix by inverting until the solution turns "milky".

5. Follow purification steps 12 through 32 of the large scale protocol.

Difficulties with restriction endonuclease digestion of DNA are usually associated with the presence of polysaccharides, phenolic, or other related secondary metabolites, in the preparation. If the amount of these impurities is small, they can be removed by precipitating the DNA in the presence of 2 M ammonium acetate (Cruse and Amorese, 1987, Perbal, 1988). If this treatment does not overcome the problem, the following procedure for the removal of contaminating macromolecules can be applied.

Recovery procedure 3

1. Transfer the DNA solution to a 25 ml Corex tube using wide-bore pipette tip. Dilute the DNA preparation with TE to final volume of 2 ml.

2. Adjust the NaCl concentration of the DNA solution to 0.7 M by adding the appropriate amount of 5 M NaCl.

3. Add 0.1 volume of CTAB-NaCl (CDA-NaCl) solution and mix both viscous liquids gently together by inverting the tube several times.

4. Incubate the tightly closed tube at 65 °C for 20 minutes.

5. Add an equal volume of CIA, mix thoroughly by quickly inverting 3 to 4 times.

6. Centrifuge at 6000 rpm (4000 x g) for 10 minutes at 4 °C and transfer the top, aqueous phase into a Corex tube with a wide-bore pipette. Repeat the CIA extraction two more times or until the white powdery interphase, containing contaminating macromolecules has been removed.

7. Transfer the aqueous phase into a sterile 50 ml Erlenmeyer flask and record the volume. Place the wide-bore end of the pipette at the bottom of the flask and slowly release the solution.

8. Add 2.5 volume of 95% ethanol using the technique described in step 17 of the large scale protocol.

9. Spool, wash and dissolve the DNA as described in steps 27 to 30 of the large scale protocol.

A large amount of RNA contamination, in most cases, results from insufficient time given for RNase treatment in step 21 of the large scale protocol. Repeating RNA digestion will overcome this problem.

Recovery procedure 4

1. Transfer the DNA solution to a 25 ml Corex tube using wide-bore pipette. Dilute the DNA preparation with TE to final volume of 2 ml.

2. Add 50 µl of RNase A and 5 µl of RNase T1. Cap the tube, mix gently and incubate at 37 °C for 30 minutes.

3. Add 8 ml TE buffer, 5 ml of phenol-8HQ and 5 ml of CIA. Close the tube and mix by holding the tube between thumb and index finger and very quickly invert back and forth until the solution turns "milky". This should not take more than 5-6 inversions.

4. Follow the large scale protocol from step 23.

Recovery procedure 5   The presence of small amounts of low molecular weight DNA in the preparation usually results from shearing the DNA during isolation. DNA is mechanically sheared when pipettes with narrow openings are used or by the DNA solution running down the side of the tube. Inverting the tube too slowly during organic extraction will also result in substantial shearing of DNA molecules. To remedy this problem the purification should be started once more and care taken to handle the DNA solution carefully.

The presence of a large amount of small molecular weight DNA in the preparation also can result from insufficient inhibition of DNase activity. This can be remedied by repeating the purification procedure with a higher concentration of EDTA in the lysis buffer.

Problems with DNA recovery; large scale procedure

Low yields, i.e., isolating less than 10% -20% of the expected amount of DNA can result from:

*   Inadequate lysis of the cells. This problem will be noticeable after the addition of the lysis buffer since normal lysis results in a drastic increase in viscosity of the solution. Most frequent causes of poor cell lysis are: Buffer I pH lower than 8.5, or too short time of lysozyme treatment. Lysozyme has a very low activity at a pH below 8.0. Also, some commercial preparations of the enzyme have very low specific activity. Moreover, some bacterial strains require much longer treatment time with lysozyme. This is particularly true for the bacterial strains selected for efficient electroporation.

*   Too high concentration of EDTA in the lysis buffer. In this case a low yield of nucleic acid will be apparent during the first precipitation with ethanol. Lowering the concentration of EDTA in the lysis buffer will remedy this problem.

- Contamination of one of the stock solutions with DNase. The low DNA yield, in this case, will be usually noticeable only after treatment with RNase's. Preparation of fresh stock solutions is the best remedy.

Problems with DNA purity; small scale procedure

Poor quality of DNA isolated using xanthogenate generally is manifested by the inhibition of restriction enzyme activity or by the inability to perform PCR with the DNA. The problem is usually associated with residual amounts of xanthogenate salt in the DNA due to inadequate washing with 70% ethanol at step 9 of the small scale protocol. Even a very small amount of this salt will inactivate proteins. The remaining xanthogenate salt can be removed using the following procedure:

1. Add TE buffer to the DNA to a combined volume of 100 µl.

2. Add 50 µl of 7.5 M ammonium acetate (half the volume) and mix well by inverting the tube several times.

3. Add 300 µl of 95% ethanol (2 x the total volume) and mix well. Centrifuge for 15 minutes at room temperature.

4. Wash and dissolve the pelleted DNA as described in steps 7 through 11 of the small scale protocol.

Recovery procedure

Problems with DNA recovery; small scale procedure

If DNA yield is low, the following can increase it:

- Freezing the cell pellet in liquid nitrogen and grinding it to fine powder using a glass homogenizer (Jhingan, 1992).

- Increasing the concentration of EDTA in xanthogenate buffer.

- Decreasing the concentration of sodium xanthogenate to 6.25 mM in the buffer.

References

Crouse J and Amorese D 1987 Ethanol precipitation: Ammonium acetate as an alternative to sodium acetate. BRL Focus 9 (2): 3-5.

Hodge R 1994 In: Protocols for nucleic acids analysis by nonradioactive probes. Method in Molecular Biology Volume 28. Ed. Isaac PG. Humana Press. Totowa, New Jersey.

Jhingan AK 1992 A novel technology for DNA isolation. Methods in Molecular and Cellular Biology 3:1-22.

Murray MG and Thompson WF 1980 Rapid isolation of high molecular weight DNA. Nucleic Acids Res. 8:43221-4325

Perbal BV 1988 A Practical guide to molecular cloning. 2nd edition. John Wiley & Sons Inc.

Isolation of Plasmid DNA

STEFAN SURZYCKI

Introduction

There are many published protocols for isolation and purification of plasmid DNA. The theoretical basis for these procedures is described in Chapter 1. In most instances, plasmid DNA, isolated by these procedures, is of sufficient quality and quantity for cloning and PCR analysis. However, the plasmid DNA obtained by many of these methods is a poor template for double-stranded dideoxy DNA sequencing, largely due to inconsistency of template purity and yield. To obtain plasmid DNA of good quality for DNA sequencing, many sequencing protocols recommend the use of plasmid templates prepared by cesium chloride centrifugation or column chromatography. Although these methods give satisfactory results, they are time consuming, expensive and not suitable for preparation of templates from multiple samples.

This chapter describes two protocols for large-scale plasmid isolation, and two protocols for minipreps of plasmid DNA. All four methods yield DNA that is an excellent template for dideoxy dsDNA automated sequencing, as well as for cloning and PCR. The procedures are a modification of the alkaline lysis procedure devised by Brinboim and Doly (Brinboim and Doly, 1979; Brinboim, 1983). The alkaline lysate is neutralized with ammonium acetate rather than potassium or sodium salts, and if necessary, the remaining proteins can be removed by precipitation with 1.87 M ammonium acetate. This allows purification of plasmid DNA without the use of toxic organic solvents, CsCl centrifugation or column chromatography. The procedures are quick and yield plasmid DNA of purity comparable to the CsCl prepared DNA. Long procedures are suitable for DNA isolation from both large and small plasmids.

▦ Outline

A schematic outline of the procedure is shown in Figure 1.

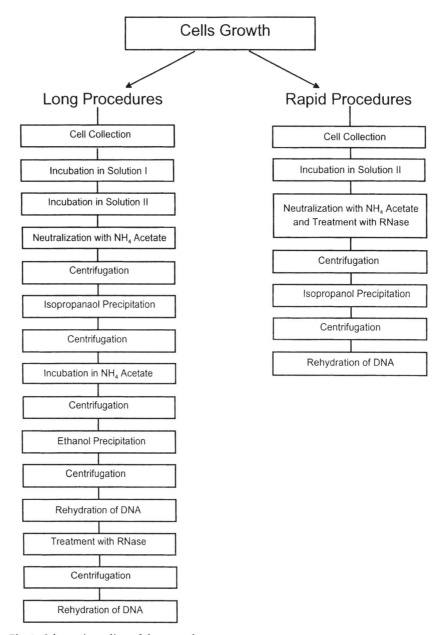

Fig. 1. Schematic outline of the procedures

Materials

- Corex 25 ml centrifuge with tubes Teflon-lined caps (Corex # 8446-25)
- 50 ml Oak Ridge polypropylene centrifuge tubes with caps. (For example Nalgene® # 21009).Only polypropylene plastic tubes should be used because they are resistant to phenol and chloroform.
- BioVortexer test tube mixer (Biospec Products # 1083-ST)
- Ribonuclease A, DNase free (Ambion # 2272). Ambion is a source of inexpensive RNase supplied in storage buffer with 50% glycerol at concentration of 1 mg/ml. The enzyme can be stored indefinitely at - 20 °C.
- Ribonuclease T1, DNase free (Ambion # 2280).
- Ambion is a source of inexpensive RNase T1. The enzyme is supplied in a storage buffer with 50% glycerol at concentration of 20,000 units/ml and can be stored indefinitely at -20 °C.
- Lysozyme (Sigma Co. # L6876) Lysozyme powder should be stored at -20 °C.

Media

- Terrific Broth Medium (TB)

Bacto-tryptone	12 gm
Yeast Extract	24 gm
Glycerol	4 ml
Water to	900 ml
Phosphate Stock Solution	100 ml

Mix first three ingredients in water and autoclave for 20 minutes, cool to room temperature and add the Phosphate Stock Solution to final volume of 1000 ml. Store at 4 °C.
- **H15 Growth Medium**
  - 5% Yeast Extract
  - 1% Casamino Acids
  - 1 % RNA (Torulla yeast Sigma # R 6625)
  - 10 μg/ml RNase A (Sigma # R 4875)
  - 150 mM MOPS
  - 272 mM Trisma base
  - 2% Glucose

Add the first four ingredients to 80 ml of deionized water and stir until the solution is no longer cloudy. This will take about 15 minutes. Add MOPS and Tris simultaneously and stir until dissolved. Adjust pH to 7.6

with concentrated HCl. Add 20 ml of 20% glucose stock solution and sterilize by filtration. Store at 4 °C. If medium is going to be used immediately, sterilization is not necessary.
- **Phosphate Stock Solution**
  - 0.72 M Potassium phosphate dibasic (KH_2PO_4)
  - 0.17 M Potassium phosphate monobasic (K_2HPO_4)

Dissolve in distilled water and autoclave for 20 minutes. Store at 4 °C.

Solutions
- **Solution I**
  - 25 mM Tris HCl, pH 8.5
  - 10 mM Na_2EDTA, pH 8.5
  - 0.9% glucose

Add the appropriate amount of 1 M Tris HCl and 0.5 M EDTA stock solutions to double distilled or deionized water. Check pH of the solution. The solution should be pH 8.5 because below this pH, lysozyme activity is drastically reduced. Sterilize by filtration and store at 4 °C. Do not autoclave this solution.
- **Solution II**
  - 200 mM NaOH
  - 1% SDS

Prepare fresh from stock solutions of 10 N NaOH and 10% SDS. Do not refrigerate.
- **SDS Stock Solution, 10% (w/v)**

Add 10 g of powder to 70 ml of distilled water and dissolve by stirring slowly. Add water to a final volume of 100 ml and sterilize by filtration through a 0.22 µm filter. Store at room temperature.

Safety Note: SDS powder is a nasal and lung irritant. Weigh the powder carefully and wear a face mask.
- **Lysozyme Solution**
  - 40.0 mg Lysozyme
  - 1.0 ml Solution I

Place lysozyme powder into a 1.5 ml microfuge tube and add 1 ml of cold buffer I. Vortex to dissolve the lysozyme. Place the tube into an ice bucket. Lysozyme must be prepared fresh before use.
- **70% Ethanol**

Add 25 ml of double distilled H_2O to 70 ml of 95% ethanol. Never use 100% ethanol because it contains an additive that can inhibit activity of most enzymes. Store at -20 °C.

- **7.5 M Ammonium Acetate Solution**
  Dissolve 57.8 g of ammonium acetate in 70 ml of double distilled or deionized water. Stir until salt is fully dissolved, but do not heat to facilitate dissolving. Fill up to 100 ml and sterilize by filtration. Store tightly closed at 4 °C. The solution can be stored for one to two months at 4 °C. Long term storage is possible at -70 °C.
- **1.87 M Ammonium Acetate Solution**
  Add 5 ml of 7.5 M ammonium acetate to 15 ml of cold double distilled or deionized water. Prepare this solution just before use and keep on ice.

- **TE Buffer**   Buffers
  - 10 mM Tris HCl, pH 7.5 or 8.0
  - 1 mM Na_2EDTA
  Sterilize by autoclaving and store at 4°C.
- **Loading Dye Solution (Stop Solution)**
  - 20% Ficoll 400
  - 5 M Urea
  - 0.1 M Sodium EDTA, pH 8
  - 0.01% Bromophenol blue
  - 0.01% Xylene cyanol
  Prepare at least 10 ml of the solution. Dissolve the appropriate amount of Ficoll powder in water by stirring at 40 to 50 °C. Add the stock solution of EDTA, powdered urea and dyes. Aliquot about 100 µl into microfuge tubes and store at -20 °C.

Procedure

Large scale procedure

1. Inoculate cells into 30 ml of Terrific Broth medium supplemented with   Large scale
   the appropriate antibiotic. Grow cells overnight at 37 °C with vigorous   protocol
   shaking. To obtain high yield of the large plasmids or low copy number
   plasmids, cell can be grown in H15 medium recently formulated for this
   purpose (Duttweiler and Gross, 1998).

Note: If medium other than TB or H15 is used, a larger volume of bacteria should be grown. When TB or H15 medium is used, the concentration of cells per milliliter at stationary phase, is approximately 5 to 15 times higher than concentrations reached in other media (Duttweiler and Gross, 1998).

Note: To provide a large surface-to-volume ratio for aeration, culture cells in an Erlenmeyer flask with the volume of medium no greater than **one-quarter** the total flask volume. Orbital shaker speed should be no less than 200 to 300 rpm.

2. Transfer the cells into a 50 ml Oak Ridge centrifuge tube. Collect the cells by centrifugation for 5 minutes at 5000 rpm (3000 x g) at 4 °C. Discard the supernatant. Centrifuge the cells for a few seconds and remove the remaining medium using a capillary tip with a Pipetman.

3. Gently resuspend the cell pellet in 3 ml of solution I. The pH of solution I must be at least 8.5.

4. Add 100 µl of a freshly-prepared lysozyme stock solution. Vortex for a few seconds and incubate for 15 to 20 minutes at room temperature.

5. Meanwhile prepare solution II and place it on ice until it becomes "cloudy". It will take only 1 to 2 minutes to achieve this. Do not allow a heavy precipitate to form.

6. Add 6 ml of solution II (two times the total volume), close the tube and mix by inverting 4-6 times. At this step, cells should lyse and the solution will become very viscose. Place the tube on ice and incubate for 10 minutes.

7. Add 4.5 ml of ice cold 7.5 M ammonium acetate (half the volume). Mix gently by inverting the tube 5-6 times, return the tube to the ice bucket and continue incubation for additional 10 minutes.

8. Centrifuge the tube in a swinging bucket rotor (e.g., Sorvall HB-4) for 10 minutes at 10 000 rpm (16 000 x g) at 4 °C.

9. Transfer the supernatant into a 25 ml Corex centrifuge tube. Be careful not to disturb the gelatinous pellet during this procedure. Discard the pellet.

10. Add 8.1 ml of isopropanol (0.6 x the total volume) to the supernatant. Mix by inverting the tube several times and incubate for 10 minutes at room temperature.

11. Centrifuge at 10 000 rpm (16 000 x g) for 10 minutes at **room temperature** using swinging bucket rotor.

12. Discard the alcohol. Centrifuge the tubes briefly (10-20 seconds) and remove the remaining alcohol with a capillary tip using a Pipetman.

13. Add 6 ml of 1.87 M ammonium acetate solution to the pellet (2 x of the original solution I volume). Resuspend the pellet by vortexing or using a Bio-Vortex mixer. The pellet is completely resuspended when the solution becomes uniformly "milky". Incubate the suspension on ice for 10 minutes.

Note: To obtain a better yield of plasmid or when preparing a large plasmid, incubate the solution overnight at 4 °C. This will increase the yield of plasmid by approximately 50%. However, the yield will not be increased substantially over that obtained in 10 minutes if incubation is carried out for only 1 to 3 hours.

14. Centrifuge for 5 minutes at 10 000 rpm (16 000 x g) at room temperature using swinging bucket rotor. Transfer the supernatant to a fresh 25 ml Corex tube.

15. Add 12 ml of 95% ethanol at room temperature (2 x of the total volume). Mix well by inverting the tube several times.

16. Centrifuge for 15 minutes at 10 000 rpm (16 000 x g) at room temperature. Discard the supernatant.

17. Slowly add 5 ml of cold 70% ethanol to the tube and keeping the tube at a 45° angle gently rinse the pellet and sides of the tube. Discard the 70% ethanol and repeat the wash one more time. Be careful not to disturb the pellet at the bottom of the tube during ethanol washing. To collect the ethanol remaining on the side of the tube, centrifuge the tube for 30 seconds. Remove ethanol from the bottom of the tube using a P200 micropipettor outfitted with a capillary tip.

18. Dissolve the pellet in 300 µl of TE buffer and transfer it to a 1.5 ml microfuge tube.

19. Add 5 µl of RNase A and 1 µl of RNase T1 to the tube. Mix well by pipetting up and down several times. Incubate the tube in a 37 °C water bath for 30 minutes.

20. Add 150 µl of 7.5 M ammonium acetate (half of the total volume). Mix well by inverting the tube 4 to 5 times.

21. Add 900 µl of 95% ethanol to the tube (2 x of the total volume). Mix well by inverting the tube 4 to 5 times.

22. Place the tube in the centrifuge and orient the attached end of the lid pointing away from the center of rotation. Centrifuge the tube at maximum speed for 10 minutes at room temperature (See Figure 2, Chapter 2).

23. Remove the tube from the centrifuge and open the lid. Gently inverting the tube, touch the lip of the tube to the rim of an Erlenmeyer flask. Hold the tube to drain the ethanol. You do not need to remove all of the ethanol from the tube. Place the tubes back into the centrifuge in the same orientation as before.

Note: When pouring off ethanol do not invert the tube more than once because this could dislodge the pellet.

24. Wash the pellet with 700 μl of cold 70% ethanol. Holding a P1000 Pipetman vertically, slowly deliver the ethanol to the side of the tube opposite the pellet. Hold the Pipetman as shown in Figure 3, Chapter 2. **Do not start the centrifuge**, in this step the centrifuge rotor is used as a "tube holder" which keeps the tubes at an angle, convenient for ethanol washing. Remove the tube from the centrifuge by holding the tube by the lid. Pour off the ethanol as described in step 23.

25. Place the tube back into the centrifuge and repeat the 70% ethanol wash one more time.

Note: This procedure makes it possible to quickly wash a large number of pellets without centrifugation and vortexing. Vortexing and centrifuging the pellet are time consuming and frequently lead to substantial loss of the material.

26. After the last wash, place the tube back into the microfuge, making sure that the tube side containing the pellet faces away from the center of rotation. Without closing the lid, start the centrifuge for 2-3 seconds collecting the remaining ethanol at the bottom of the tube. Remove all ethanol with a P200 micropipettor outfitted with a capillary tip.

Note: Never **dry the DNA pellet** in a vacuum. This is an unnecessary step that may make dissolving the DNA pellet very difficult if not impossible.

27. Add 20-30 μl of TE or water to the tube and dissolve the pelleted DNA. Gently pipette the buffer up and down directing the stream of the buffer towards the pellet. If the pellet does not dissolve in several minutes, place the tube into a 60 °C to 65 °C water bath and incubate for 10-20 minutes mixing occasionally.

28. Determine the concentration of DNA by measuring absorbance at 260 nm. Initially use a 1:100 dilution of the DNA in PBS. The absorbance reading should be in the range of 0.1 to 1.5 OD_{260}. A solution of 50 µg/ml has an absorbance of 1 in a 1-cm path cuvette. Determine the purity of the DNA by measuring the absorbance at 280 nm and 234 nm. Calculate the 260/280 and 260/234 ratios. See Chapter 1 for a detailed discussion about the validity of this method and calculation of DNA concentration in the presence of a small amount of protein contamination.

DNA concentration should never be measured in water or TE buffer. For discussion of determination of DNA concentration and its dependence on ionic strength and pH see Chapter 1.

Rapid large scale procedure

Plasmid prepared by this procedure can be used in sequencing, PCR, restriction enzyme digestion and yeast and mammalian cells transformation. The procedure is also suitable for isolating very large plasmids. The yield is about 180 µg of a large copy number plasmid from 20 ml of cells, that is about 5 to 10% less than the yield using a more involved large scale procedure. However, the procedure can be completed in about 30 minutes.

1. Collect 20 ml of cells following the first two steps of the long large scale procedure. **Do not resuspend the cell pellet in solution I.**

   Rapid large scale protocol

2. Add 8 ml of freshly prepared **solution II** to the cell pellet. Stir with a wooden applicator to break up the pellet completely. Incubate 10 minutes on ice.

3. Add 6 ml of 7.5 M ammonium acetate (0.75 x the volume) and 90 µl of RNase A (9 mg). Mix by inverting the tube gently 3 - 5 times. Incubate on ice for 10 minutes.

4. Centrifuge at 10 000 rpm (16 000 x g) in a swinging bucket rotor for 5 minutes at room temperature.

5. Transfer the supernatant to a clean 25 ml Corex tube with a 10 ml pipette. Avoid breaking the pellet or transferring flocculent precipitate that may be present at the top of the liquid.

6. Add 8 ml of isopropyl alcohol (0.6 x the volume), mix gently by inverting the tube several times and incubate at room temperature for 10 minutes.

7. Centrifuge the tube for 10 minutes in a swinging bucket rotor (e.g., Sorvall HB-4) at 10 000 (16 000 x g) at room temperature. Pour off the supernatant.

8. Slowly add 5 ml of cold 70% ethanol to the tube and keeping the tube at a 45° angle, gently rinse the pellet and sides of the tube. Discard the 70% ethanol and repeat the wash two more times. Be careful not to disturb the pellet at the bottom of the tube during ethanol washing. To collect ethanol remaining on the side of the tube, centrifuge the tube for 30 seconds. Remove ethanol from the bottom of the tube using capillary tip and a micropipettor.

9. Add 20-30 μl of TE or water to the tube and rehydrate the pelleted DNA. Gently pipette the buffer up and down directing the stream of the buffer towards the pellet. If the pellet does not dissolve in several minutes, place the tube into a 60 °C to 65 °C water bath and incubate for 10-20 minutes mixing occasionally.

10. Determine the concentration of DNA by measuring absorbance at 260 nm. Initially use a 1:100 dilution of the DNA in PBS. The absorbance reading should be in the range of 0.1 to 1.5 OD_{260}. A solution of 50 μg/ml has an absorbance of 1 in a 1-cm path cuvette. Determine the purity of the DNA by measuring the absorbance at 280 nm and 234 nm. Calculate the 260/280 and 260/234 ratios. See Chapter 1 for a detailed discussion about the validity of this method and calculation of DNA concentration in the presence of a small amount of protein contamination.

Note: DNA concentration should never be measured in water or TE buffer. For discussion of determination of DNA concentration and its dependence on ionic strength and pH see Chapter 1.

Small scale procedures

Two protocols are described. The first protocol, long small scale protocol, can be used for isolation of plasmid DNA when more material is required (4-6 μg) or from low copy number plasmids. The second protocol, rapid small scale protocol, requires only 400 μl of cells to isolate 2-3 μg of plasmid in approximately 30 minutes. The DNA obtained by both procedures is suitable for manual or automatic sequencing and for restriction enzyme analysis and PCR work.

1. Inoculate cells into 5 ml of Terrific Broth or H15 medium supplemented with the appropriate antibiotic. Use a 50 ml Erlenmeyer flask to grow the cells. Grow cells overnight at 37 °C with vigorous shaking.

2. Add 1.5 ml of cells to a microfuge tube and collect the cells by centrifugation for 5 minutes at room temperature. Discard the supernatant. Centrifuge the cells for a few seconds and remove the remaining medium using a capillary microtip and a P200 Pipetman.

3. Gently suspend the cell pellet in 150 µl of solution I by pipetting up and down. Do not vortex the cells.

4. Add 10 µl of a freshly prepared lysozyme stock solution of 40 mg/ml and mix gently by inverting the tube several times. Incubate the tube at room temperature for 10 minutes.

5. Meanwhile, prepare solution II and place it on ice until it becomes "cloudy". It will take only 1 to 2 minutes to achieve this. Do not allow a heavy precipitate to form.

6. Add 300 µl of solution II (2 x volume), close the tube and mix by inverting 4-6 times. Place the tube on ice and incubate for 5 minutes.

7. Add 225 µl of ice cold 7.5 M ammonium acetate (half the total volume). Invert the tube 5 times and incubate for 5 minutes on ice.

8. Centrifuge the tube at full speed for 10 minutes in a microfuge at room temperature.

9. Remove and discard the gelatinous pellet using a toothpick. Add 405 µl of isopropanol (0.6 x volume) to the supernatant. Mix well and incubate at room temperature for 10 minutes.

10. Centrifuge for 10 minutes at room temperature to collect the plasmid.

11. Remove the tube from the centrifuge and open the lid. Gently inverting the tube, touch the lip of the tube to the rim of an Erlenmeyer flask. Hold the tube to drain the isopropanol. You do not need to remove all of the isopropanol from the tube.

12. Place the tube back into centrifuge in the same orientation as before. Without closing the tube lid start centrifuge for 2-3 seconds and collect the remaining isopropanol in the bottom of the tube. The centrifuge should not reach maximum speed. A speed of 500 to 1000 rpm is sufficient to collect all isopropanol from the sides of the tube. Remove all isopropanol from the bottom of the tube using a P200 micropipette with a capillary tip.

Long small scale protocol

13. Add 300 μl of 2.0 M ammonium acetate (2 x the solution II volume) to the pellet. Suspend the pellet by vortexing or using a Bio-Vortex mixer. The solution will become uniformly "milky" in appearance when fully resuspended. Draw the suspension in and out of a yellow tip to suspend the granular precipitate. Incubate the suspension on ice for 10 minutes.

Note: To obtain a better yield of plasmid or when preparing low copy number plasmid, incubate the solution overnight at 4 °C. This will increase the yield of plasmid by approximately 50%. However, the yield will not be increased substantially over that obtained in 10 minutes if the incubation is carried out for only 2 - 3 hours.

14. Centrifuge for 5 minutes at room temperature. Transfer the supernatant to a fresh tube with a P1000 Pipetman.

15. Add 750 μl of 95% ethanol (2.5 x volume) and mix well by inverting the tube 4-5 times.

16. Return the tube to the centrifuge with the lid attachment oriented away from center of the rotor. This is necessary to mark the position of the invisible DNA pellet. Centrifuge for 15 minutes at room temperature.

17. Remove the tube from the centrifuge and open the lid. Gently inverting the tube, touch the lip of the tube to the rim of an Erlenmeyer flask. Hold the tube to drain the ethanol. You do not need to remove all of the ethanol from the tube.

18. Place the tubes back into the centrifuge in the same orientation as before. **Do not start the centrifuge.** In this step, the centrifuge rotor is used as a "tube holder" that keeps the tubes at an angle convenient for ethanol washing.

19. Wash the pellet with 700 μl of cold 70% ethanol. Holding a P1000 Pipetman vertically, deliver the alcohol to the side of the tube opposite the pellet, i.e., the side facing the center of the rotor. Hold the Pipetman as shown in the Figure 3, Chapter 2. Withdraw the tube from the centrifuge holding it by the lid. Gently inverting the tube, touch the lip to the rim of an Erlenmeyer flask. Hold the tube to drain the ethanol. Place the tube back into the centrifuge and repeat ethanol wash one more time.

20. Place the tube back into centrifuge in the same orientation as before. Without closing the lid, start centrifuge for 2-3 seconds and collect the remaining ethanol to the bottom of the tube. The centrifuge should not reach maximum speed. A speed of 500 to 1000 rpm is sufficient to collect all ethanol from the sides of the tube. Remove all ethanol from the bottom of the tube using a P200 micropipette fitted with a capillary tip.

21. Dissolve the DNA pellet (invisible) in 10-15 µl of sterile deionized water. This operation will be successful only if you know the position of the pellet on the side of the tube. Gently pipette the liquid up and down, directing the stream of liquid toward the pellet.

Note: Centrifugation using microfuge with fixed-angle rotors will deposit most of the DNA pellet on the side of the tube rather than on its bottom. To dissolve all plasmid DNA it is necessary to direct the stream of the solution **toward the bottom lower 2/3 side of the tube.**

22. Determine the concentration of DNA by measuring absorbance at 260 nm. Initially use a 1:100 dilution of the DNA in PBS. The absorbance reading should be in the range of 0.1 to 1.5 OD$_{260}$. A solution of 50 µg/ml has an absorbance of 1 in a 1-cm path cuvette. Determine the purity of the DNA by measuring the absorbance at 280 nm and 234 nm. Calculate the 260/280 and 260/234 ratios. See Chapter 1 for a detailed discussion about the validity of this method and calculation of DNA concentration in the presence of a small amount of protein.

Note: DNA concentration should never be measured in water or TE buffer. For discussion of determination of DNA concentration and its dependence on ionic strength and pH see Chapter 1.

Rapid small scale protocol

1. Inoculate cells into 2 ml of Terrific Broth or H15 medium supplemented with the appropriate antibiotic. Use 20 ml tube to grow the cells. Grow cells overnight at 37 °C with vigorous shaking.

2. Transfer 400 µl of cells into a 1.5 ml microfuge tube and collect the cells by centrifugation at **room temperature** for 3 minutes. Discard supernatant. Re-centrifuge the cells for a few seconds and remove the remaining medium using a Pipetman (P200) fitted with capillary microtip.

3. Add 400 µl of freshly prepared **solution II** to the cell pellet. Break the pellet into small pieces with a toothpick. Close the tube and mix by inverting 4-6 times. Place tube on ice and incubate for 5 minutes.

4. Meanwhile prepare the ammonium acetate-RNase solution by adding 10 µl of RNase A stock solution to 1 ml of 7.5 M ammonium acetate. Place the solution on ice.

5. Add 300 μl (0.75 volume) of ice cold ammonium acetate-RNase solution to the tube and mix by gently inverting the tube 5-10 times. Place the tube back on ice and incubate for 5 minutes.

6. Centrifuge the tube at **room temperature** for 5 minutes.

7. Pour off the supernatant into a fresh 1.5 ml microfuge tube. Be careful not to transfer the gelatinous pellet.

8. Add 420 μl (0.6 x total volume) of isopropanol to the supernatant. Mix well by inverting the tube several times. Incubate at room temperature for no more than 10 minutes.

9. Place the tube in the centrifuge, with the attached side of the lid away from the center of rotation. Centrifuge the tube at maximum speed for 10 minutes at room temperature (See Figure 2, Chapter 2).

10. Remove the tube from the centrifuge and open the lid. Gently inverting the tube, touch the lip to the rim of an Erlenmeyer flask. Hold the tube to drain the isopropanol. You do not need to remove all of the isopropanol from the tube. Place the tubes back into the centrifuge in the same orientation as before.

Note: When pouring off alcohol do not invert the tube more than once because this may dislodge the pellet.

11. Place the tubes back into the centrifuge in the same orientation as before. **Do not start the centrifuge**, in this step, the centrifuge rotor is used as a "tube holder" which keeps the tube at an angle convenient for ethanol washing.

12. Wash the pellet with 700 μl of cold 70% ethanol. Holding a P1000 Pipetman vertically, slowly deliver ethanol to the tube side of the tube opposite the pellet, i.e., side facing the center of the rotor. Hold the Pipetman as shown in Figure 3, Chapter 2. Remove the tube from the centrifuge by holding it by the lid. Gently inverting the tube, touch the lip to the rim of an Erlenmeyer flask. Hold the tube to drain the ethanol. You do not need to remove all of the ethanol. Place the tubes back into the centrifuge in the same orientation as before and wash with 70% ethanol one more time.

Note: This procedure makes it possible to quickly wash a large number of pellets without centrifugation and vortexing. Vortexing and centrifuging the pellet are time consuming and frequently lead to substantial loss of plasmid DNA.

13. After the last wash, place the tube into the centrifuge, making sure that the side of the tube with the pellet is away from the center of rotation. Without closing the tube lid, start the centrifuge for 2-3 seconds to collect remaining ethanol at the bottom of the tube. The centrifuge should not reach maximum speed. A speed of 500 to 1000 rpm is sufficient to collect all ethanol remaining on the sides of the tube. Remove all ethanol using a P200 micropipette fitted with a capillary tip.

Note: Never **dry the DNA pellet** in a vacuum. This is an unnecessary step that can make rehydration of the DNA very difficult, if not impossible.

14. Add 10-15 µl of TE or sterile deionized water to the tube and dissolve the pelleted DNA. Gently pipette the liquid up and down directing the stream of the liquid toward the pellet.

Note: Centrifugation using microfuge with fixed angle rotors holding 24 tubes will deposit most of the DNA pellet on the side of tube rather than on the bottom. To dissolve all plasmid DNA when these centrifuges are used, it is necessary to direct the stream of the dissolving solution **toward the bottom lower 2/3 side of the tube.**

15. Determine the concentration of DNA by measuring absorbance at 260 nm. Initially use a 1:100 dilution of the DNA in PBS. The absorbance reading should be in the range of 0.1 to 1.5 OD_{260}. A solution of 50 µg/ml has an absorbance of 1 in a 1-cm path cuvette. Determine the purity of the DNA by measuring the absorbance at 280 nm and 234 nm. Calculate the 260/280 and 260/234 ratios. See Chapter 1 for a detailed discussion about the validity of this method and calculation of DNA concentration in the presence of a small amount of protein.

Note: DNA concentration should never be measured in water or TE buffer. For discussion of determination of DNA concentration and its dependence on ionic strength and pH see Chapter 1.

Results

Typical absorption spectra of plasmid DNA purified using fast protocol is shown in Figure 2. The 260/280 ratio and 260/234 ratio are 1.9 and 2.0, respectively. The purity of the DNA obtained is comparable to DNA obtained using cesium chloride centrifugation. Plasmid yield for all three procedures is between 50% to 75% of the expected amount and is similar to that obtained by the CsCl purification procedure (Heilig et al., 1994). For example,

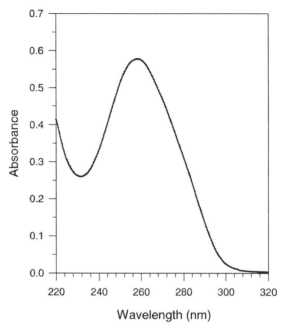

Fig. 2. Absorption spectrum of plasmid DNA purified by the rapid large scale protocol. Plasmid DNA was purified from 30 ml of cells. The sample was diluted 10 x in PBS and scanned using a UV spectrophotometer. The 260/280 absorbance ratio was 1.89 and 260/234 absorbance ratio was 2.0.

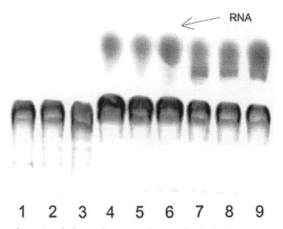

Fig. 3. Gel electrophoresis of plasmid prepared using both the long and rapid procedures. Lanes 1 and 2 have 1 μl plasmid DNA prepared from 400 μl of cells by the rapid protocol. Lane 3 has 1 μl of plasmid DNA prepared from 30 ml of cells using the rapid protocol. Lanes 4, 5, 6, 7, 8 and 9 have 5 μl of plasmid DNA prepared by the small scale, long procedure.

the total amount of DNA isolated from plasmid pUC 19 with a 500 bp insert using 30 ml cells and the large scale protocol was 300 µg. Because pUC 19 is a high copy number plasmid (Brent and Irvin, 1994), the expected amount of DNA from 30 ml cells is between 60 to 400 µg (Heilig et al., 1994). The 300 µg constitute about 75% of the expected yield.

Results of agarose gel electrophoresis of 1 µl of plasmids isolated using fast and long procedures are presented in Figure 3. The preparations do not contain chromosomal DNA. The majority of plasmid isolated is a super-coiled monomer. The weak second band represents supercoiled dimer.

Troubleshooting

Problems encountered in both the large and small scale protocols can be classified into two groups: problems resulting from inadequate purity of the product or problems resulting in inadequate plasmid yield.

Problems with plasmid purity

Poor purity of plasmid can manifest itself by:

- A low 260/280 ratio.
  This indicates the presence of proteins in the preparation. This usually is a result of too short an incubation time with neutralized solution II. It is also possible to introduce protein contaminates into the preparation by not removing the entire pellet after centrifugation. However, for most applications, removing all proteins is not necessary. Plasmid prepara-tions having a 260/280 ratio as low as 1.4 give satisfactory results in clon-ing, ligation, restriction enzyme and PCR experiments as well as in man-ual and automatic sequencing.

- Contamination with bacterial chromosomal DNA.
  This contamination can only be detected by gel electrophoresis and ap-pears as a faint, high-molecular weight DNA band. Plasmid DNA, con-taminated with bacterial chromosomal DNA, cannot be used in any sub-sequent applications. This impurity usually results from mixing solu-tion II with the cell lysate too vigorously. There is no immediate remedy for this problem. A new plasmid preparation should be made, taking care to mix solution II with the bacterial suspension very gently. Also mixing during the neutralization step should only be done by gentle inversion.

- Contamination with RNA.
  This contamination appears as a diffuse band running together or slightly ahead of the bromophenol blue dye. For most procedures, some contamination with RNA is not deleterious. These include restriction enzyme and PCR experiments. If necessary RNA contamination can be removed by treating the plasmid preparation with RNase again.

Problems with plasmid yield

Poor yield of the plasmid in all three protocols usually results from:

- Inadequate lysis of the cells after the addition of solution II.
  Lysis of the cells in the rapid protocol is always adequate due to the very high concentration of NaOH and SDS, but it may be a problem in the two other procedures. To remedy this problem, it is necessary to begin a new preparation and increase the concentration and/or time of lysozyme treatment. Also, be sure that the cell pellet is completely resuspended before adding the lysozyme solution.

- Low concentration of ammonium acetate solution used for neutralization step.
  Upon prolonged storage, the molarity of 7.5 M ammonium acetate decreases due to the volatility of both ions. This solution should be made freshly every two months and stored, tightly closed, at 4 °C. To remedy the problem, prepare fresh ammonium acetate solution and repeat the procedure once more. A good indication of this problem is the presence of a large quantity of RNA and a small amount of plasmid on agarose gels.

References

Brent R and Irwin N (1994) Vector derived from plasmids. In: Current Protocols in Molecular Biology. Vol. 1. Ed. Janssen K. Current Protocols Publisher.

Brinboim HC (1983) A rapid alkaline extraction method for the isolation of plasmid DNA. In: Method Enzymol, Vol. 100. Academic Press, London, pg. 243-255.

Brinboim HC and Doly J (1979) A rapid alkaline extraction procedure for screening recombinant plasmid DNA. Nucleic Acid Res. 7:1513-1523.

Duttweiler HM and Gross DS (1989) Bacterial growth medium that significantly increases the yield of recombinant plasmid. BioTechniques 24:438-444.

Heilig JS, Lech K and Brent R (1994) Large-scale preparation of plasmid DNA. In: Current Protocols in Molecular Biology. Vol. 1. Ed. Janssen K. Current Protocols Publisher.

Isolation and Purification of RNA

STEFAN SURZYCKI

Introduction

Obtaining pure RNA is an essential step in the analysis of patterns of gene expression and understanding the mechanism of gene expression. Thus, isolation of pure, intact RNA is one of the central techniques in molecular biology and represents an important step in Northern analysis, nuclease protection assays, RNA mapping, RT-PCR, cDNA library construction and in vitro translation experiments.

Two strategies of RNA isolation are usually employed: isolation of total RNA and isolation of mRNA. A typical eukaryotic cell contains about 10 to 20 pg of RNA, most of which is localized in the cytoplasm, whereas a prokaryotic cell contains 0.02 to 0.05 pg of RNA (Davis et al., 1994). About 80 to 85 percent of eukaryotic RNA is ribosomal RNA, while 15 to 20 percent is composed of a variety of stable low molecular weight species such as transfer RNA and small nuclear RNA. Usually about 1 to 3% of the cell RNA is messenger RNA (mRNA) that is heterogeneous in size and base composition. Almost all eukaryotic mRNAs are monocystronic and contain a post-transcriptionally added, poly-adenylic acid (poly A) tract at their 3' terminal. This 3' poly-A tail permits separation and isolation of mRNA from all other classes of RNA present in the cell. In prokaryotes many mRNAs are polycistronic and none contain poly A tracts. This adds to the difficulty of mRNA purification from prokaryotic cells.

This chapter describes methods used for total RNA isolation, whereas methods used in purification of eukaryotic mRNA are described in Chapter 7.

Purification of total RNA

Isolation of total RNA is most frequently used when pure RNA is required for experiments. This is because these techniques are less laborious and require less time to perform than isolation of mRNA. Various techniques for purification of total RNA are now in use and their application depends on the nature of RNA required. For example, if RNA is going to be used for quantitative RT-PCR, intactness of the purified RNA is not critical, while intact RNA is required for cDNA library preparation or Northern blot analysis. Complete removal of DNA contamination is critical if RNA is to be used in RT-PCR but is not important in in vitro translation. The techniques described here yield pure intact RNA good for any applications.

The physical and chemical properties of RNA and DNA are very similar, thus, the basic procedures used in RNA purification are similar to those of DNA described in Chapter 1. All of the RNA purification methods incorporate the following steps:

a. disruption of cells or tissue,

b. effective denaturation of nucleoprotein complexes and removal of proteins,

c. concentration of RNA molecules and,

d. determination of purity and integrity of isolated RNA.

In addition, methods must include procedures that remove co-purified DNA from the preparation. In contrast to DNA purification, guarding against physical shearing of RNA molecules is not necessary because RNA molecules are much smaller and much more flexible than DNA molecules. Therefore, in RNA protocols, strong physical forces during cell and tissue breakage are frequently used and the use of wide-mouth pipettes is not required. However, RNA isolation is much more difficult than DNA purification largely due to the sensitivity of RNA to degradation by internal and external ribonucleases. These enzymes are omnipresent and are very stable molecules that do not require any cofactors for their function. A crucial aspect of any procedure for RNA purification is fast and irreversible inactivation of endogenous RNases and protection against contamination with exogenous RNase during the isolation procedure. To these ends, all extraction buffers include powerful RNase inhibitors and all solutions and equipment used are treated to remove exogenous RNases.

Elimination of RNases

The most commonly used inhibitors included in extraction buffers to inhibit endogenous RNase are:

- Strong protein denaturation agents. To these belong guanidinium hydrochloride and guanidinium isothiocyanate used at a concentration of 4 M. These chaotropic agents can quickly inactivate endogenous RNases and contribute to denaturation of nucleoprotein complexes. To irreversibly denature RNase by these compounds, a high concentration of 2-mercapthoethanol is also included.

- Vanadyl-ribonucleoside complexes. Oxovanadium IV ions form complexes with any ribonucleoside and bind to most RNases inhibiting their activity (Berger, and Birkenmeir, 1979). A 10 mM solution is used during cell breakage and is added to other buffers used in RNA isolation. Complexes can be used with deproteinizing agents (phenol or CIA) and with chaotrophic agents. However, the compound is expensive and difficult to remove from purified RNA. Any residual amount of this compound will inhibit many enzymes used in subsequent RNA manipulations.

- Aurintricarboxylic acid (ATA). This compound binds selectively to RNase and inhibits its activity. ATA is usually incorporated into extraction buffers used for bacterial RNA preparations (Hallick et al, 1997). The inhibitor can affect certain enzymes and should not be used if RNA will be needed for primer extension or S1 nuclease analysis.

- Macaloid. This naturally occurring clay (sodium magnesium lithofluorosilicate) being negatively charged, strongly absorbs all RNase. The macaloid and bound RNase are removed from the preparation by centrifugation (Marcus and Halvorson, 1967).

- Protein RNase inhibitors such as RNasin. A protein originally isolated from human placenta, inhibits RNase by non-competitive binding. It cannot be used in extraction buffers containing a strong denaturant. It is usually included in solutions used in later stages of purification or in buffers used in storage or subsequent RNA manipulations. Its presence is critical in in vitro transcription and translation assays and during cDNA synthesis.

The most frequent sources of exogenous RNase contamination are one's hands, and bacteria and fungi present on airborne dust particles. To remove exogenous RNase contamination, the most frequently used inhibitors are:

- Diethyl pyrocarbonate (DEPC). DEPC causes enzyme inactivation by denaturing proteins. Inactivation of RNase is irreversible. The compound is used for removing RNase from solutions and glassware used in RNA preparation. DEPC should be used with care because it is highly flammable and a suspected strong carcinogen.

- RNaseZap® or RNaseOff solutions. These commercially available reagents destroy RNases on contact very effectively. The decontamination solutions are not toxic and can be used to remove RNase from all surfaces and equipment. The compositions of these reagents are trade secrets.

Moreover to prevent the contamination of equipment and solutions with RNase, the following precautions should be taken:

- Gloves should be worn always. Because gloves can be easily contaminated with RNase they should be changed frequently.

- Tubes should be kept closed.

- Whenever possible disposable, certified RNase free, plasticware should be used.

- RNase-free tips and microfuge tubes should be sterilized. Regular microfuge tubes and tips usually are not contaminated with RNase and they do not require special treatment if they are used from unopened bags. When autoclaving them, it is necessary to be careful not to contaminate them when loading into containers. Use gloves when loading.

- All glassware should be treated with 0.1% DEPC water solution and autoclaved to remove DEPC. It also is possible to inactivate RNase by baking glassware at 180°C for at least 2 hours or overnight. Alternatively RNase can be easily and efficiently eliminated from glassware, countertops, pipettors and plastic surfaces using RNaseZap® solution. We strongly recommend its use.

- All solutions should be made with DEPC-treated water or sterilized MilliQ water.

Outline

Three methods of RNA isolation are described: a guanidinium hot-phenol method, a high-salt lithium chloride method and a method using a commercially available reagent, TRI-Reagent™.

The guanidinium hot-phenol procedure is a modification of a method first described by Chirgwin (Chirgwin et al., 1979) and Chomczynski and Sacchi (1987). This single-step extraction procedure takes advantage of the characteristic of RNA, under acidic conditions, to remain in the aqueous phase containing 4 M guanidine thiocyanate, while DNA and proteins are distributed into the phenol:chloroform, organic phase. Distribution of DNA into the organic phase is particularly efficient if DNA molecules are small. The method, therefore, uses a procedure to fragment DNA into molecules not larger than 10 kb. To facilitate cellular disruption, a high concentration of Sarcosyl is used and the pH of the solution is carefully controlled at pH 5.5 or lower to limit RNA degradation (Noonberg et al., 1995). This method can be used to isolate total RNA from a variety of prokaryotic and eukaryotic cells. The efficiency of this method is very high (80-90%), affording purification of a large quantity of high quality RNA. Using this protocol, RNA was successfully purified from many bacterial species and lower eukaryotic cells such as yeast and Chlamydomonas. This protocol can be used to isolate total RNA from animal or plant cells and tissues that are very rich in RNase, such as pancreatic cells or old leaves (Asubel et al., 1987). Any modification of the protocol for specific types of cells and tissues are confined to the preparation of material prior to RNA isolation, rather then to the protocol itself, making the application of this method very convenient.

The **high-salt lithium chloride** method can be used to isolate RNA from some plant tissues especially rich in various secondary products such as anthocyanins, phenolic compounds, polysaccharides and latex. It was shown that it is very difficult to isolate pure RNA from such plants using chaotropic agents (Shultz, et al., 1994; Bugos, et al., 1995). The procedure presented is an adaptation of a protocol of Bugos et al. (1995). The procedure involves cell breakage in low pH, high salt buffer in the presence of RNase inhibitors. Protein and DNA are removed by acidic phenol:CIA extraction and RNA is recovered by lithium chloride precipitation.

The **TRI-Reagent**™ method is a single-step method of RNA isolation using a monophasic solution of phenol and guanidine isothiocyanate (TRI-Reagent™) combined with precipitation of RNA by isopropanol in the presence of high-salt (Chomczynski and Mackey, 1995). The method is particularly useful for fast isolation of RNA from numerous small samples and can be used with all types of cells and tissues.

At the end of this chapter a brief description of commercially available kits for total RNA purification is given.

A schematic outline of the procedure is given in Figure 1.

Materials

– 50 ml Oak Ridge polypropylene centrifuge tubes with caps (for example Nalgene® # 21009)

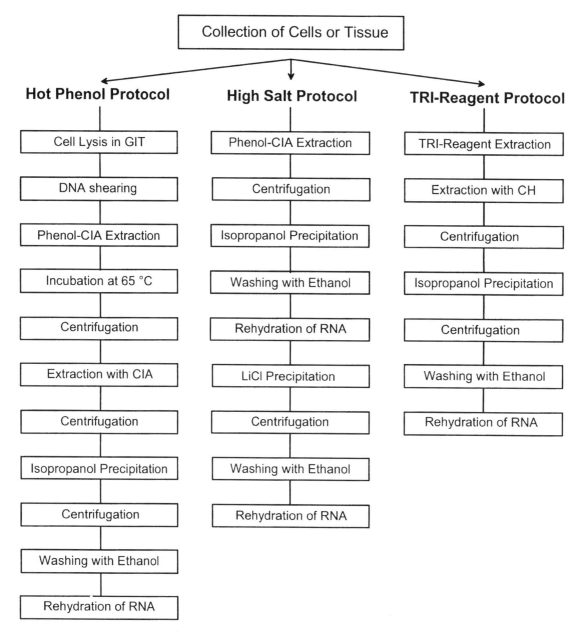

Fig. 1. Schematic outline of the procedures

- Corex 25 ml centrifuge tubes with Teflon-lined caps (Corex # 8446-25)
- 15 ml and 50 ml sterilized polypropylene conical centrifuge tubes (for example Corning # 25319-15; 25330-50)
- Phenol, redistilled, water-saturated. (Ambion # 9712).
  Redistilled phenol is commercially available. There are a number of companies that supply high quality redistilled phenol. Water-saturated phenol is preferable to the crystalline form because it is easier and safer to use. Water-saturated phenol can be stored indefinitely in a tightly closed, dark bottle at -70 °C.
- Diethyl pyrocarbonate (DEPC) (Sigma # D 5758)
- RNaseZap® Solution (Ambion # 9780 or equivalent)
- Sarcosyl (N-laurylsarcosine, sodium salt, Sigma Co. # L 5125)
- 8-hydroxyquinoline free base (Sigma Co. # H 6878)

Note: Do not use hemisulfate salt of 8-hydroxyquinoline.

- 2-mercaptoethanol (Sigma Co. # M 6250)
- Antifoam C (Sigma Co. # A 8001)
- Guanidinium isothiocyanate (Fluka # 50981)
- Glass beads 0.4-0.5 mm (Sigma Co. # G 1634)
  To prepare glass beads, soak them for 1 hour in concentrated nitric acid, wash beads with deionized water and dry them by baking in 180 °C oven overnight.
- TRI-Reagent™ (Molecular Research Center, Cincinnati, OH, # TR - 118)

- **DEPC Treated Water**   Solutions
  Add 0.1 ml DEPC to 100 ml of deionized water. Shake vigorously to dissolve DEPC in water. Incubate 2 hours or overnight at 37 °C. Autoclave to inactivate DEPC. DEPC must be completely inactivated before using any solution treated with it because it will carboxylate RNA. Carboxylated RNA will inhibit reverse transcriptase, DNA:RNA hybridization and in vitro translation. Use DEPC treated water for preparation of the solutions that cannot be treated with DEPC directly. Primarily these are solutions containing primary amines, such as Tris or ammonium acetate that will react with it.
  Safety Note: When using DEPC, wear gloves and use a fume hood because DEPC is highly flammable and a suspected carcinogen.

Note: Deionized water from a MilliQ RG apparatus (Millipore Co. # ZD5111584) can be used directly in all applications instead of DEPC-treated water because it does not contain RNase.

- **PBS Solution, pH 7.4**
  - 137 mM NaCl
  - 2.7 mM KCl
  - 4.3 mM $Na_2HPO_4 \cdot 7 H_2O$
  - 1.4 mM KH_2PO_4

  After the addition of each salt, completely dissolve it in distilled or deionized water. Adjust pH to 7.4 with 1 N HCl. Add DEPC to a final concentration of 0.01% and shake vigorously to dissolve DEPC. Incubate for at least 2 hours. Autoclave for 20 minutes. Store at 4 °C.

- **Phenol 8-HQ Solution**

  Water-saturated, twice distilled phenol is equilibrated with an equal volume of 0.1 M of sodium borate. Sodium borate should be used rather than the customary 0.1 M Tris solution because of its superior buffering capacity at pH 8.5, its low cost, and its antioxidant properties. In addition sodium borate removes oxidation products from phenol during the equilibration procedure. Mix water-saturated phenol with an equal volume of 0.1 M sodium borate in a separatory funnel. Shake until the solution turns milky. Wait for the phases to separate and then collect the bottom phenol phase. Add 8-hydroxyquinoline to the phenol at a final concentration of 0.1% w/v. Phenol 8-HQ can be stored in a dark bottle at 4 °C for several weeks. This solution can be stored for several years at -70 °C.

  Safety Note: Phenol is rapidly absorbed by, and is highly corrosive to, the skin. It initially produces a white softened area, followed by severe burns. Because of the local anesthetic properties of phenol, skin burns may not be felt until there has been serious damage. Gloves should be worn when working with this chemical. Because some brands of gloves are soluble or permeable to phenol, they should be tested before use. If phenol is spilled on the skin, flush off immediately with a large amount of water and treat with a 70% solution of PEG 4000 in water (Hodge, 1994). Used phenol should be collected into a tightly closed glass receptacle and stored in a chemical hood until it can be properly disposed of.

- **Chloroform:Isoamyl Alcohol Solution (CIA)**

  Mix 24 volumes of chloroform with 1 volume of isoamyl alcohol. Because the chloroform is light sensitive, the CIA solution should be stored in a brown glass bottle, preferably in a fume hood.

  Safety Note: Handle chloroform with care. Mixing chloroform with other solvents can involve a serious hazard. Adding chloroform to a solution containing strong base or chlorinated hydrocarbons could result in an explosion. Prepare the CIA solution in a hood because isoamyl alcohol vapors are poisonous. Used CIA can be collected in the same bottle as the phenol solution and discarded together.

- **0.5 M EDTA Stock Solution, pH 8.5**
  Weigh an appropriate amount of EDTA (disodium ethylenediamine-tetraacetic acid, dihydrate) and add it to distilled water while stirring. EDTA will not go into solution completely until the pH is greater than 7.0. Adjust the pH to 8.5 using a concentrated NaOH solution, or by adding a small amount of pellets, while mixing and monitoring the pH. Adjust the solution to its final volume with water and sterilize by filtration. EDTA stock solution can be stored indefinitely at room temperature.
- **1 mM EDTA Solution**
  Prepare this solution using 0.5 M EDTA stock and RNase-free water. Sterilize by autoclaving and store at -20 °C to prevent contamination.
- **3 M Sodium Acetate pH 5.2**
  Weigh 24.6 g of sodium acetate and add 50 ml deionized water. Dissolve the salt and adjust pH with glacial acetic acid to pH 5.2. Add water to 100 ml and DEPC to 0.1%, mix very well and incubate for 2 hours or overnight at 37 °C. Autoclave for 30 minutes to remove DEPC and store at 4 °C.
- **Sarcosyl Stock Solution, 20% (w/v)**
  Dissolve 40 g of Sarcosyl in 100 ml of RNase-free water. Adjust to a final volume of 200 ml. Sterilize by filtration through a 0.22 μm filter. Store at room temperature.
- **GIT Solution**
  - 4 M guanidinium isothiocyanate
  - 25 mM sodium acetate pH 5.2
  - 1 mM EDTA
  - 0.05% Antifoam C

  Prepare 100 ml of solution at a time. Add 47.2 g of guanidinium isothiocyanate to 50 ml water followed by 0.85 ml of a 3 M solution of sodium acetate, and 1 μl of antifoam C. Fill up to 100 ml with water and heat to facilitate dissolving. The solution is light sensitive and can be stored at room temperature in a dark bottle for a month.
- **8 M Lithium Chloride Solution**
  - 8 M lithium chloride

  Dissolve 33.9 g of lithium chloride in 100 ml of deionized water. Add DEPC to 0.1%, mix very well and incubate for 2 hours or overnight at 37 °C. Autoclave for 30 minutes to remove DEPC and store at 4 °C.
- **Homogenization Buffer for Plant Material**
  - 100 mM Tris-HCl, pH 9.0
  - 200 mM NaCl
  - 15 mM EDTA
  - 0.5% Sarcosyl.

Sterilize solution by filtration and store at 4 °C. Add 2-mercaptoethanol to a concentration of 100 mM just before use.

- **High-Salt Precipitation Solution**
  - 1.2 M NaCl
  - 0.8 M sodium citrate

Prepare solution in RNase-free water and store at 4 °C.

Procedure

Preparation of cell material

Bacterial cells
1. To prepare RNA from bacteria, use mid-exponential growth phase cultures. Consult Table 1 for the number of cells required to isolate the desired amount of RNA. For isolation of RNA from 10^9 bacteria, prepare a 50 ml Erlenmeyer flask containing 10 ml of TB (Terrific Broth) medium. Inoculate the Erlenmeyer flask with 0.1 ml from an overnight culture of bacteria.

Note: To provide a large surface-to-volume ratio for aeration, culture cells in an Erlenmeyer flask filled with no more than **one-quarter** the total flask volume. The orbital shaker speed should be no less than 200-300 rpm.

2. Grow the cells at 37 °C to a density of 1.0 at 600 nm. This should take approximately 2 to 3 hours.

3. Collect cells by pouring them on ice. Fill a 25 ml Corex tube half full with crushed ice and add 10 ml of cells. Centrifuge in a fixed angle rotor (Sorvall SS-34 or equivalent) at 5000 rpm (3000 x g) for 5 minutes at 4 °C.

4. Discard the supernatant and add 10 ml of hot (65 °C) GIT solution. Transfer the tube to a 65 °C water bath and go to step 1 of the RNA isolation protocol.

Yeast cells
1. To prepare RNA from yeast, use mid-exponential growth phase cultures. Consult Table 1 for the number of cells required to isolate the desired amount of RNA. For isolation of RNA from about 10^8 cells, grow 10 ml of yeast on the desired medium to mid-exponential phase ($A_{600} = 1.0$).

Note: To obtain consistent yields of RNA, purification should not be started from cells approaching or at stationary phase.

2. Collect cells by pouring them on ice. Fill a 25 ml Corex tube half full with crushed ice and add 10 ml of cells. Centrifuge at 7000 rpm (6000 x g) for 3 minutes at 4 °C using a Sorvall SS-34 fixed angle rotor or the equivalent.

3. Discard the supernatant and add 10 ml of hot (65 °C) GIT solution. Transfer the tube to a 65 °C water bath and go to step 1 of the RNA isolation protocol.

Monolayer cell cultures

1. Grow cells as a monolayer in the desired medium. For monolayers, a confluent cell culture from one large bottle (75 cm^2 area) will yield approximately 50 - 80 mg of cells, that is, about 5 x 10^7 cells. For a large scale RNA preparation, use 1-2 large bottles. Remove medium from monolayer bottle and add 10 ml of PBS to the bottle, tilt the bottle back and forth to rinse the cells, then withdraw the PBS. Repeat this washing procedure one more time. Washing with PBS removes serum left over from the medium that frequently contains large amounts of RNases.

2. Add approximately 5 ml of hot (65 °C) GIT solution **directly** to the cell monolayer in each bottle. Mix the solution with pipettes to ensure that all cells are released from the surface and lysed. Transfer the lysate to a 25 ml Corex tube. If two bottles are used, transfer the cell lysate from both bottles to one 25 ml Corex tube. Otherwise add 5 ml of GIT solution to 5 ml of cell lysate. Transfer the tube to a 65 °C water bath and go to step 1 of the RNA isolation protocol.

Cell suspension cultures

1. Collect 1 to 2 x10^8 cells for a single preparation. The suspension cultures at saturation will contain 1-2 x 10^6 cells/ml. Use 100 ml of the cell culture. Pour 50 ml of the cells into two 50 ml Oak Ridge centrifuge tubes and centrifuge at 2000 rpm (500 x g) for 5 minutes at 4 °C using a Sorvall SS-34 fixed angle rotor or the equivalent. Discard the supernatant.

2. Wash the cell pellet with 25 ml of ice-cold PBS. Resuspend the cell pellet by gentle pipetting up and down. Centrifuge the cells at 500 x g for 5 minutes at 4 °C and discard the supernatant. Wash the cells with ice cold PBS one more time. Resuspend each cell pellet in 5 ml of hot (65 °C) GIT. Collect the cell lysate from both tubes to one 25 ml Corex tube. Transfer the tube to a 65 °C water bath and go to step 1 of the RNA isolation protocol.

Animal and plant tissue

Harvest solid tissue material from animal or plants as described below.

1. Place a 250 ml plastic beaker into an ice bucket and pour liquid nitrogen into it. Harvest fresh and soft tissue in pieces less than 1 cm^3, using sterile scissors or a razor blade, and immediately freeze them with liquid nitrogen. Tissues harvested this way can be stored at -70 °C for several years.

2. Place a mortar into a second ice bucket and pour about 30 ml of liquid nitrogen into it. Cover the ice bucket with a lid and cool the mortar for 5-10 minutes.

3. Weigh between 100 to 1000 mg of frozen tissue as fast as possible, taking care that the tissue does not thaw in the process. Consult Table 1 for the amount of animal or plant tissues needed to isolate the desired quantity of RNA. Place a plastic weighing boat on the balance and tare it. Pick frozen tissue fragments from the liquid nitrogen with forceps and place them in the weighing boat. Transfer the frozen tissue to the frozen mortar. For preparation of RNA from plant material, if possible choose young leaves or shoots.

4. Add a few pieces of dry ice to the mortar. The volume of dry ice used should be roughly 3 times the volume of tissue.

5. Grind the tissue with the mortar and pestle until it appears as a white powder. Transfer the powder into 25 ml Corex tube and add 10 ml of hot (65 °C) GIT solution. Add the solution slowly to allow the dry ice to evaporate. Transfer the tube to a 65 °C water bath and go to step 1 of the RNA isolation protocol.

RNA isolation procedures

Hot-phenol protocol

Prepare cells or tissue as described in section Preparation of cell material. For all centrifugations described in this procedure, unless otherwise directed, use a Sorvall SS-34 rotor or the equivalent.

1. As fast as possible, add 1 ml of 20% Sarcosyl (2% final concentration) and 120 μl of 2-mercapthoethanol (1.2% final concentration) to the cell lysate. During this operation keep the tube at 65 °C. Close the Corex tube with a Teflon-lined cap and vortex it for 1 min. Be sure that the temperature of the mix does not drop during vortexing. Vortex in short 10 - 20 seconds intervals and place the tube back into the hot water bath to keep the temperature at 65 °C.

Note: The ratio of GIT to cells should be 1 ml solution to 100 mg of tissue or 1×10^7 cells. This is approximately 10 volumes of GIT to one volume of the cells. Ten milliliters of GIT solution is sufficient for the isolation of RNA from 1g of tissue or 2-3 10^8 cells.

Note: For preparation of RNA from bacteria or yeast, add an equal volume of acid-washed glass beads to the cells to facilitate cell breakage (Kormanec and Farkašovsk, 1994).

2. Place the tube back into a 65 °C water bath. Draw the lysate forcefully up and down through a 19 g needle using a 20 ml syringe. This will shear genomic DNA. Continue passing the solution through the needle until it is no longer viscous. This should not take more than 3 - 5 passes.

3. Add 5 ml (0.5 volume) of hot (65 °C) phenol, close the tube with a cap and vortex it for 10 to 30 seconds. Place the tube back into the 65 °C water bath.

4. Add 1 ml (0.1 volume) of 3 M Na acetate, pH 5.2 and mix by vortexing briefly. Transfer a few drops of the mixture into 15 ml of distilled water and measure the pH with pH paper. The pH of the solution should be between pH 5.2 and 5.5. If the pH is higher, add a few drops of glacial acetic acid to the lysate, mix by vortexing and measure the pH again. Continue until the mixture reaches the correct pH.

5. Add 5 ml (0.5 volume) of CIA, close the tube and vortex briefly to mix solutions.

6. Place tube back into the 65 °C water bath and incubate for 15 minutes. During incubation, vortex the tube frequently to keep the phases mixed. Alternate 1 minute of vortexing with 2 minutes of incubation to keep the temperature from dropping substantially below 65 °C.

7. Cool the tube to room temperature and centrifuge for 10 minutes at 10 000 rpm (12 000 x g) to separate phases.

8. Collect the aqueous phase with a pipette and transfer it into a fresh 25 ml Corex tube. Add an equal volume of CIA (about 11 ml); close the tube and vortex it vigorously to mix the phases.

9. Centrifuge the tube at 10 000 rpm (12 000 x g) for 5 minutes at 4 °C.

10. Transfer the aqueous phase to a fresh 25 ml Corex tube. When transferring, measure the volume of the liquid with the pipette. Add 0.1 volume (about 1.1 ml) of 8 M lithium chloride. Precipitate RNA by the addition of an equal volume of isopropanol (about 12 ml). Incubate the solution at room temperature for 5 to 10 minutes.

Note: Precipitation with isopropanol helps to remove residual DNA because the lithium salt of small molecular weight DNA is soluble in isopropanol, whereas RNA salts are not.

11. Centrifuge 10 minutes at 10 000 rpm (16 000 x g) using a swinging bucket rotor (e.g., Sorvall HB-4 rotor). Discard the supernatant.

Note: Centrifugation in a swinging bucket rotor is recommended to collect RNA at the bottom of the tube. This facilitates dissolving the RNA pellet.

12. Wash the pellet 2 times with cold 70% ethanol as follows: Add 5 ml of cold 70% ethanol to the tube and wash the pellet by carefully rolling the tube at a 45° angle in the palm of your gloved hand. Be careful not to dislodge the RNA pellet from the bottom of the tube during this procedure. **Never vortex the tube.** Discard the ethanol and drain the tube by inverting it over paper towels for a few minutes. Repeat the wash one more time.

13. Dissolve the RNA pellet in 100-200 μl of RNase free 1 mM EDTA. Store at -70 °C.

14. Determine the concentration of RNA by measuring absorbance at 260 nm. Initially use a 1:100 dilution of the sample in PBS. The absorbance reading should be in the range 0.1 to 1.5. Calculate the concentration of RNA using the equation:

$$N = \frac{A_{260}}{\varepsilon_{260}}$$

Where ε_{260} is the RNA absorption coefficient, N is RNA concentration in μg/ml and A_{260} is the absorbance reading (corrected for dilution). The absorption coefficient for total RNA is usually taken to be $0.025\ \mu g^{-1}\ cm^{-1}$ giving a solution of 40 μg/ml an absorbance of 1 (e.g., 1/0.025 = 40 μg/ml).

15. To determine the purity of the RNA, measure absorbance at 260 nm, 280 nm and 234 nm and calculate the 260/280 and 260/234 ratios. See Chapter 1 for a detailed discussion about the validity of this method and calculation of nucleic acid concentration in the presence of a small amount of protein.

High-salt protocol   This method is used for isolation of RNA from plant material that is recalcitrant to isolation of RNA by the hot-phenol procedure. Consult Table 1 for the amount of material needed to isolate the required quantity of RNA. For all centrifugations described in this procedure, unless otherwise directed, use a Sorvall SS-34 rotor or the equivalent.

1. Place a 250 ml plastic beaker into an ice bucket and pour liquid nitrogen into it. Harvest plant tissue, by cutting it into pieces less than 1 cm^3 size and immediately freeze them in liquid nitrogen. Tissues harvested this way can be stored at -70 °C for several years.

2. Place a mortar into a second ice bucket and pour about 30 ml of liquid nitrogen into it. Cover the ice bucket with a lid and cool the mortar for 5-10 minutes.

3. Quickly weigh the appropriate amount of frozen tissue, taking care that the tissue does not thaw. Use at least 1 g of material for RNA isolation. Place a plastic weighing boat on the balance and tare it. Pick frozen tissue fragments from the liquid nitrogen with forceps and place them in the weighing boat. Transfer the frozen tissue to the frozen mortar.

4. Add a few pieces of dry ice to the mortar. The volume of dry ice used should be roughly 3 times the volume of tissue.

5. Grind the tissue with the mortar and pestle until it appears as a white powder. Transfer the powder into a 50 ml Oak Ridge tube and add 10 ml of homogenization buffer. Add the buffer slowly to allow the dry ice to evaporate. Add as fast as possible add 10 ml of 8-HQ-phenol and 2 ml of CIA. Cap the tube and mix by vortexing for 1 to 2 minutes.

6. Add 0.7 ml of 3 M sodium acetate (pH 5.2) and continue vortexing for 1 minute.

7. Cool the tube on ice and centrifuge at 10 000 rpm (12 000 x g) for 10 minutes at 4 °C.

8. Transfer the aqueous phase to a 25 ml Corex centrifuge tube and add an equal volume of isopropanol. Mix by inverting the tube several times.

9. Collect RNA by centrifugation at 10 000 rpm (16 000 x g) in a swinging bucket rotor (e.g., Sorvall HB-4) for 10 minutes at 4 °C. Discard the supernatant. Centrifugation in a swinging bucket rotor is recommended to collect RNA at the bottom of the tube. This facilitates dissolving the RNA pellet.

10. Wash pelleted RNA two times with cold 70% ethanol. Add 5 ml of cold 70% ethanol to the tube and wash the pellet by carefully rolling the tube at a 45° angle in the palm of your gloved hand. Be careful not to dislodge the RNA pellet from the bottom of the tube during this procedure. Discard the ethanol and drain the tube by inverting it over a paper towel for a few minutes. Repeat the wash one more time.

11. Resuspend the RNA in 900 μl of RNase-free water. Remove insoluble material by centrifugation at 10 000 rpm (16 000 x g) for 5 minutes at 4 °C using a swinging bucket rotor.

12. Transfer the solution to a microfuge tube and adjust the volume to 900 μl with RNase-free water if needed.

13. Add 300 μl of 8 M LiCl to precipitate RNA. The final concentration of LiCl should be 2 M. Incubate on ice for 3 hours or overnight at 4 °C.

14. Collect precipitated RNA by centrifugation. Place the tube into the microfuge, orienting the attached end of the tube lid away from the center of rotation. Centrifuge the tube at maximum speed for 10 minutes at room temperature (See Figure 2, Chapter 2).

15. Remove the tube from the centrifuge. Pour off the ethanol by holding the tube by the open lid and gently inverting it, touch the lip of the tube to the rim of an Erlenmeyer flask. Hold the tube to drain the ethanol. You do not need to remove all the ethanol. Place the tubes back into the centrifuge in the same orientation as before.

Note: When pouring off ethanol do not invert the tube more than once because this could loosen the pellet from the side of the tube.

16. Wash the pellet with 700 μl of cold 70% ethanol. Holding a P1000 Pipetman vertically, slowly deliver ethanol to side of the tube away from the pellet. Hold the Pipetman as shown in Figure 3, Chapter 2. **Do not start the centrifuge,** in this step the centrifuge rotor is used as a "tube holder" that keeps the tube at an angle convenient for ethanol washing. Withdraw the tube from the centrifuge by holding the tube by the lid. Remove ethanol, place the tube back into the centrifuge and repeat the 70% ethanol wash one more time.

Note: This procedure makes it possible to quickly wash the pellet without centrifugation and vortexing. Vortexing and centrifuging the pellet is time consuming and frequently leads to substantial loss of material.

17. Place the tube into the microfuge, making sure that the side containing the pellet faces away from the center of rotation. Start the centrifuge until it reaches 500 rpm (1-2 seconds) to collect the remaining ethanol at the bottom of the tube. Remove all ethanol using a capillary tip.

18. Resuspend the RNA pellet in 200-300 μl of 1 mM EDTA. Measure RNA concentration and store RNA at -70 °C.

19. Determine the concentration of RNA by measuring absorbance at 260 nm. Initially use a 1:100 dilution of the sample in PBS. The absorbance reading should be in the range 0.1 to 1.5. Calculate the concentration of RNA using the equation:

$$N = \frac{A_{260}}{\varepsilon_{260}}$$

Where ε_{260} is the RNA absorption coefficient, N is RNA concentration in $\mu g/ml$ and A_{260} is the absorbance reading (corrected for dilution). The absorption coefficient for total RNA is usually taken to be $0.025\,\mu g^{-1}\,cm^{-1}$ giving a solution of $40\,\mu g/ml$ an absorbance of 1 (e.g., $1/0.025 = 40\,\mu g/ml$).

20. To determine the purity of the RNA, measure absorbance at 260 nm, 280 nm and 234 nm and calculate the 260/280 and 260/234 ratios. See Chapter 1 for a detailed discussion about the validity of this method and calculation of nucleic acid concentration in the presence of a small amount of protein.

The procedure described here is a modification of the procedure recommended by Molecular Center Inc. The procedure is carried out in 1.5 ml microfuge tubes. It can be used with 10 mg of plant or animal tissue prepared by grinding in liquid nitrogen (see section Preparation of cell materials) or with 1×10^6 eukaryotic cells or 1×10^7 bacterial cells. This procedure can be scaled up to prepare a large quantity of RNA.

TRI-reagent™ protocol

1. Add 10 mg of powdered tissue to a microfuge tube. Let the liquid nitrogen evaporate for a few minutes. To isolate RNA from bacterial or yeast cells, the appropriate amount of cells should be collected by centrifugation. Consult Table 1 for the required amount of cells needed to isolate the desired amount of RNA. Washing the cells is not recommended because it can lead to RNA degradation.

2. Add 700 μl of TRI Reagent™. Close the tube and mix by vortexing. Incubate for 5 minutes at room temperature.

Note: The amount of TRI Reagent™ should be no less than 9 volumes of reagent to one volume of material.

Note: When preparing RNA from yeast or bacteria, add an equal volume of acid-washed glass beads and continue vortexing for 1 minute to facilitate cell breakage. To prepare glass beads soak them for 1 hour in concentrated nitric acid, wash acid off with deionized water and dry beads by baking in 180 °C oven overnight.

3. Add 140 µl (0.2 volume) of chloroform (not CIA) and mix by vortexing for 15 seconds. Incubate at room temperature for 2 to 15 minutes.

4. Centrifuge for 15 minutes at room temperature. After centrifugation, the mixture will be separated into two phases, a bottom phase containing chloroform and an upper, aqueous phase containing RNA.

5. Transfer the aqueous phase to a fresh microfuge tube and precipitate RNA by the addition of 175 µl of isopropanol (0.25 volume of the original volume of TRI Reagent™) and 175 µl of high salt precipitation solution. Mix by inverting the tube several times and incubate at room temperature for 5 to 10 minutes.

6. Place the tube into the centrifuge, orienting the attached end of the tube lid, away from the center of rotation. Centrifuge for 10 minutes at room temperature to collect precipitated RNA.

7. Remove the tube from the centrifuge. Remove supernatant using a P200 Pipetman.

8. Wash the pellet with 1 ml of 75% ethanol. Add ethanol to the tube and mix by inverting several times.

9. Place the tube into a microfuge and centrifuge for 5 minutes at room temperature. Remove ethanol with a P200 Pipetman. Repeat ethanol washing one more time.

10. Place the tube into the centrifuge, making sure that the side containing the pellet faces away from the center of rotation. Start the centrifuge until it reaches 500 rpm (1-2 seconds). This will collect ethanol from the sides of the tube. Remove ethanol using a P200 Pipetman equipped with capillary tip.

11. Dissolve the pelleted RNA in 25 – 50 µl of RNase free 1 mM EDTA. To facilitate dissolving RNA, place the tube into a 65 °C water bath for 10 to 15 minutes. Store the RNA sample at -70 °C.

12. Determine the concentration of RNA by measuring absorbance at 260 nm. Initially use a 1:100 dilution of the sample in PBS. The absorbance reading should be in the range 0.1 to 1.5. Calculate the concentration of RNA using the equation:

$$N = \frac{A_{260}}{\varepsilon_{260}}$$

Where ε_{260} is the RNA extinction coefficient, N is RNA concentration in μg/ml and A_{260} is the absorbance reading (corrected for dilution). The absorption coefficient for total RNA is usually taken to be 0.025 μg^{-1} cm^{-1} giving a solution of 40 μg/ml an absorbance of 1 (e.g., 1/0.025 = 40 μg/ml).

13. To determine the purity of the RNA, measure absorbance at 260 nm, 280 nm and 234 nm and calculate the 260/280 and 260/234 ratios. See Chapter 1 for a detailed discussion about the validity of this method and calculation of nucleic acid concentration in the presence of a small amount of protein.

Results

RNA was isolated from 700 mg of corn leaves using the guanidinium hot-phenol procedure. Plant tissue was collected in liquid nitrogen and processed as described on page 130. The total amount of RNA isolated was 183 μg. The yield of the total RNA is consistent with the expected yield pre-

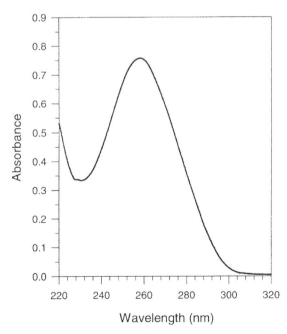

Fig. 2. Absorption spectrum of RNA purified from corn leaves using the hot-phenol procedure. RNA was diluted 40 x in PBS and scanned using a UV spectrophotometer. The 260/280 absorbance ratio was 2.0 and the 260/234 absorbance ratio was 2.0.

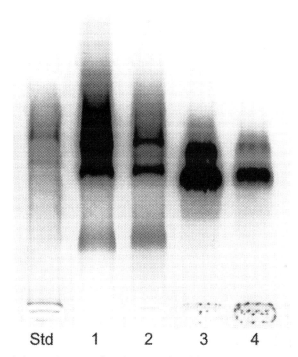

Fig. 3. Native gel electrophoresis of total RNA isolated from corn leaves. Lanes 1 and 2, RNA isolated using the hot-phenol procedure. Lanes 3 and 4 RNA isolated by the TRI Reagent procedure. Lanes 1 and 3 contain 10 µg of RNA. Lanes 2 and 4 contain 5 µg of RNA.

Table 1. Expected yields of total RNA and mRNA obtained from different sources using guanidinium hot-phenol method

Source	Sample size	Total RNA (mg)	mRNA (µg)
Bacteria	4 - 8 x 10^9 cells[a]	0.5 - 1.0	10 - 20
Yeast	2 - 3 x 10^8 cells[b]	0.1 - 0.3	5 - 15
Animal cell cultures	5.0 x 10^7 cells	0.25 - 1.0	12 - 50
Animal tissue (muscle)	100 mg	0.010 - 0.150	0.5-8.0
Animal tissue (liver)	100 mg	0.5 - 0.8	25 - 40
Plant tissue	1.0 g	0.3 - 0.5	15 - 25
Arabidopsis	1.0 g	0.2 - 0.3	10 - 25

[a] 10 ml cells from mid-exponential phase of growth ($A_{600} = 1.0$).

[b] 10 ml of cells from mid-exponential phase of growth ($A_{600} = 1.0$)

sented in Table 1. The 260/280 absorbance ratio was 2.0 and 260/234 absorbance ratio was 2.0, indicating pure RNA. Figure 2 presents the absorption spectrum of 40 x diluted RNA sample. Figure 3 presents native agarose gel electrophoresis of the sample. The presence of mRNAs is indicated by the diffused staining pattern above and below the ribosomal bands.

Isolation of RNA from 700 mg of corn leaves was also carried out using TRI Reagent^{TM}. The amount of RNA prepared was 170 mg giving a yield comparable to that obtained using the hot-phenol procedure. The absorption spectrum of this RNA is presented in Figure 4. The 260/280 absorbance ratio was 2.0 and 260/234 absorbance was 1.5 indicating some contamination with proteins. Native gel electrophoresis of this RNA is presented in Figure 3. Intensity of stain indicates that the ratio of large to small ribosomal RNA is approximately 2:1 but mRNA is less evident in this preparation.

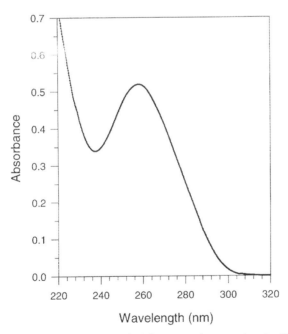

Fig. 4. Absorption spectrum of RNA purified from corn leaves using the TRI-Reagent procedure. RNA was diluted 20 x in PBS and scanned using a UV spectrophotometer. The 260/280 absorbance ratio was 2.0 and the 260/234 absorbance ratio was 1.5.

Troubleshooting

Problems met in purification of RNA are low yield, low purity or degradation of RNA. Low yield of RNA usually results from the following causes listed according to the frequency of their occurrence:

- Inadequate homogenization or lysis of the samples
  If frozen tissue is used as a starting material, this will result in inadequate grinding in liquid nitrogen. In this case, the tissue will not appear as a homogenous white powder. When the hot-phenol protocol is used, insufficient lysis of the cells or tissue will be noticeable after the addition of Sarcosyl and 2-mercapthoethanol in step 1 and before DNA breakage. At this stage the lysate should be viscous.

- Presence of secondary metabolites that interfere with RNA solubility. This problem is most frequently encountered with plant tissue. If the hot-phenol procedure was used, change to the high-salt protocol, which was especially developed for plant material with high levels of secondary metabolites.

- Incomplete solubilization of the final RNA pellet
  To ensure full solubilization of the pellet never dry the RNA pellet under a vacuum. Complete removal of ethanol from the pellet, is not necessary and drying the pellet will interfere with RNA solubility. Difficulties with solubilization of the pellet usually indicate the presence of impurities (See note above).

- Contamination of a stock solution with RNases
  This would result in low yield and poor quality of RNA. Preparation of fresh stock solutions is the best remedy.
  To test for the presence of RNase in stock solutions, the RNaseAlert™ kit from Ambion can be used (Ambion Inc., # 1960). This dipstick-based kit is a convenient and inexpensive way to test reagents for RNase contamination. The detection procedure is very sensitive and can be carried out in a very short time (10 minutes).

Inadequate purity of RNA will be apparent from a low 260:280 ratio (<1.7), low 260:234 ratio (<1.8) or difficulty in dissolving RNA. This can result from (listed according to the frequency of occurrence):

- Too small volume of reagents used during homogenization. The volume of the reagent should always be more than 9 times the volume of the sample. This is particularly important for TRI Reagent™. Increase the ratio of reagent to cell volume.

- Contamination of the sample with secondary metabolites. These impurities will also make RNA difficult to dissolve. To remove these impurities follow recovery procedure 1.

- Contamination of the sample with protein. To remove protein follow recovery procedure 1. However some loss of RNA (up to 40%) may occur.

- Low 260:234 ratio but satisfactory 260:280 ratio indicates contamination with guanidine isothiocyanate or 2-mercaptoethanol rather than with proteins. To remedy this problem precipitate RNA again using the procedure described in recovery procedure 1 step 5.

RNA samples also can be contaminated with residual DNA. This makes it impossible to correctly determine RNA concentration or perform RT-PCR but usually does not interfere with Northern blot hybridization or isolation of poly(A)$^+$ RNA. Removal of DNA is critical if the sample will be used for RT-PCR. To remove DNA, follow recovery procedure 2.

Degradation of RNA can be detected using native agarose gel electrophoresis (see Chapter 12 for the protocol). Nondegraded total RNA should give sharp bands of large and small ribosomal RNA in a ratio of approximately 2:1 when stained with ethidium bromide. Plant material should contain two bands for each class of ribosomal RNA. The smaller rRNA bands originate from chloroplast ribosomes. The chloroplast large rRNA subunit (23S) is particularly sensitive to RNase digestion and its absence in an RNA preparation is a very sensitive indicator of RNA degradation. If RNA is degraded, the most common cause is contaminated stock solutions, and this can be tested as described above.

However, frequently when gel electrophoresis results indicate RNA degradation, the cause is not contamination of stock solutions but contamination of electrophoresis buffers and gel boxes. The source of RNase, in this case, is usually from electrophoresis of DNA samples in the same gel boxes. Since RNase treatment is frequently used in DNA isolation procedures, gel electrophoresis of DNA can cause electrophoresis tray and gel box contamination. Therefore, using the same electrophoresis equipment for DNA and RNA electrophoresis is not recommended. Decontaminate the gel tray and box and use them exclusively for RNA gel electrophoresis. To decontaminate the gel apparatus and gel casting trays, treat them with 0.2 N NaOH for 15 minutes and rinse before use with RNase-free water. RNaseZap® solution can be used instead alkaline of treatment.

Recovery procedure 1

The following procedure will remove contaminating proteins or other impurities from an RNA sample:

1. Add an RNase-free solution of 1 mM EDTA to the sample to a final volume of 500 µl. If insoluble material is present remove it by centrifugation for 10 minutes at room temperature.

2. Transfer the supernatant to a fresh 1.5 ml microtube and add 250 µl of 8-HQ phenol and 250 µl of CIA. Mix by vortexing for 1 minute.

3. Centrifuge for 5 minutes at room temperature to separate phases. Pipette the aqueous phase to a fresh tube and add an equal volume of CIA. Mix by vortexing for 1 minute.

4. Separate phases by centrifugation for 5 minutes at room temperature. Collect the aqueous phase in a clean tube.

5. Add 8 M LiCl to a final concentration of 2 M to precipitate RNA. Incubate on ice for several hours or overnight at -20 °C.

6. Recover precipitated RNA by centrifugation for 15 minutes at room temperature. Wash the pellet twice with cold 70% ethanol using the procedure described in steps 14 to 17 of the high salt protocol. Resuspend RNA in the desired volume of RNase-free water or 1 mM EDTA and determine the purity of the sample by measuring the 260:280 ratio as described in step 14 and 15 of the hot-phenol procedure.

Recovery procedure 2

To remove DNA contamination from RNA, three methods can be used: acid phenol:chloroform extraction, lithium chloride reprecipitation and DNase digestion. We prefer to use the last method that is a little more laborious but ensures best results (Smith, 1998).

1. Add RNase-free water to the RNA sample to final volume of to 50 µl.

2. Prepare DNase treatment solution on ice. DNase I enzyme used should be RNase-free (Ambion # 2222 or the equivalent).

TE buffer	39.7 µl
100 mM $MgCl_2$	10 µl
10 mM DTT	10 µl
DNase (2.5 mg/ml)	0.2 µl
RNasin (25 u/µl)	0.1 µl

3. Add 50 μl of this reagent to 50 μl of RNA sample. Mix by pipetting up and down and incubate 15 minutes at 37 °C.

4. Stop the DNase reaction by adding 1 μl of 0.5 M EDTA.

5. Precipitate RNA with LiCl as described in steps 5 and 6 of recovery procedure 1.

References

Asubel FM, Brent R, Kingston RG, Moor DD, Seidman JG, Smith JA and Struhl K (1987) Current protocols in molecular biology. Wiley & Sons Interscience. New York.

Berger SL, Birkenmeir CS (1979) Biochemistry 18:5143-5149.

Bugos RC, Chiang VL, Zhang X.-H, Campbell ER, Podilla GK, Campbell WH (1995) RNA isolation from plant tissues recalcitrant to extraction in guanidine. BioTechniques 19:734-737.

Chirgwin J, Przybylska A, MacDonald R, Rutter W (1979) Isolation of biologically active ribonucleic acid from sources enriched in ribonuclease. Biochemistry 18:5294-5299.

Chomczynski P, Sacchi N (1987) Single-step method of RNA isolation by acid guanidinium thiocyanate-phenol-chloroform extraction. Anal Biochem 162:154-159.

Chomczynski P (1993) A reagent for the single-step simultaneous isolation of RNA, DNA and proteins from cell and tissue samples. BioTechniques 15:532-537.

Chomczynski P, Mackey K (1995) Modification of the TRI Reagent™ procedure for isolation of RNA from polysaccharide- and proteoglycan-rich sources. BioTechniques 19:942-945.

Davis L, Kuehl M, Battey J (1994) Basic method on molecular biology. 2nd edition. Appelton & Lang, Norwalk, CT.

Hallick RB, Chelm BK, Gray PW, Orozco EM (1977) Use of aurintricarboxylic acid as an inhibitor of nucleases during nucleic acid isolation. Nucleic Acids Res. 4:3055-3064.

Kormanec J, Farkašovsk M (1994) Isolation of total RNA from yeast and bacteria and detection of rRNA in Northern blots. BioTechniques; 17:840-841.

Marcus L, Halvorson HO (1967) Resolution and isolation of yeast polysomes. In: Method in Enzymology 12A:498-502. Grossman L, Moldave K, Eds. Academic Press 1967. New York and London.

Noonberg SB, Scott GK, Benz C (1995) Effect of pH on RNA degradation during guanidinium extraction. BioTechniques 19:731-733.

Smith TJ (1998) Methods to remove DNA contamination from RNA samples. Ambion® TechNotes™ from The RNA Company™ 5:14-15. Ambion Inc. 1998.

Schultz DJ, Craig R, Cox-Foster DL, Mumma RO, Medford JI (1994) RNA isolation from recalcitrant plant tissue. Plant Molecular Biology Reporter. 12:310-316.

Suppliers

Commercially available kits for purification of total RNA

Many total-RNA isolation kits are commercially available. The majority of the kits use a single-step procedure. These kits can be divided into three groups.

The first group uses the method of RNA purification described here as the hot-phenol procedure. These kits include all necessary reagents to carry out the protocol. Thus, use of the kit saves time needed to make all of the components, but at a substantially higher price.

The second group of kits uses solutions identical or similar to TRI Reagent^{TM} to isolate total RNA. The protocol described here uses a kit from Molecular Research Center Inc.; the company that first introduced this methodology. Their recent kit incorporates a modification of this procedure that significantly increases the purity of isolated RNA (Chomczynski and Mackey, 1995).

The third group uses procedures to limit exposure to dangerous chemicals such as organic solvents. Cells or tissues are homogenized in the presence of chaotropic agents to inhibit endogenous RNases and denature RNA-protein complexes. RNA is separated from proteins and DNA using a silica matrix that specifically binds nucleic acids in the presence of a high concentration of chaotropic salts. First, protein is removed by extensive washing, second DNA is eluted followed by the elution of RNA. These kits are well suited to isolate small amounts of RNA from multiple small samples. However, in our hands, the isolated RNA is usually contaminated with proteins and DNA, and large scale procedures are time consuming. These kits are useful for applications where purity of the RNA isolated is not critical.

Isolation of PolyA$^+$ RNA

STEFAN SURZYCKI

Introduction

Extraction, purification and analysis of mRNA are essential in understanding gene expression. Usually 1 to 3% of cellular RNA is present as mRNA, the remaining being ribosomal RNA and various species of small stable RNAs. For some applications, such as construction of cDNA libraries or analysis of low-abundance mRNAs, isolation of mRNA instead of total cellular RNA is essential.

A vast majority of eukaryotic mRNAs contain a poly(A) tract at their 3' end, whereas these tracts are not present in other cellular RNAs. The length of these tracts is quite uniform for all mRNA species and consists of 200 to 250 nucleotides in higher eukaryotes and about 50 to 100 nucleotides in lower eukaryotes (e.g., yeast or green algae). This feature is used to separate mRNA from the remaining nonpolyadenylated RNA in cells. In practice, the poly(A) tracks of mRNA are hybridized, at high salt concentration, to poly dT or poly U oligonucleotides, coupled to a solid matrix. After extensive washing to remove unbound RNA, the retained mRNA is eluted from the solid phase. This procedure gives a 20 to 50-fold enrichment of mRNA that is usually sufficient for Northern analysis or ribonuclease protection assays. For preparation of cDNA libraries, usually a second round of purification is applied to remove at least 90% of rRNA species.

This technique was introduced by Aviv and Leder (1972) who used oligo(dT) bound to cellulose to isolate mRNA. Since then, only small modifications have been made to this procedure, mainly to improve its efficiency. Some of these changes involve the introduction of different matrices to which oligo(dT) is bound. The new matrices used are: (a) latex beads that have a large surface to bind oligo(dT). This increases the poly(A)$^+$ RNA binding capacity of the matrix. (b) Magnetic particles that permit easy collection of bound mRNA without centrifugation or column elution. These new matrices make it possible to isolate mRNA directly from cell lysates. Another improvement to the original procedure is substitution of poly(U) for oligo(dT) as the binding material (More and Sharp, 1984). Use of poly(U) allows for more efficient capture of mRNA with short poly(A) residues due to the high stability of DNA:RNA hybrids.

Methods of mRNA isolation

Three protocols for mRNA isolation are presented. Two are useful for the isolation of large amounts of mRNA, primarily from previously isolated total RNA. The third procedure can be used for fast isolation of mRNA from multiple samples directly from cell lysates.

The first procedure uses gravity-run, oligo(dT) cellulose-packed columns. This method is ideally suited to purify a large quantity of mRNA from total RNA. Other advantages of the procedure are: high quality of the product obtained and low cost of material used. A disadvantage of the procedure is that it is relatively time consuming, increasing the possibility of accidental RNase contamination and degradation of the product. The method should be used with no less than 1 to 5 mg of total RNA as starting material. Direct isolation of mRNA from cell lysates does not give satisfactory results using this method and is not recommended.

The second method uses oligo(dT)-cellulose packed into spin columns. The advantage of this method is its speed and convenience, especially when used with commercially prepared spin columns. The method can be used with about 1 mg or less of total RNA as starting material. The use of spin columns is ideal for second round purification of mRNA after the first round isolation of poly(A)$^+$ RNA using gravity columns. Direct isolation of mRNA from cell lysates is also possible, but satisfactory results are obtained only when lysates are not very rich in protein. Disadvantages of this method are: low capacity in the first round of purification (1 mg of total RNA), difficulty in removing rRNA from the sample and somewhat higher cost. Careful calibration of centrifuge speed and close attention to manufacturer-recommended centrifugation time is critical for the successful application of this procedure.

The third method uses oligo(dT) bound to the surface of magnetic particles. The method is useful for rapid isolation of mRNA directly from cell lysates of multiple samples. The protocol can be scaled up and used for rapid isolation of mRNA from total RNA. The purified product is well suited for Northern blot analysis and RT-PCR. The main disadvantage of this method is its relatively high cost, since it requires the purchase of magnetic separators.

A crucial aspect of all mRNA purification procedures, even more so than in most RNA techniques, is protection against contamination with exogenous RNases. To isolate intact messenger by binding its poly(A) tail to the matrix, no breaks between the 5' and 3' end of mRNA molecules can be tolerated. This results in separation of mRNA from its poly(A) end and as a consequence, the ability to isolate intact mRNA molecules is lost. To guard against exogenous RNase contamination, all necessary precautions and methods of RNase removal from reagents and equipment should be followed. These methods are described in detail in Chapter 6. Moreover, when working with low concentrations of RNA, nonspecific binding of RNA to the sides of columns and tubes could be substantial. Using plastic tubes and tips during the entire procedure is essential. Silanizing plasticware is not necessary and we do not recommend it due to the high toxicity of the reagent. Most commercially available plastic tubes and tips do not bind RNA.

A list and short description of commercially available kits for mRNA isolation is presented at the end of this chapter.

Outline

A schematic outline of the procedures is shown in Figure 1.

Materials

- 25 ml Corex centrifuge tubes with Teflon-lined caps (Corex # 8446-25)
  Tubes should be made RNase-free by baking at 180 °C for at least 2 hours or overnight. For other methods of removing RNase from glassware see Chapter 6.
- 15 ml sterilized, polypropylene centrifuge tubes (for example Corning # 25319-15)
- Oligo(dT)-Cellulose (Type 7, Pharmacia # 27-5543-02 or equivalent).
- Spin columns (4 ml Select-D column, Eppendorf 5' # 0-300750 or equivalent)
  These are the least expensive DEPC-treated columns available that can be used for column chromatography without autoclaving.
- Oligo(dT) cellulose spin column kit (polyA Spin kit, New England Bio-Labs, Inc. or equivalent)
- Oligo(dT) magnetic beads (mRNaid oligo(dT$_{30}$) kit, Bio 101 # 2510-200 or equivalent). Magnetic stand is included with the kit.

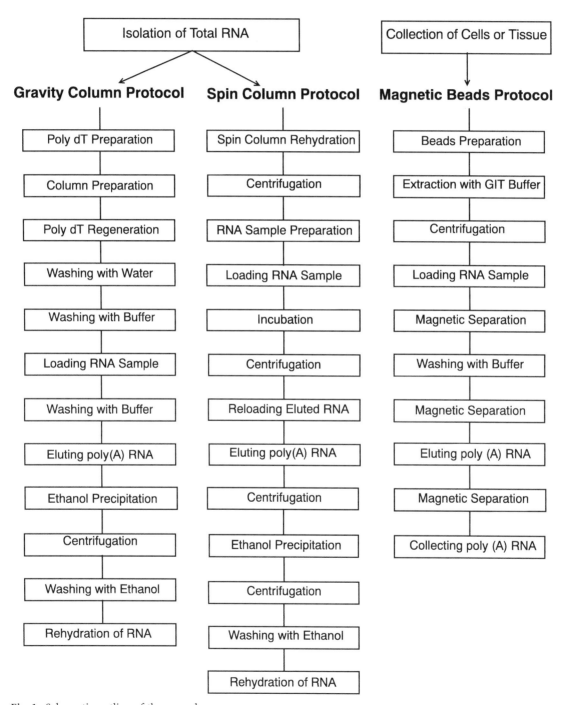

Fig. 1. Schematic outline of the procedures

- 2-mercaptoethanol (Sigma Co. # M 6250)
- Antifoam C (Sigma Co. # A 8011)
- Guanidinium isothiocyanate (Fluka # 50981)
- Diethyl pyrocarbonate (DEPC) (Sigma # D 5758)
- RNaseZap Solution (Ambion # 9780)
- Sarcosyl (N-lauroylsarcosine, sodium salt, Sigma Co. # L 5125)

- **DEPC Treated Water**

  Add 0.1 ml DEPC to 100 ml of deionized water. Shake vigorously to dissolve DEPC in water. Incubate 2 hours or overnight at 37 °C. Autoclave to inactivate DEPC. DEPC must be completely inactivated before using any solution treated with it because it will carboxylate RNA. Carboxylated RNA will inhibit reverse transcriptase, DNA:RNA hybridization and in vitro translation. Use DEPC treated water for preparation of the solutions that cannot be treated with DEPC directly. Primarily these are solutions containing primary amines, such as Tris or ammonium acetate.
  Safety Note: When using DEPC, wear gloves and use a fume hood because DEPC is highly flammable and a suspected carcinogen.

Note: Deionized water from a MilliQ RG apparatus (Millipore Co. # ZD5111584) can be used directly in all applications instead of DEPC-treated water because it does not contain RNase.

- **0.5 M EDTA Stock Solution, pH 8.5**
  Weigh an appropriate amount of EDTA (disodium ethylenediamine-tetraacetic acid, dihydrate) and add it to distilled water while stirring. EDTA will not go into solution completely until the pH is greater than 7.0. Adjust the pH to 8.5 using a concentrated NaOH solution, or by adding a small amount of pellets, while mixing and monitoring the pH. Adjust the solution to its final volume with water and sterilize by filtration. EDTA stock solution can be stored indefinitely at room temperature.
- **1 mM EDTA Solution**
  Prepare this solution using 0.5 M EDTA stock and RNase-free water. Sterilize by autoclaving and store at -20 °C to prevent contamination.
- **Loading Buffer**
  - 20 mM Tris-HCl pH 7.6
  - 0.5 M NaCl
  - 1.0 mM EDTA, pH 8.5
  - 0.1% SDS

  Prepare enough for a single application using RNase-free stock solutions and water.

- **Washing Buffer**
  - 20 mM Tris-HCl, pH 7.6
  - 100 mM NaCl
  - 1 mM EDTA, pH 8.5
  - 0.1% SDS

  Prepare from RNase-free stock solutions. Do not store, use immediately.
- **Elution Buffer**
  - 20 mM Tris-HCl, pH 7.6
  - 1 mM EDTA, pH 8.5
  - 0.05% SDS

  Prepare before use from stock solutions and RNase-free water.
- **Regenerating Solution**
  - 0.1 N NaOH
  - 5 mM EDTA

  Prepare before use from 10 N NaOH stock. Do not store.
- **3 M Sodium Acetate pH 5.5**

  Weigh 24.6 g of sodium acetate and add 50 ml deionized water. Dissolve the salt and adjust pH with glacial acetic acid to pH 5.5. Add water to 100 ml and DEPC to 0.1%, mix very well and incubate overnight or for 2 hours at 37 °C. Autoclave for 30 minutes to remove DEPC and store at 4 °C.
- **GIT Solution**
  - 4 M guanidinium isothiocyanate
  - 25 mM sodium acetate pH 5.2
  - 1 mM EDTA
  - 0.05% Antifoam C

  Prepare 100 ml of solution at a time. Add 47.2 g of guanidinium isothiocyanate to 50 ml water followed by 0.85 ml of 3 M sodium acetate, and 1 µl of antifoam C. Fill up to 100 ml with water and heat to facilitate dissolving. The solution is light sensitive and can be stored at room temperature in a dark bottle for a month.

Procedure

Gravity column protocol

1. Weigh 0.15 g of oligo(dT) cellulose powder. Transfer it to a 15 ml sterile, plastic centrifuge tube and add 5 ml of washing buffer. Mix by inverting several times. The amount of material will be enough to build a single column of 0.5 ml packed volume. This column can be loaded with 1 to 10 mg of total RNA and can absorb about 650 µg of poly(A)$^+$ RNA.

2. Open the bottom of a select-D column and fit it with a laurel lock or tygon tubing with a clamp. Pour the oligo(dT)-cellulose slurry into the column and drain washing buffer slowly by opening the bottom clamp. Make sure that the final volume of the oligo(dT) cellulose bed is at least 0.5 ml.

3. Wash the column two times with 3 ml of regenerating solution. Let it stand at room temperature the first time for 20-30 minutes. This will sterilize oligo(dT) cellulose. If oligo(dT) cellulose was used previously this will restore its binding capacity.

4. Wash the column with 4 ml portions of sterilized water until the pH drops to 8. This usually requires only two to three washes.

5. Add 4 ml of washing buffer and wash the column slowly. Check the pH of eluent with pH paper. It should be between 7.4 to 7.8. Continue washing with 4 ml portions of loading buffer until the pH reaches the desired value.

6. Add 1 to 10 mg of total RNA sample to a microfuge tube. Fill up to 900 µl with eluting buffer and add 5 µl of 10% SDS. Heat the sample at 65 °C for 5 minutes. Add 100 µl of 5 M NaCl and cool to room temperature for 2 minutes.

7. Apply 1 ml of sample to the column. Add sample to the top and let it pass completely into the column. Collect any eluent into a 15 ml plastic centrifuge tube. Stop the flow from the column and allow the RNA to bind to the column for 5 minutes. Heat the eluate collected to 65 °C for 10 minutes. Cool quickly on ice and reapply to the column.

Note: The rate of elution from the column during loading and subsequent washes should be about 1 drop per 5-10 seconds. Adjust the flow to this rate using a clamp.

8. Wash the column with two 4 ml portions of washing buffer. Collect eluate in a single 15 ml plastic centrifuge tube. This fraction contains nonpolyadenylated RNA and can be saved for other uses if desired.

9. Continue washing the column three times with 4 ml portions of washing buffer. The total washing volume should be approximately 10 times column volume.

10. Elute poly(A)⁺ RNA with two, 2 ml portions, of elution buffer. Collect both eluates into a single 15 ml plastic centrifuge tube. The total volume of elution buffer should always exceed 2 column volumes.

Note: The eluted mRNA will be contaminated with small amounts of rRNA. Ribosomal RNA should be removed if you are going to use mRNA for preparation of a cDNA library or RT-PCR. A second application of the RNA to a new column will remove 90% of the ribosomal RNA. Adjust the first eluted sample to 0.5 M NaCl with 5 M NaCl and reapply it to an oligo(dT) cellulose column equilibrated with binding buffer. Alternatively, the first eluate can be purified a second time using a spin column rather than gravity column.

Note: Used oligo(dT) cellulose columns can be stored at -20 °C and regenerated for further use as described in step 3 of this protocol.

11. To precipitate RNA, add 0.4 ml of 3 M sodium acetate (1/10 of volume) and 11 ml (2.5 volume) of 95% ethanol. Transfer the solution to a 25 ml Corex tube and leave it at -70 °C until the fraction freezes (about 30 to 60 minutes) or overnight at -20 °C.

12. Collect precipitated RNA by centrifugation for 20 minutes at 10 000 x g (16 000 rpm) in a swinging bucket rotor (Sorvall HB-4 or equivalent). Drain the tube by inverting it over Kimwipes for 5 minutes.

13. Add 5 ml of cold 70% ethanol to the tube and wash the pellet by carefully rolling the tube at a 45° angle in the palm of your hand. Be careful not to dislodge the RNA pellet from the bottom of the tube. Discard ethanol and drain the tube well by inverting it over a paper towel for a few minutes. Repeat the wash one more time.

14. Add 50 µl of RNase-free 1 mM EDTA and dissolve RNA by gently pipetting up and down. More buffer may be necessary if the RNA does not dissolve. Transfer the RNA to a microfuge tube and store at -70 °C.

15. Determine the concentration of RNA by measuring absorbance at 260 nm. Initially use a 1:10 dilution of the RNA in RNase-free PBS. For correct calculation of RNA concentration, the absorbance reading should be in the range of 0.1 to 1.5 OD_{260}. A solution of 40 µg/ml has an optical density of 1 in a 1 cm path cuvette.

Note: RNA concentration should never be measured in water or TE buffer. This will result in incorrect determination of RNA concentration (Wilfinger et al., 1998). For a discussion of determination of nucleic acid concentration and its dependence on ionic strength and pH see Chapter 1.

16. Determine RNA purity by measuring the absorbance at 280 nm and 234 nm. Calculate the 260/280 and 260/234 ratios. See Chapter 1 for a detailed discussion about the validity of this method and calculation of nucleic acid concentration in the presence of a small amount of protein.

17. Check the integrity of the RNA by native agarose gel electrophoresis as described in Chapter 12.

Spin column protocol

The procedure described here essentially follows the recommended protocol for the polyA Spin mRNA Isolation Kit from BioLabs® Inc. Similar kits are available from other manufacturers. The spin column has the capacity to isolate poly(A)⁺ RNA from 1 mg of total RNA.

1. To prepare a spin column, add 200 µl of loading buffer to dry cellulose. Mix well and centrifuge for 10 seconds. Discard column eluent. Prewarm the elution buffer in a 65 °C water bath.

2. Pipette the desired amount of total RNA into a microfuge tube and add loading buffer to a final volume of 450 µl. Add 50 µl of 5.0 M NaCl and mix well by inverting the tube several times.

3. Incubate the RNA sample for 5 minutes in a 65 °C water bath, then quickly cool it on ice for at least 3 minutes.

4. Apply the RNA sample to a spin column. Close the column with the cap provided, and mix by inverting several times. Oligo(dT) cellulose should be thoroughly mixed with the sample, forming a uniform white solution.

5. Incubate at room temperature for 5 minutes, occasionally mixing by inverting the column.

6. Centrifuge for 10 seconds to collect liquid. Remove the spin column reservoir and transfer the eluent back into the original microfuge tube.

7. Incubate collected eluate for 5 minutes at 65 °C, then cool quickly on ice. Add the sample back into a spin column and repeat steps 4 to 6. This eluent contains nonpolyadenylated RNA, it can be stored at - 70 °C for further use if desired.

8. Wash the spin column with loading buffer. Add 400 µl of loading buffer and resuspend the cellulose matrix by inverting the column several times.

9. Collect the eluent by centrifugation for 10 seconds at room temperature. Remove the column reservoir and discard the eluent. Repeat steps 8 and 9 three times. Collected eluent can be discarded because it contains little or no RNA.

10. Wash the spin column once with washing buffer. Add 400 µl of washing buffer and mix well with the cellulose matrix. Centrifuge for 10 seconds at room temperature and discard the eluent.

11. Place the spin column into a clean microfuge tube. Add 200 µl of Elution buffer prewarmed to 65°C. Resuspend cellulose matrix by inverting the column several times. Incubate for 2 minutes with occasional agitation.

12. Centrifuge for 10 seconds and collect eluent into a fresh microfuge tube. The eluent contains poly(A)$^+$ RNA. Repeat elution of mRNA one more time with 200 µl of prewarmed elution buffer. Join both eluents.

Note: If very pure mRNA is desired, repeat the poly(A$^+$) RNA purification using a new spin column. Adjust the salt concentration of the first eluent to 0.5 M NaCl and repeat steps 3 to 12 one more time.

13. Add 44 µl of 3 M sodium acetate and 20 µl of glycogen carrier, supplied with the kit, to the eluent. Mix well by inverting several times.

14. Precipitate mRNA by adding 1 ml of 95% ethanol. Mix well and leave at -20 °C for 30 minutes.

Note: At this point mRNA can be stored as an ethanol precipitate at -70 °C until needed.

15. Collect precipitated RNA by centrifuging for 10 minutes at room temperature. Place the tube into the centrifuge, orienting the attached end of the lid away from the center of rotation.

16. Remove the tube from the centrifuge. Pour off the ethanol by holding the tube by the open lid and gently lifting the end, touch the tube to the edge of an Erlenmeyer flask. Hold the tube to drain the ethanol. You do not need to remove all the ethanol. Place the tubes back into the centrifuge **in the same orientation as before.**

17. Wash the pellet (often not visible) with 1 ml of 75% ethanol. Holding a P1000 Pipetman vertically, slowly deliver ethanol to the side of the tube opposite the pellet. Hold the Pipetman as shown in Figure 3, Chapter 2. **Do not start the centrifuge**, in this step the centrifuge rotor is used as a "tube holder" that keeps the tube at an angle convenient for ethanol washing. Withdraw the tube from the centrifuge by holding the tube by the lid. Remove ethanol, place the tube back into centrifuge and repeat the 75% ethanol wash one more time.

Note: This procedure makes it possible to quickly wash the pellet without centrifugation and vortexing.

18. Place the tube into a centrifuge, making sure that the side containing the pellet faces away from the center of rotation. Start the centrifuge until it reaches 500 rpm (1 - 2 seconds). This will collect ethanol from the sides of the tube at the bottom. Withdraw the ethanol using a capillary tip.

19. Dissolve the pelleted RNA in 20 - 50 µl of RNase-free, 1 mM EDTA. To facilitate dissolving RNA, place the tube into a 65 °C water bath for 5 to 10 minutes. Store the RNA sample at -70 °C.

20. Determine the concentration of RNA by measuring absorbance at 260 nm. Initially use a 1:10 dilution of the RNA in RNase-free PBS. For correct calculation of RNA concentration, the absorbance reading should be in the range of 0.1 to 1.5 OD_{260}. A solution of 40 µg/ml has an absorbance of 1 in a 1 cm path cuvette.

Note: RNA concentration should never be measured in water or TE buffer. This will result in incorrect determination of RNA concentration (Wilfinger et al, 1998). For a discussion of determination of nucleic acid concentration and its dependence on ionic strength and pH see Chapter 1.

21. Determine RNA purity by measuring the absorbance at 280 nm and 234 nm. Calculate the 260/280 and 260/234 ratios. See Chapter 1 for a detailed discussion about the validity of this method and calculation of nucleic acid concentration in the presence of a small amount of protein.

Magnetic beads protocol

This method essentially follows the protocol described for mRNaid kit from Bio 101 Inc. This kit uses magnetic beads that have been coupled to an oligo(dT_{30}) matrix. Direct coupling of oligo(dT) is preferable to that using streptavidin-labeled particles and oligo(dT) bound to biotin. The kit is inexpensive and easy to use. It includes a magnetic stand and is useful for fast preparation of mRNA directly from lysed samples. The yield and purity of isolated mRNA is not high, but it is sufficient for RT-PCR or single Northern analysis. If higher yields and purity of RNA are desired, magnetic kits from other manufactures can be used (for example, Straight A's mRNA from Novagen or Biomag mRNA kit from PerSeptive Biosystems), but at substantially higher costs.

1. Prepare magnetic beads. Add 100 µl of magnetic beads to a microfuge tube. Add 300 µl of ice cold loading buffer and wash the beads by inverting the tube several times. Place the tube into the single tube matrix collector supplied with the kit and collect the beads on the side of the tube. Remove and discard the supernatant. Transfer the tube from the magnetic stand to an ice bucket. Add 100 µl of loading buffer and resuspend magnetic beads by inverting the tube several times.

Note: 100 µl of the matrix is sufficient for isolation of poly(A)$^+$ RNA from about 100 µg of total RNA.

2. Prepare tissue or cell lysate. Grind animal or plant tissue as described in Chapter 6. Use between 20 to 50 mg of tissue or 1 to 5 x 10^6 animal cells. Consult Table 1 in Chapter 6 for amounts of other material needed to prepare a lysate containing 100 µg of total RNA.

3. Collect cells or ground tissue in a microfuge tube and add 1 ml of hot (65 °C) GIT solution. Mix the tube by vortexing briefly and add 20 µl of 2-mercapthoethanol and 100 µl of 5 M NaCl. Mix by vortexing. Continue vortexing until the lysate becomes less viscous.

Note: Final concentration of NaCl at this step should be 0.5 M.

4. Centrifuge at room temperature for 5 minutes to collect cell debris.

5. Collect the supernatant and add it to the tube with magnetic beads. Mix well and incubate 30 to 60 minutes on ice with occasional mixing.

6. Place the tube into the magnetic stand and collect the beads on the side of the tube. Remove the supernatant using a Pipetman. Return the tube to an ice bucket. This supernatant will contain nonpolyadenylated RNA that can be stored if desired.

7. Resuspend the beads in 100 µl of ice cold loading buffer. Keep the tube on ice during this procedure.

8. Wash the pellet three times with 300 µl of ice cold loading buffer, each time separating the beads magnetically as described above. Discard supernatants.

9. Add 10 to 25 µl of hot (60 °C to 80 °C) elution buffer to elute mRNA. Incubate for 2 to 5 minutes at 60 °C to 80 °C.

10. Use the magnetic stand to pull the beads to the side of the tube. Transfer the supernatant to a clean microfuge tube. The supernatant contains polyadenylated RNA that is ready for further use.

Note: More than 90% of bound RNA should be eluted in this step. To remove remaining mRNA, the elution procedure can be repeated one more time.

Results

The gravity column procedure permits isolation of poly(A)+ RNA from as much as 5 mg of total RNA. The spin column and magnetic bead procedures, although faster to perform, have much lower capacity. For most preparations, one should expect recovery of at least 3 to 5% of total input RNA when using single selection, and about 1 to 1.5% of the original starting material after the second selection. This corresponds to RNA that is about 50% poly(A)+ RNA after a single selection and greater than 90% poly(A)+ after the second selection. The main contaminant remaining in the preparation is rRNA which cannot be entirely removed by any procedure so far designed (Milburn, 1996). The amount of mRNA isolated directly from cells using the magnetic beads protocol will be about 5 to 10 times lower than the amount of mRNA expected on the basis of total RNA present in the cells (see Table 1, Chapter 6 for amounts of RNA per cell).

A single round of selection yields mRNA of sufficient quality for most Northern analysis and ribonuclease protection assays and in vitro translation. A second selection is necessary if mRNA is to be used for cDNA library construction or RT-PCR, especially if analysis of a rare transcript is desired. In gel electrophoresis, the selected RNA should appear as a smear from 20 kb down with the greatest intensity in the 5- to 20-kb range for higher eukaryotes and somewhat smaller size (1 to 5 kb) for lower eukaryotic cells, respectively. If the second round of selection is successful, there should be little or no visible rRNA bands present when up to 5 μg of sample is electrophoresed.

Troubleshooting

Problems met in the purification of poly(A)+ RNA are: low yields and low quality of product. The yield largely depends on the quality of total RNA from which polyadenylated RNA is isolated. The quality of isolated mRNA depends on two factors:

a. the presence of intact, full length mRNA, and

b. degree of contamination with nonpolyadenylated RNA species, mainly ribosomal RNA.

Low yield and poor quality of mRNA usually results from (given in the order of their occurrence):

- Poor quality of total RNA used for selection of mRNA
  The total RNA sample should be analyzed on a native agarose gel and visualized by ethidium bromide staining before using the sample for mRNA purification. Two ribosomal RNA bands should be present with the upper band (large ribosomal subunit RNA) about twice as intense as the lower band (small subunit ribosomal RNA). The mRNA should be visible as a smear extending from slightly above the large ribosomal subunit to slightly below the small ribosomal subunit. Absence of rRNA bands or an RNA smear indicates poor quality of total RNA. Total RNA should be reisolated and used for mRNA purification.

- Overestimation of the concentration of total RNA
  This most frequently results from DNA present in the sample. This contamination is impossible to detect by absorbance at 260 nm alone. However, the degree of DNA contamination can be evaluated when native gel electrophoresis of total RNA is performed (see above). The gel should contain standard RNA of a known concentration. The intensity of ethidium bromide staining of the standard RNA should be compared to the intensity of the unknown RNA sample. Lower sample intensity than expected from the amount of RNA estimated from absorbance at 260 nm indicates DNA contamination. DNA contamination can be removed using recovery procedure 2 described in Chapter 6.

- Incomplete elution from oligo(dT) cellulose columns
  This will happen when loading buffer was not completely removed from the column before application of elution buffer. This problem is mostly encountered when small elution volumes are used. Repeat the elution with an additional volume of hot elution buffer or RNase-free water.

- Inadequate precipitation of RNA from the eluted sample
  This usually results from too low salt concentration in the sample. Add the correct amount of 3 M sodium acetate to the eluent before addition of ethanol. Incubate the tube at -20 °C for at least a half hour or overnight. Addition of a carrier such as glycogen can aid in precipitation of RNA. However, we did not find that carrier, as claimed, helps in RNA precipitation. All gains in recovery of precipitated RNA in the presence of the carrier can be attributed to making the pellet visible. This helps in redissolving the RNA completely. Follow the precipitation procedure as described in steps 11 to 13 of the gravity column procedure or 13 to 18 of spin column procedure to precipitate the RNA completely.

- Contamination of stock solutions or columns with RNase
  To test for the presence of RNase in stock solutions, the RNaseAlert kit from Ambion can be used (Ambion # 1960). The dipstick-based kit is a convenient and inexpensive way to test reagents for RNase contamination. The detection procedure is very sensitive and can be carried out in a short time (10 minutes).
  Oligo(dT) cellulose can be easily contaminated with RNase during weighing and column preparation. Always wash the column with regenerating solution as described in step 3 of the gravity column procedure. Spin column and magnetic beads are supplied RNase-free and do not require this treatment.

References

Aviv H, Leder P (1972) Purification of biologically active globin messenger RNA by chromatography on oligo-thymidilic acid-cellulose. Proc Natl Acad Sci USA 69:1408-1412.

Milburn S (1996) Purify mRNA rapidly with high yield. Ambion TechNotes 3:1-8.

Moore CL Sharp PA (1984) Site-specific polyadenylation in a cell-free reaction. Cell 36:581-591.

Wilfinger WW, Mackey K, Chomczynski P (1997) Effect of pH and ionic strength of the spectrophotometric assessment of nucleic acid purity. BioTechniques 22:474-480.

Suppliers

Commercially available kits for poly(A)⁺ RNA isolation

Ambion, Inc. (www.ambion.com)
Two kits are offered. Poly(A)Pure® and MicroPoly(A)Pure®. These kits are designed for the rapid purification of mRNA directly from cells or tissues. Kits utilize guanidinium thiocyanate solution during homogenization and batch binding of mRNA to oligo(dT). Washing and collection of mRNA is performed using spin columns. PolyA tails as short as eight bases can be isolated.

Amersham Pharmacia Biotech. (www.apbiotech.com)
Two kits, QuickPrep and QuickPrep Micro, are available for direct isolation of mRNA. These kits work with a wide variety eukaryotic cells and tissues. Guanidinium thiocyanate buffer is used to lyse the cells or tissues. The Quick-Prep kit uses an oligo(dT) slurry for isolation of mRNA from 5×10^7 cells or 0.5 g tissue. QuickPrep Micro uses spin columns containing pre-packed oligo(dT) cellulose to isolate mRNA from 1×10^7 cells or 100 mg of tissue.

For isolation of mRNA from total RNA preparations, the company offers a mRNA purification kit containing pre-packed oligo(dT) columns.

Amresco, Inc. (www.amresco-inc.com)
Amresco, Inc. offers a Rapid mRNA Purification kit that uses oligo(dT) cellulose columns for isolation of mRNA from total RNA.

Bio 101 Co. (www.bio101.com)
The mRNAid kit uses oligo(dT)$_{30}$ bound to magnetic particles. This kit can be used to isolate mRNA directly from a cell lysate or from total RNA preparations.

Biometra Co. (www.biometra.com)
This company offers a mRNA isolation kit that uses oligo(dT) chains covalently bound to magnetic beads. A long chain linker is used to reduce steric interactions between particles and mRNA. Very pure poly(A)$^+$ RNA can be obtained in an exceptionally small elution volume. The source of mRNA purification is a total RNA preparation.

CPG, Inc. (www.cpg-biotech.com)
Their MPG Direct mRNA purification kit uses magnetic particle technology to isolate mRNA directly from cells or tissues. Streptavidin magnetic particles are used to bind oligo(dT). A high density of oligo(dT)$_{25}$ is provided for poly(A) tail binding. The binding capacity of these particles is greater than any other magnetic particles to which oligo(dT) is bound directly.

Clontech, Inc. (www.clontech.com)
Clontech provides mRNA Separator kit for isolation of poly(A)$^+$ RNA from total RNA preparations. This kit uses oligo(dT) cellulose spin columns.

Cruachem, Inc. (www.cruachem.com)
Cruachem's ISOLATE mRNA Pure Prep kit uses magnetic particles with covalently bound oligo(dT)$_{25}$ with long linker arms to prevent steric hindrance with mRNA binding. This kit is used to isolate poly(A)$^+$ RNA from a total RNA preparation.

Dynal, Inc. (www.dynal.no)
Dynal is the company that introduced magnetic beads technology. Dynabeads Oligo(dT)$_{25}$ are uniform supermagnetic polymer beads with a very high binding capacity. Two kits are offered. The mRNA Direct Kit is used for direct isolation of mRNA from cells or tissues lysates. The Dynabeads mRNA Purification Kit can be used for purification of mRNA from total RNA preparations.

Invitrogen, Inc. (www.invitrogen.com)
Two kits are offered Fast Trac® and Micro-Fast Trac®. Both kits use oligo(dT) cellulose for isolation of mRNA directly from cells or tissues.

Life Technologies BRL (www.lifetech.com)
This company offers a number of mRNA isolation kits for direct or indirect isolation of mRNA. All kits are based on oligo(dT) cellulose technology in a filter-syringe format. Either TRIZOL reagent or phenol guanidine is used for cell or tissue lysis.

Novagen, Inc. (www.novagen.com)
This company offers several kits for direct isolation of mRNA from cells and tissues. Paramagnetic bead technology is used in all of these kits with proprietary lysis and washing buffers. Each kit is optimized for specific starting material e.g., plant tissues, animal tissues, frozen or fresh materials, etc. These kits give the most consistent results with exceptional RNA purity.

PerSpective Biosystem Co. (www.pbio.com)
BioMag mRNA purification kit consists of supermagnetic particles that are covalently attached to oligo(dT)$_{20}$. This kit is used for direct isolation of mRNA from cells, plant and animal tissue or from a total RNA preparation. Yields of mRNA isolated are high when using this kit.

Promega Co. (www.promega.com)
Promega offers a number of mRNA isolation kits all based on magnetic bead technology. Magnetic particles have streptavidin attached to them. The mRNA is hybridized in solution to biotinylated oligo(dT) that in turn is captured on beads by streptavidin-biotin interaction. This system allows both direct isolation of mRNA from lysates and isolation of mRNA from total RNA preparations.

Roche Molecular Biochemicals, formerly Boehringer Mannheim (biochem.roche.com)
Several kits are offered all based on oligo(dT) cellulose technology. Starting material can be cells, tissues or total RNA preparations. mRNA is hybridized with biotinylated oligo(dT) that in turn is captured in streptavidin-coated microfuge tubes. Kits work best with total RNA preparations.

Qiagen Co. (www.qiagen.com)
Company offers a series of mRNA isolation kits that use spherical latex particles with very large surface area for binding oligo(dT). The size of the particles is such that they remain in solution during hybridization but can be quickly pelleted by centrifugation. A high yield of very pure mRNA is isolated using these kits. Kits can be used for direct isolation of mRNA from cells and tissues or from total RNA preparations.

Sigma Chemical Co. (www.sigma.sial.com)
Kits offered by this company can be used for isolation of mRNA from cells or tissues. Technology used is oligo(dT) cellulose columns.

Stratagene, Inc. (www.stratagen.com)
Poly A Quick mRNA Isolation Kit uses push columns filled with oligo(dT) cellulose for fast isolation of mRNA from total RNA. Using quanidinium thiocyanate and β-mercaptoethanol to disrupt cells. These kits can be used for direct isolation of mRNA.

Agarose Gel Electrophoresis of DNA

STEFAN SURZYCKI

Introduction

This chapter outlines the theory and practice of agarose gel electrophoresis. The description includes a detailed discussion of all essential factors influencing the optimal separation of DNA bands in agarose gels. Protocols for running standard agarose gels and high resolution agarose gels are given, as well as a method for recovery of DNA fragments from the gel.

Principle of electrophoresis

When a molecule is placed in an electric field it will migrate to the appropriate electrode with a velocity or free electrophoretic mobility (M_0) described by the following formula:

$$M_0 = \frac{E}{d} \frac{q}{6\pi R\eta} \tag{1}$$

Where: E is the potential difference between electrodes measured in volts (V); q is the net charge of the molecule; d is distance between electrodes (cm); η is viscosity of the solution; R is Stock's radius of the molecule, and E/d is field strength.

Since, under physiological conditions, phosphate groups in the phospho-sugar backbone of DNA (RNA) are ionized, these polyanions will migrate to the positive electrode (anode) when placed in an electric field. Due to the repetitive nature of the phospho-sugar backbone, double-stranded DNA molecules have a net charge to mass ratio approximately the same. This results in approximately the same free electrophoretic mobility (M_0) of DNA molecules, irrespective of their size. However, by adjusting the viscosity of the electrophoretic medium, one can accentuate the effects of friction on the mobility of the molecules. If viscosity is very large, the mobility of the molecules subjected to electrophoresis will depend largely on their shape and size. Equation 1 simplifies to:

$$M_0 = \frac{E}{d} \frac{1}{R} \tag{2}$$

To increase the viscosity of an electrophoretic medium, specific support matrixes are used. These include agarose and polyacrylamide. Varying the pore size using various agarose concentrations or different crosslinking ratios of polyacrylamide alters the viscosity of these materials. The mobility of DNA molecules is profoundly influenced by the size and shape of the molecules, as well as, by the size of the matrix pores. Using these gels, DNA molecules are fractionated by their size and conformation in a relatively fast and inexpensive way.

Principle of agarose gel electrophoresis of DNA

At one point or another, virtually every experiment with nucleic acids requires the use of agarose gel electrophoresis. The technique is simple, rapid to perform, and adaptable to analytical or preparatory applications.

Agarose is a polysaccharide consisting of basic agarobiose repeat units of 1,3-linked β-D-galactopyranose and 1,4-linked 3,6-anhydro-α-L-galactopyranose. These units form long chains of approximately 400 repeats, reaching a molecular weight of approximately 120,000 daltons. Long polymer chains contain small amounts of charged residues, consisting largely of pyruvate and sulfate. These residues are responsible for agarose properties that are important in its use in gel electrophoresis. Most important is the phenomenon of electroendosmosis (EEO) (Adamson, 1976; Himenz, 1977). During electrophoresis only hydrated positive ions, normally associated with the fixed anionic groups of agarose (pyruvate or sulfate residues), can move toward the cathode. Water is therefore pulled with these positive ions toward the negative electrode and negative molecules, like DNA migrating toward the positive electrode, are slowed down. For DNA electrophoresis, agarose with the lowest possible EEO is recommended for maximum separation of DNA molecules.

The electrophoretic migration rate of DNA through agarose gels depends on the following parameters:

- Size of the DNA molecules.

- The concentration of agarose gel.

- Voltage applied.

- Conformation of the DNA.

- The buffer used for electrophoresis.

On first approximation, DNA molecules travel through a gel at a rate inversely proportional to the logarithm of their molecular weights or number of base pairs (Helling et al., 1974). Thus, a plot of mobility against log of the size should give a straight line for all DNA sizes. However, this is true for a narrow range of DNA size. A better linear relationship between mobility and DNA size is obtained in plots of DNA base pair number (DNA size) versus 1/mobility (Sealey and Southern, 1982).

Mobility of DNA fragments

The useful linear range of mobility depends on the gel concentration used and voltage applied. A DNA fragment of a given size migrates at different rates in gels containing different concentrations of agarose. A model for gel structure predicts that the log of the mobility of different DNA molecules (M) as a function of gel concentration (C) should result in a straight line with different slopes called retardation coefficients (K_r) and intercepts named free mobility (M_o), mobility of DNA molecules at zero concentration of agarose. This can be expressed mathematically by the following equation:

$$\text{Log M} \; = \; \log M_0 - CK_r \tag{3}$$

Using gels of different agarose concentrations, it is possible to resolve a wide range of DNA fragment sizes, provided that the voltage gradient applied to the gel is chosen correctly. Normally the migration rate of DNA fragments is directly proportional to the voltage applied. With increased voltage, however, large DNA molecules migrate at a rate proportionally faster than small molecules. Consequently, the field strength applied to most gels should be between 0.5 V/cm and 10 V/cm. In general, higher resolution is achieved at a low voltage gradient, especially if higher molecular weight DNA is used. Small DNA molecules should be run at higher-voltage gradients to prevent their diffusion during electrophoresis. The amount of DNA in a sample will also affect its apparent mobility. Overloaded bands will appear to move faster than bands with the correct amount of DNA. Therefore, the amount of DNA loaded into each sample well should be similar when comparing the mobility of DNA fragments. The useful separation ranges of various gel concentrations do overlap, and different electrophoretic conditions can shift their useful range. Data presented in Table 1 can serve as a starting point for choosing a gel concentration and voltage gradient for agarose electrophoresis of DNA.

The DNA shape (conformation) also influences the migration rate. Plasmid DNA can exist in three conformational states: closed circular supercoiled molecules (form I); nicked circular relaxed molecules (form II) and linear molecules (form III). Introducing a single-stranded nick converts form I molecules into form II molecules. Introducing a double-stranded cut to

DNA conformation

Table 1. Suggested agarose concentration and field strength for optimal resolution of DNA fragments

Agarose	DNA Fragment Size (bp)[a]			
Concentration (% w/v)	SeaKem®	MetaPhor®	SeaPlaque®	Field strength (V/cm)[b]
0.60-0.75	1 000-23 000	n/a	500-25 000	0.5-1.0
0.80	800-10 000	n/a	500-20 000	0.5-1.0
1.00	400-8 000	n/a	300-20 000	1.0-2.0
1.20-1.25	300-7 000	n/a	200-12 000	1.0-2.0
1.50	200-4 000	n/a	150-6 000	2.0-3.0
1.75	150-2 000	n/a	100-3 000	2.0-3.0
2.00	100-3 000	150-800	50-2 000	2.0-3.0
3.00	n/a	100-600	n/a	2.0-3.0
4.00	n/a	50-25	n/a	2.0-3.0
5.00	n/a	2-130	n/a	2.0-3.0

[a] Modified from date presented in FMC BioProducts Catalog.

[b] Data from Davis et al., 1980.

supercoiled molecules (form I) converts them into linear molecules (form III). Thus, forms II and III are the result of the action of nucleases during plasmid purification and should not be present in preparations of plasmid DNA isolated using an alkaline procedure (for explanation see Chapter 1).

The relative mobility of the three forms is dependent primarily on the agarose concentration and, to a lesser extent, on the strength of the current applied and the ionic strength of the buffer. Under most conditions, supercoiled DNA (form I) migrates faster than linear DNA (form III) of the same size. The circular DNA (form II) migrates faster than supercoiled DNA in TAE buffer and slower than supercoiled DNA in TBE buffer.

The presence of ethidium bromide in the gel can drastically change the mobility of DNA due to changes in the number and direction of superhelical twists in form I DNA. As the concentration ethidium bromide increases the negative superhelical turns in form I DNA molecules are progressively removed and their rate of migration is slowed. Incorporation of more ethidium bromide into these molecules introduces positive superhelical turns into molecules resulting in an increase in their electrophoretic mobility.

As a general rule, in TAE buffer, plasmid DNA migrates in 0.8-1% gels in the following order (from slower to faster molecules, or reading from the well down): linear DNA (form III) supercoiled DNA (form I) and, relaxed circular DNA (form II). In TBE buffer, this order would be as follows: relaxed circular DNA (form II), linear DNA (form III) and supercoiled DNA (form I).

Several different buffers are used for agarose gel electrophoresis. These are: Tris-acetate EDTA buffer (TAE), Tris-borate EDTA buffer (TBE) and Tris-phosphate EDTA buffer (TPA).

Electrophoresis buffers

Tris-acetate EDTA buffer (TAE; 40 mM Tris-base, 20 mM acetic acid 1 mM EDTA) buffer is the most frequently used buffer for DNA electrophoresis. This buffer has rather low buffering capacity that may require recirculation of the buffer upon prolonged electrophoresis if a two-tank apparatus is used. The buffer is particularly well suited for submarine gel electrophoresis, permitting short running times at relatively high voltage gradients without excessive heating. The ratio of voltage applied to current (mA) is approximately one, for a wide variety of gel sizes and buffer volumes when this buffer is used (Perbal, 1988). The tracking dye, bromophenol blue,

Table 2. Migration of DNA fragments in relation to bromophenol blue dye

Agarose concentration (%)	Size of DNA fragment co-migrating with dye in TAE buffer (bp)			Size of DNA fragment co-migrating with dye in TBE buffer (bp) (%)		
	SeaKem[®]	MetaPhor[®]	SeaPlaque[®]	SeaKem[®]	MetaPhor[®]	SeaPlaque[®]
0.60-0.75	1 000	n/a	500	720	n/a	250
0.80	900	n/a	500	700	n/a	200
1.00	500	n/a	350	400	n/a	180
1.20-1.25	370	n/a	200	260	n/a	100
1.50	300	n/a	150	200	n/a	70
1.75	200	n/a	100	110	n/a	50
2.0	150	70	60	70	40	30
3.0	n/a	40	n/a	n/a	35	10
4.0	n/a	35	n/a	n/a	30	n/a
5.0	n/a	30	n/a	n/a	15	n/a

[a] Table was assembled using data presented in FMC BioProducts Catalog.

will travel in this buffer, at a rate of about 1 cm/hr at 1 to 10 V/cm field strength. Thus, this marker dye co-migrates with the smallest DNA molecules at each agarose concentration (see Table 2).

Tris-borate EDTA buffer (TBE; 89 mM Tris-base, 89 mM boric acid 2 mM EDTA) has very high buffering capacity. It can be used when DNA less than 12 000 bp is electrophoresed but gives superior results to TAE in electrophoresis of DNA fragments less than 1000 bp. DNA mobility in this buffer is approximately two times slower than in TAE buffer. This is due to the lower porosity of agarose gel when agarose polymerizes in the presence of borate. Ionic strength of TBE is high, resulting in 4:1 ratio of voltage to current (mA) during electrophoresis for a wide variety of gel sizes and buffer volumes (Perbal, 1988). In general, DNA bands are sharper when TBE buffer is used but time of electrophoresis is considerably longer.

An important advantage of TBE is that bacteria or fungi will not grow in a 1x buffer solution so it is possible to store the buffer at room temperature. A considerable disadvantage is that borate interacts with the hydroxyl groups on the agarose polysaccharide, forming a non-covalent tetrahydro-borate complex. This complex interferes with the action of chaotropic agents used to dissolve agarose gels in procedures of DNA fragment recovery. Thus, TBE buffer cannot be used when a silica-binding procedure for DNA fragment recovery is used.

Tris-phosphate buffer (TPE; 89 mM Tris-base, 23 mM H_3PO_4, and 2.5 mM EDTA). This buffer has high buffering capacity, comparable to that of TBE buffer. DNA mobility in this buffer, however, is similar to that in TAE buffer due to a similar pore size formed during polymerization. The buffer has a high ionic strength resulting in a voltage (V) to current (mA) ratio similar to that obtained in TBE buffer. This buffer has an advantage over Tris-borate buffer, in that gels can be dissolved in concentrated chaotropic agents, permitting recovery of DNA fragments using a silica powder binding procedure.

Gel size Agarose gel electrophoresis is commonly carried out using submerged horizontal slab gels (submarine gels). The best separation between DNA bands in such a system is achieved in gels approximately 20 cm long, 15 cm wide and about 4 mm thick. Electrophoresis should be continued until the bands of interest migrate 1/2 to 2/3 the length of the gel. To obtain maximum resolution of many bands electrophoresis should be continued until tracking dye (for example, bromophenol blue) has moved 70% to 80% the length of the gel. The size of the sample well also can affect the resolution of DNA bands. The optimal length of the sample wells for a large gel is between 0.5 to 1 cm long, and optimal width of the well is 1 to 2.0 mm. The sample well bottom should be 0.5 to 1.0 mm above the gel bottom. Minigels usually have sample wells 0.2 to 0.5 cm long and 1 mm (or less) wide. The distance between sample wells should be less then half of the length of the well to ensure comparable mobility of the DNA band across the gel. Large gels should be run using at least 1 liter of electrophoresis buffer to prevent heating of the gel slab during electrophoresis. Most of the commercially available submarine electrophoresis gel boxes fulfill the above requirements. An article evaluating the performance of all commercially available agarose gel systems has been published and is an excellent guide for buying new equipment (Baldwin, 1995).

Sample concentration

The amount of DNA loaded into one well can vary considerably without affecting the mobility of the DNA. For standard large gels (see description above), the DNA load can vary from 1 to 10 ng of DNA per band. Concentrations above 100 ng per band should be avoided, since this will overload an analytical gel. The total amount of DNA loaded per well should not exceed 10 μg. In general, one should use decreasing amounts of DNA per well as the running voltage of the gel is increased.

Sample loading solutions

DNA samples are prepared for electrophoresis by the addition of loading dye solution. The composition of loading dye solution plays an essential role in obtaining sharp DNA bands. This solution serves three vital functions: it is used to terminate enzymatic reactions before electrophoresis (stop solution), it provides density for loading the sample into the well, and it provides a way to monitor the progress of electrophoresis.

Most loading dye solutions contain EDTA to stop enzymatic reactions. However, this is frequently not sufficient to fully dissociate DNA-protein complexes, the presence of which will affect not only mobility of DNA fragments but also can cause an excessive smearing and widening of the bands. To remove these complexes, all loading dye solutions should contain a protein denaturing agent. Urea, at concentration of 5 M, is the best protein denaturing agent used in the loading dye solution because it does not interact with agarose or affect DNA mobility.

Glycerol or sucrose, at concentrations of 5% to 10%, is commonly used to provide density. However, using these low molecular weight compounds results in U-shaped DNA bands, due to sample streaming up the side of the well before beginning electrophoresis (Sealy and Southern, 1983). This effect is particularly pronounced when electrophoresis is carried out at low field strength. To increase the sharpness of the bands and prevent their U-shape appearance, Ficoll 400 at concentration of 15% to 20%, should be used to provide density.

A number of different dyes are in use as tracking dyes. Bromophenol blue (dark blue) and xylene cyanol FF (turquoise) are two most frequently used dyes. Tables 2 and 3 present the relationship between migration of these dyes and DNA fragments at various agarose gel concentrations in two of the most commonly used gel electrophoresis buffers. These dyes cannot be used in alkaline agarose gels because they decompose. Bromocresol green is used in such gels. Different tracking dyes are also used such as Tartrasine (orange) or Cresol red (violet). These dyes usually are used in PCR reaction buffers because they do not affect the activity of Taq DNA polymerase. This permits direct loading of PCR products on the gel without the addition of loading dye solutions.

Table 3. Migration of DNA fragments in relation to xylene cyanol dye [a]

Agarose concentration (%)	Size of DNA fragment co-migrating with dye in TAE buffer (bp)			Size of DNA fragment co-migrating with dye in TBE buffer (bp)		
	SeaKem®	MetaPhor®	SeaPlaque®	SeaKem®	MetaPhor®	SeaPlaque®
0.60-0.75	9 200	n/a	4 000	7,100	n/a	2 850
0.80	8 000	n/a	3 000	6 000	n/a	2 000
1.00	6 100	n/a	2 300	4 000	n/a	1 700
1.20-1.25	4 100	n/a	1 500	2 250	n/a	1 000
1.50	2 600	n/a	1 000	1 900	n/a	700
1.75	2 00	n/a	700	1 400	n/a	500
2.0	1 500	480	550	1 000	310	400
3.0	n/a	200	n/a	n/a	140	n/a
4.0	n/a	120	n/a	n/a	85	n/a
5.0	n/a	85	n/a	n/a	60	n/a

[a] Table was assembled using data presented in FMC BioProducts Catalog.

Gel staining   To visualize DNA, agarose gels are usually stained with ethidium bromide. This is the most rapid, sensitive and reproducible method currently available for staining single-stranded DNA and double-stranded DNA (Sharp et al. 1973). Ethidium bromide binds to double-stranded nucleic acid by intercalation between stacked base pairs. The mobility of linear DNA in gel electrophoresis is reduced by about 15%. Ultraviolet irradiation of ethidium bromide at 302 nm and 366 is absorbed and re-emitted as fluorescence at

590 nm. Similarly, energy absorbed by DNA irradiated at 260 nm is transmitted to intercalated dye and re-emitted as fluorescence at 590 nm. The intensity of fluorescence of dye bound to DNA is much greater than that of free dye suspended in agarose. This results in very low background and a high intensity of fluorescence from DNA bands. The best staining results are obtained by incorporating ethidium bromide into the gel at a concentration of 0.5 μg/ml. This permits direct observation of the progress of electrophoresis and limits the amount of ethidium bromide-contaminated liquid waste. The gels can be stained after electrophoresis by immersing the gel in a solution containing 0.5 μg/ml of ethidium bromide for about 0.5 to 1 hour. This procedure is not recommended because it is less sensitive than incorporating ethidium bromide into the gel and generates a large volume of ethidium bromide-contaminated liquid.

The most sensitive photographs of ethidium-bromide-stained DNA are obtained when DNA is illuminated with ultraviolet light at 254 to 260 nm rather than by ethidium bromide direct illumination at 300 nm.

Agarose gels also can be stained using silver stain. Silver stained agarose gels will yield sensitivity similar to ethidium bromide staining. The best results are obtained with the use of Silver Stain Plus kit manufactured by Bio-Rad Laboratories. However, this staining takes a long time and cannot be used to observe DNA during electrophoresis or when DNA fragments are going to be isolated from the gel.

Photographing gels

Photographs of the gels provide not only a permanent record of the experiment, but also permit analysis of the data and visualization of DNA bands not visible to the unaided eye. Polaroid cameras, equipped with appropriate filters, are usually used for this purpose. The most commonly used fast Polaroid film type 667 (ASA 3000) can record a DNA band containing 2-4 ng when loaded into a 1 cm well. A more sensitive method of recording gel results is the use of a computer imaging system equipped with a charge-couple device (CCD) digital camera. The sensitivity of the computer imaging system is approximately ten times greater than the sensitivity of photography. This permits visualization and recording of as little DNA per band as 0.1 ng. The commercially available instruments for digital recording are rather expensive. However, construction of such equipment, using an inexpensive camera ($100), was recently described that, in our hands, gave results comparable to those of the expensive commercial instruments (Scott et al. 1996).

Elution of DNA fragments from agarose gels

Many satisfactory protocols for purifying DNA fragments from agarose gels have been developed. They can be classified into four categories: elution of DNA fragments by electrophoresis, elution by using low-melting-point agarose or enzymatic digestion of the agarose polymer, dissolving agarose by chaotropic salts and binding DNA fragments to silica particles. The most important consideration for choosing any of these protocols is speed of technique, yield of the DNA fragments and purity of the DNA obtained.

Electroelution

The first procedure, introduced in the 1970s, is easily reproducible for a wide variety of DNA fragment sizes and gives pure DNA (Wienand et al., 1978). A major disadvantage of its use is that it is rather laborious and time consuming. Eluted fragments, frequently, require further purification to remove contaminating material co-eluted from agarose. Organic solvent extraction or a matrix that specifically binds DNA fragments is used to remove these impurities. This method is most frequently applied when it is necessary to isolate large fragments of DNA (more than 10 000 kb) that are inefficiently recovered by most other methods.

An interesting modification of the electroelution method that obviates the requirement for additional purification of eluted DNA fragments is electroelution of DNA during electrophoresis directly into DEAE paper (Girvitz et al., 1980; Dretzen et al., 1981). An incision in the gel is made ahead of the DNA band to be eluted in which DE81 paper is inserted. Electrophoresis is continued until the entire fragment is absorbed onto the paper. The DNA fragment is eluted from the paper with a high salt buffer.

Low-melting-point agarose

This method uses a low-melting-temperature agarose to recover DNA fragments (Wieslander, 1979). The section of agarose containing the desired DNA fragment is melted at 65 °C and DNA is recovered either by repeated organic solvent extraction or absorption to silica powder. One advantage of this method is that it permits recovery of large DNA fragments. A second, is the ability to perform enzymatic reactions such as ligation, restriction digestion and labeling of DNA fragments without removing the agarose. The disadvantage of this method is its rather high cost and low resolving power. Moreover, the recovery of large DNA fragments from low melting-point agarose is less reproducible than the other two methods.

Agarose-digesting enzyme

An alternative procedure for dissolving the gel is the use of enzymes that digest agarose (Burmeister and Lehrach, 1989). Numerous modifications of the original protocol have been introduced that eliminate the need for further purification with organic solvents or silica powder. DNA is recovered by ethanol precipitation and ready for subsequent manipulations. Large molecular weight DNA can be recovered with satisfactory efficiency. Disadvantages of the enzymatic protocol are the high cost and the time required to completely digest agarose, particularly when high concentration gels are used. Also, digestion of agarose from some sources leaves an ethanol-perceptible, water-insoluble precipitate that interferes with subsequent procedures such as ligation and restriction digestion.

Silica powder

The method was originally described by Vogelstein and Gillespie (1979) and is now the most frequently used method for recovering DNA fragments from agarose gels. It takes advantage of the fact that, at high concentration of chaotropic salts, DNA binds selectively to silica particles. Salts most frequently used are either sodium iodide or guanidine thiocyanate. These salts not only promote DNA binding to the silica particle but also can dissolve agarose gels. Unbound agarose, salts, proteins, nucleotides, oligonucleotides are washed away with a buffered ethanol solution. The bound DNA is eluted in a small volume of TE buffer or water. Using this method, it is possible to recover DNA fragments between 0.5 to 10 Kb in size. A modification of the original method permits the recovery of DNA fragments in the 100 to 500 bp ranges (Smith et al., 1995). Numerous kits that employ this method are commercially available. Some commercial vendors offer such kits and a brief description of each kit is given at the end of this chapter.

Gel staining for recovery

To recover DNA fragments from an agarose gel using any of the procedures described, the DNA must first be visualized on the gel. However, when DNA is going to be recovered from agarose gels, ethidium bromide staining and visualization of the DNA bands by UV illumination should not be used. Irradiation of DNA by short wave UV light, in the presence of ethidium bromide, results in its rapid degradation (Nikogosyan, 1990). It has been shown that viability of lambda DNA decreases ten times per one second irradiation at 260 nm (Davis, et. al., 1980). Gel illuminators at wavelength of 300 to 320 nm are commonly assumed not to cause detectable single-stranded breaks, even after a few minutes exposure (Brunk and Simpson, 1977). Long-wave illuminators are, therefore, used to visualize DNA fragments for excision. However, some limited cross-linking of DNA strands was reported to occur during this illumination that eventually interfered with efficient ligation, cloning and priming in PCR experiments (Cariello et al., 1988; Hartman, 1991, Mäueler et al., 1994). It was recently shown that only 1 % of DNA remains intact after 20 to 45 seconds of UV-B light illumination (Gründemann and Schömig, 1996). Thus, when DNA bands are visualized for their subsequent excision from the gel, even the use of long wave UV illuminators can cause DNA damage when such illumination is not kept to a minimum.

Solutions to the DNA-damage problem have recently been proposed. The first method avoids illumination of agarose gels with UV light altogether by staining the gel with crystal violet or nile blue (Rand, 1996: Adkins and Burmeister 1996). As little as 5 ng of DNA per band can be visualized by incorporating crystal violet or nile blue into the gel and electrophoresis buffer at a concentration of 10 μg/ml. The second method uses cytidine or guanosine incorporated into the gel running buffer to protect DNA against UV damage (Gründemann and Schömig, 1996). In this method the gel is stained using a standard concentration of ethidium bromide and is illuminated by UV light to visualize DNA bands. Thus, the sensitivity of the visualization of DNA bands provided by ethidium bromide is not sacrificed with this procedure. The protection of DNA, using this method, is substantial because more than 99% of the irradiated DNA is available for ligation, cloning and PCR.

Outline

A schematic outline of the procedure is shown in Figure 1.

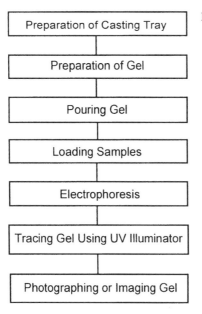

Fig. 1. Schematic outline of the procedure.

Materials

- Agarose powder SeaKem® LE (FMC BioProducts # 50001) or equivalent
- MetaPhor® agarose (FMC BioProducts # 50181)
- SeaPlaque® low melt agarose (FMC BioProducts # 50101, or equivalent)
- Ethidium bromide (Sigma Co. # E 8751)
- Ficoll 400 (Sigma Co. # F 4375)
- GELase (Epicentre Technologies # G09050), or β-agarase (New England Biolab # 392S)
  Epicentre Technologies GELase is a mixture of agarase digesting enzymes and can be used with regular and low melting agarose, as well as with denaturing gels.
- Elu-Quick DNA recovery kit (Schleicher and Schuell # 74450)
- Gel electrophoresis apparatus (minimum 13 cm x 20 cm gel size) with power supply (e.g., Owl Scientific # A1)

Solutions

- **0.5 M EDTA Stock Solution, pH 8.5**
  Weigh an appropriate amount of EDTA (disodium ethylenediamine-tetraacetic acid, dihydrate) and add it to distilled water while stirring. EDTA will not go into solution completely until the pH is greater than 7.0. Adjust the pH to 8.5 using a concentrated NaOH solution or by adding a small amount of pellets while mixing and monitoring the pH. Bring the solution to a final volume with water and sterilize by filtration. The EDTA stock solution can be stored indefinitely at room temperature.
- **50 x TAE Electrophoresis Buffer**
  - 2 M Trisma base
  - 1 M acetic acid
  - 50 mM Na_2 EDTA
  Weigh 242 g of Trisma base and add to 800 ml of double distilled or deionized water. Add 57.1 ml of glacial acetic acid and 100 ml of 0.5 M EDTA stock solution (pH 8.5). Dissolve the powder by continuous stirring for half an hour and add water to final volume of 1 liter. Do not autoclave. Store at room temperature tightly closed.
- **Ethidium Bromide Stock Solution (10 mg/ml)**
  - 100 mg ethidium bromide
  - 10 ml water.
  Dissolve the powder in the water by stirring under a chemical hood. Store at room temperature in a tightly closed, dark bottle.
  Safety Note: Ethidium bromide is a powerful mutagen and suspected carcinogen. It is particularly toxic when inhaled in powder form. Take all possible precautions not to inhale it during preparation of

the stock solution. Weigh powder under a chemical hood and wear gloves.

– **Loading Dye Solution**
  – 15% Ficoll 400
  – 5 M Urea
  – 0.1 M Sodium EDTA, pH 8
  – 0.01% Bromophenol blue
  – 0.01% Xylene cyanol

Prepare at least 10 ml of the solution. Dissolve an appropriate amount of Ficoll powder in double distilled or deionized water by stirring at 40 - 50 °C. Add stock solution of EDTA, powdered urea and dyes. Aliquot about 100 µl into microfuge tubes and store at -20 °C.

Procedure

Large scale agarose gel electrophoresis

This procedure describes the use of large agarose gels (25 cm x 13 cm x 0.4 cm) for separation of DNA fragments. DNA band separation in such gels is sufficient to detect a single copy gene in a large genome. These gels require longer electrophoresis times but give better resolution of DNA fragments than smaller gels. The well length is 6.5 mm and well width is 1.5 mm, dimensions optimal for obtaining sharp DNA bands with as little as 0.2 µg or as much as 10 µg of DNA per well. In addition, the large buffer volume limits heating during gel electrophoresis. If other gel systems are used, the amount of DNA should be adjusted accordingly.

Standard protocol

1. Seal the opened ends of the gel casting tray with tape. Regular labeling tape or electrical insulation tape can be used. Check that the teeth of the comb are about 0.5 mm above the gel bottom. To adjust this height, it is most convenient to place a plastic charge card (e.g., MasterCard) under the comb and adjust the comb height to a position where the card is easily removed from under the comb.

2. Prepare 1500 ml of 1x TAE by adding 30 ml of a 50 x TAE stock solution to a final volume of 1500 ml of deionized water.

3. Place 130 ml of the buffer into a 500 ml conical flask and add the appropriate amount of agarose. Weigh 1.0 g of agarose for a 0.8% agarose gel. Melt the agarose by heating the solution in a microwave oven at full power for approximately 3 minutes or by boiling for 10 minutes in a

boiling water bath. Swirl the agarose solution to ensure that the agarose is dissolved; that is, no agarose particles are visible. If evaporation occurs during melting, adjust the volume to 150 ml with deionized water.

Note: The efficient range of separation of DNA molecules in 0.8% SeaKem® agarose is 800 to 10 000 bp. See Table 1 for other size ranges and agarose types.

4. Cool the agarose solution to approximately 60 °C and add 5 µl of ethidium bromide stock solution. Slowly pour the agarose into the gel casting tray. Remove any air bubbles by trapping them in 10 ml pipette.
Safety Note: Ethidium bromide is a mutagen and suspected carcinogen. Contact with skin should be avoided. Wear gloves when handling ethidium bromide solution and gels containing ethidium bromide.

5. Position the comb approximately 1.5 cm from the edge of the gel. Let the agarose solidify for about 30 to 60 minutes. After the agarose has solidified, remove the comb with a gentle back and forth motion, taking care not to tear the gel.

6. Remove the tape from the ends of the gel casting tray and place the tray on the central supporting platform of the gel box.
Safety Note: For safety purposes, the electrophoresis apparatus should be always placed on the laboratory bench with the positive electrode (red) facing away from the investigator, that is, away from the edge of the bench.

7. Add electrophoresis buffer to the buffer chamber until it reaches a level approximately 0.5-1 cm above the surface of the gel.

8. Load the samples into the wells using a yellow tip. Place the tip **under** the surface of the electrophoresis buffer just **above** the well. Expel the sample slowly, allowing it to sink to the bottom of the well. Take care not to spill the sample into a neighboring well. During sample loading, it is very important not to place the end of the tip into the sample well or touch the edge of the well with the tip. This can damage the well resulting in uneven or smeared bands. The DNA size standard should be loaded in the first well from the left. This is because many computer programs that are used for calculating DNA fragment size require the size standard in this position.

Note: Samples must be loaded in sequential sample wells. Do not "skip" wells when loading fewer samples than number of wells. In this case, load DNA samples in wells at the center of the gel.

9. Place the lid on the gel box and connect the electrodes. DNA will travel toward the positive (red) electrode positioned away from the edge of the laboratory bench. Turn on the power supply. Adjust the voltage to about 1 to 5 V/cm. For example if the distance between electrodes (not the gel length) is 40 cm, to obtain a field strength of 3 V/cm, voltage should be set to 120 V. Continue electrophoresis until the tracking dye moves at least half of the gel length. It will take the tracking dye approximately 3-5 hours to reach this position on a gel 20 cm long.

Note: For best separation of large fragments of eukaryotic DNA, electrophoresis should be carried out at field strength of 1 V/cm or less. For a standard electrophoresis apparatus with 40 cm between electrodes the voltage setting should be 35 V and time of electrophoresis approximately 12 to 17 hours (overnight).

10. Turn the power supply off and disconnect the positive (red) lead from the power supply. Remove the gel from the electrophoresis chamber. Trim the gel to the desired size discarding pieces that do not contain DNA.
    Safety Note: To avoid electric shock always disconnect red (positive) lead first.

11. Place the gel on a UV illuminator, cover it with a transparent acetate sheet and draw the outline of the gel with a felt pen. Mark the position of the wells and position of visible DNA bands. Do not forget to mark the position of DNA bands in the standard lanes. It is very important that your drawing be as precise as possible. Label the contents of each well on the acetate sheet and mark the bottom left corner of the gel. This schematic drawing can be used to locate the positions of the hybridized bands on the autoradiogram of the Southern blot.
    Safety Note: Ultraviolet light can damage the retina of the eye and cause severe sunburn. Always use safety glasses and a protective face shield to view the gel. Work in gloves and wear a long-sleeved shirt or labcoat when operating UV illuminators.

12. You can photograph the gel using Polaroid film 667 and a Polaroid camera at a speed of 1 and F8 or use a computer imaging system to record the results (see Introduction, Principle of agarose gel electrophoresis of DNA for details). It is not necessary to photograph the gel with a ruler because the location of hybridized bands can be directly determined from the schematic drawing prepared in step 11.

High resolution agarose electrophoresis

This procedure uses MetaPhor® agarose that has twice the resolution capabilities of standard agarose. DNA fragments that differ by only 4 bp can be resolved using this matrix in the size range between 40 to 800 bp. MetaPhor™ agarose has approximately the resolution power of standard polyacrylamide gels. This agarose is particularly useful for separation of PCR products.

DNA fragments can be recovered easily from the gel using standard recovery techniques. The concentrations of MetaPhor® gels for various DNA sizes suggested by manufacturer are presented in Table 1 and mobility of the standard tracking dyes, bromophenol blue and xylene cyanol, are presented in Tables 2 and 3.

1. Use a minigel casting tray when working with this agarose (e.g., gel size: 7.5 cm x 7.5 cm x 0.4 cm). Preparing smaller gels limits the cost and does not affect resolution of DNA bands. Seal the ends of the gel casting tray with tape. Regular labeling tape or electrical insulation tape can be used. Check that the bottom of the comb is about 0.5 mm above the gel bottom. To adjust this height it is most convenient to place a plastic charge card (for example, MasterCard) at the bottom of the tray and adjust the comb height to a position where it is easy to remove card from under the comb.

2. Prepare 500 ml of 1 x TAE by adding 10 ml of a 50 x TAE stock solution to 490 ml of deionized water.

3. Place 30 ml of the buffer into a 250 ml Erlenmeyer flask and add the appropriate amount of agarose. For example, use 0.75 g of agarose for a standard 2.5% gel. Allow the powder to swell in the buffer for at least 25 minutes. Melt the agarose by heating the solution in a microwave oven at full power for 20 to 30 seconds at a time, until the agarose is fully dissolved. The MetaPhor™ agarose is more difficult to melt than regular agarose since it "boils over" very easily. If evaporation occurs during melting, adjust the volume to 30 ml with deionized water.

4. Cool the agarose solution to approximately 60 °C and add 1 μl of ethidium bromide stock solution. Slowly pour the agarose into the casting tray. Remove any air bubbles by trapping them in 10 ml pipette.
Safety Note: Ethidium bromide is a mutagen and suspected carcinogen. Avoid contact with skin. Wear gloves when handling ethidium bromide solution and gels containing ethidium bromide.

Agarose
electrophoresis
protocol

5. Position the sample comb at approximately 1.0 cm from the edge of the gel. Let the agarose solidify for about 30 minutes. To achieve maximum resolution, after the MetaPhor® has solidified, transfer the gel to a 4 °C refrigerator for 30 minutes.

6. Remove the comb with a gentle back and forth motion, taking care not to tear the gel. Remove the tape from the ends of the gel casting tray and place the tray on the central supporting platform of the buffer chamber. **Safety Note:** For safety purposes the electrophoresis apparatus should always be placed on the laboratory bench with the positive electrode (red) facing away from the investigator, that is, away from the edge of the bench.

7. Add electrophoresis buffer to the buffer chamber until it reaches a level approximately 0.5 to 1 cm above the surface of the gel.

8. Load the samples into the wells using a yellow tip. Place the tip **under** the surface of the electrophoresis buffer and **above the sample well** opening. Deliver the sample slowly, allowing it to sink to the bottom of the well. During loading it is very important not to place the tip into the well or touch the edge of the well with it. This can damage the well resulting in uneven or smeared bands. DNA size standards should be loaded in the first well from left. This is because many computer programs that are used for calculating DNA fragment size require the size standard in this position.

Note: Samples should always be loaded in sequential sample wells. Do not "skip" wells when loading fewer samples than number of wells. If a small number of samples is electrophoresed, place them in the sample wells at the center of the gel.

9. Place the lid on the gel box and connect the electrodes. DNA will travel toward the positive (red) electrode positioned away from the edge of the laboratory bench. Turn on the power supply. Adjust the voltage to obtain field strength about 1 to 5 V/cm. For example, if the distance between electrodes (not the gel length) is 20 cm, to obtain a field strength of 3 V/cm, the power supply should be set to 60 V. Continue electrophoresis until the tracking dye moves at least 2/3 of the gel length. It should take the tracking dye approximately 30 to 40 minutes to reach this position using approximately 70 to 90 volts for a gel 10 cm long.

10. Turn the power off and disconnect the leads from the power supply. **Safety Note**: To avoid electric shock always disconnect red (positive) lead first.

11. Remove the gel tray from the electrophoresis chamber and place it on an UV illuminator. Photograph the gel using Polaroid film 667 and a Polaroid camera at speed 1 and F stop 8 or use a computer imaging system to record results.

    Safety Note: Ultraviolet light can damage the retina of the eye and cause severe sunburn. Always use safety glasses and a protective face shield to view the gel. Work in gloves and wear a long-sleeved shirt or labcoat when working with the UV illuminator.

Recovery of DNA fragments from agarose gels

Two protocols to recover DNA fragments from agarose are given: The first protocol uses silica powder, and the second, a GELase enzyme mixture that digests agarose. The silica powder method is very fast but frequently there is poor recovery of very small or very large fragments. Moreover some DNA fragments cannot be eluted from silica powder. The GELase method is expensive and requires more time, but gives nearly 100% recovery of all sizes of DNA fragments. This method should be used if the first method fails to give satisfactory results. A schematic outline of the procedures is shown in Figure 2.

This protocol describes recovery of DNA fragments using Elu-Quick silica matrix manufactured by Schleicher & Schuell. This procedure can be used to elute DNA fragments from low-melting point agarose. This kit gives better recovery of large DNA fragments than other kits commercially available.

Silica powder protocol

1. Prepare an agarose gel as described in the main protocol. Use the TAE buffer system. It is not possible to recover DNA fragments from gels made in TBE buffer. Consult Table 1 for the concentration of gel to use for best separation of the DNA fragments to be isolated. To protect DNA fragments from damage by UV light during fragment excision, use 1 mM guanosine or cytosine in the electrophoresis buffer. Run gel electrophoresis using the conditions described in the general protocol for agarose gel electrophoresis.

2. Prepare microfuge tubes for extraction by weighing them and recording their weight.

3. Place the gel on a UV-B light transilluminator (313 nm) and, using a sharp scalpel or razor blade, cut out the slice of agarose containing the band of interest. Cut the smallest slice of agarose possible. Collect

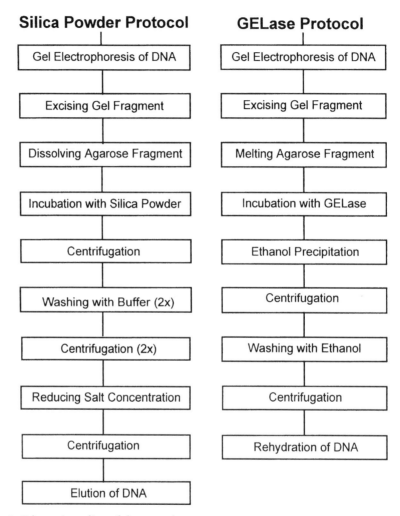

Fig. 2. Schematic outline of the procedure.

the gel slice in the pre-weighed microfuge tube. Weigh the tube with the agarose slice and calculate the volume of agarose, assuming that 100 mg of gel has 100 μl of volume.

4. Add 2.8 volumes of binding buffer. Incubate the tube containing the gel slice at 50 °C to 65 °C for 5 minutes or until the gel is completely melted.

5. Vortex the tube with silica to suspend settled particles. Add 20 μl of silica to the tube and mix by inverting several times. Twenty μl of silica sus-

pension will bind approximately 2.5 µg of DNA. Add enough binding buffer to give a ratio of 3:1 of binding buffer plus silica to the volume of agarose slice. For example, if 200 µl of agarose was dissolved in 560 µl of binding buffer (2.8 volumes) and 20 µl of silica solution was used, add 20 µl of binding buffer to obtain this ratio (560 + 20 + 20 = 600; 600/200 = 3).

6. Incubate the tube at room temperature for 10 minutes. Occasionally mix the silica suspension by gently inverting the tube. Do not vortex the tube because this will shear large DNA fragments. For better recovery of small DNA fragments incubate at 55 °C rather than room temperature.

7. Centrifuge at 10 000 rpm for 30 seconds to pellet silica particles. Using a pipette tip, remove all of the supernatant.

8. Add 500 µl of washing buffer to the tube. Gently pipette up and down to resuspend the pelleted silica. Use blue Pipetman tip with a cut off end for this procedure. The pellet should appear flocculent.

9. Centrifuge the tube at room temperature at 10 000 rpm for 30 seconds. Remove the supernatant using a P1000 Pipetman.

10. Repeat the washing procedure described in steps 8 and 9 one more time.

11. Add 500 µl of salt reduction buffer and gently resuspend the silica pellet as described in step 8. To limit DNA shearing be sure to use a cut off pipette tip.

12. Centrifuge the tube at room temperature at 10 000 rpm for 30 seconds. Remove the supernatant with a Pipetman (P200). Centrifuge the tube again at 10 000 rpm for 30 second and remove residual supernatant with a Pipetman (P200) fitted with capillary tip.

13. Add 20 to 50 µl of water or TE to the silica pellet. Gently resuspend the pellet by pipetting up and down using a tip with a cut off end.

14. Incubate at 55 °C for 10 minutes to release the DNA from the silica particles.

15. Centrifuge at room temperature at maximum speed for 5 minutes. Collect the supernatant containing the DNA fragment into a fresh microfuge tube.

Note: DNA sample solution is sometimes contaminated with a small amount of silica particles. The particles can interfere with the activity of some enzymes. In particular, ligase is adversely affected by these particles. In general, it is good practice to **briefly centrifuge the solution** before using the DNA.

GELase protocol   This protocol is a modification of the protocol recommended by the manufacturer of the enzyme, GELase. GELase can digest agarose gels that have been electrophoresed in many commonly used electrophoresis buffers, including TAE, TBE, MOPS and TPE. Ethidium bromide in the gel does not affect GELase activity. However, the activity of the enzyme is decreased when TBE buffer is used.

1. Prepare the agarose gel as described in the general protocol. Use low-melting temperature agarose such as SeaPlaque® instead of regular agarose to build the gel. Consult Table 1 for the appropriate concentration of gel to use for the fragment size to be separated. For fragments larger than 10 000 bp, use a 1% gel. To protect DNA fragments from damage by UV light during excision, use 1 mM guanosine or cytosine in the electrophoresis buffer. Run gel electrophoresis using the conditions described in the general protocol.

2. Place the gel on a UV-B light transilluminator (313 nm). Using a sharp scalpel or razor blade cut out a slice of agarose containing the band of interest. Cut the smallest slice of agarose possible to reduce the amount of enzyme needed.

3. Transfer the gel slice to a microfuge tube. Incubate it at 68 °C to 70 °C until the gel is completely melted. This process usually takes about 15 minutes. Be sure that no unmelted gel fragments remain. Measure the volume of liquid using cut off yellow tip.

4. Add 2 μl of 50 x GELase buffer per 100 μl of melted agarose. Mix well by inverting the tube 2 to 3 times. Do not pipette the solution up and down because this can shear large DNA fragments.

5. Place the tube into a 45 °C water bath and equilibrate the molten gel to 45 °C. It takes less than 5 minutes for 100 μl of gel to reach this temperature. Larger volumes may require a longer time.

6. Add 1 unit of GELase per 300 μl of molten 1-% SeaPlaque® or NuSieve® agarose gel. To avoid re-gelling, do not remove the tube from the water bath when adding enzyme. Mix enzyme with molten agarose by gently stirring with the pipette tip while holding the tube in the water bath. Incubate the reaction for 2 hours at 45 °C. Make sure that the tube is completely immersed in water to prevent the agarose from gelling.

Note: For agarose concentrations other than 1%, the units of GELase should be adjusted proportionally. Use 2 units of GELase for 300 mg of a 2% gel that was run in TAE. Also, the amount of gel digested by GELase is approximately proportional to incubation time. One unit of GELase will digest 600 mg of a 1% gel run in TAE buffer in 2 hours at 40 °C to 45 °C. For longer incubations, add a little extra enzyme to compensate for any loss of GELase activity. When digesting a large quantity of agarose, the incubation can be carried out at 40 °C for 12 hours.

7. Add 0.5 volume of 7.5 M ammonium acetate to the solution. Mix well by inverting the tube several times.

8. Add 2 volumes of 95% ethanol and mix gently by inverting the tube several times. For example, 100 μl of digested gel to which 50 μl of ammonium acetate was added will require 300 μl (150 x 2) of 95% ethanol.

9. Place the tube in a centrifuge, orienting the attached end of the tube lid pointing away from the center of rotation. Centrifuge at maximum speed for 10 to 15 minutes at room temperature (see Figure 2, Chapter 2).

10. Open the lid of the tube and remove it from the centrifuge holding it by the lid. Pour off ethanol into an Erlenmeyer flask. Invert the tube and touch the lip of the tube to the rim of the flask. Hold the tube until all ethanol has drained. Some ethanol may remain in the tube. **Place the tubes back** into the centrifuge in the same orientation as before.

Note: When pouring off ethanol do not invert the tube more than once because this may dislodge the pellet.

11. Wash the pellet with 700 μl of cold 70% ethanol. Holding the P1000 Pipetman vertically (see Figure 3, Chapter 2), slowly deliver the ethanol to the side of the tube opposite the pellet. **Do not start the centrifuge**, in this step centrifuge rotor is used as a "tube holder" that keeps the tube at an angle convenient for ethanol washing. Remove the tube from the centrifuge by holding the tube by the lid. Discard ethanol as before. Place the tube back into the centrifuge and wash with 70% ethanol one more time.

Note: This procedure makes it possible to wash the pellet quickly without centrifugation and vortexing. Vortexing and centrifuging is time consuming and frequently leads to a substantial loss of material.

12. After the last ethanol wash, place tubes back into centrifuge with the side of the tube containing the pellet facing away from the center of rotation. It is not necessary to close the tubes. Start the centrifuge for 2-3 seconds. This will collect the ethanol remaining on the sides of the tube. Remove all ethanol from the bottom of the tube using a Pipetman (P200) with a capillary tip.

Note: Never dry the DNA pellet in a vacuum. This will make dissolving the DNA very difficult if not impossible.

13. Add 20 to 30 µl of sterile water or TE buffer to rehydrate the DNA pellet. Use minigel electrophoresis to check the efficiency of DNA recovery.

Troubleshooting

Photographs of an ideal gel and gels with the most common errors in agarose gel electrophoresis are illustrated in Figures 3 to 5. Effect of gel overloading and puncturing the bottom of the well are presented in Figure 3. Figure 4A presents the effect of a poorly formed well and Figure 4B shows the effect of electrophoresis at excessively high voltage. Figure 5 illustrates a gel that was made with diluted buffer or with water.

Problems with DNA fragments recovery

Poor recovery of DNA fragments is a frequent problem in recovery of DNA fragments from gels using silica powder or GELase.

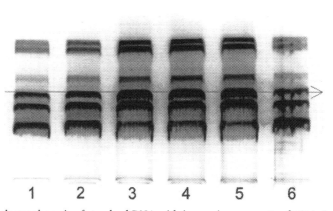

Fig. 3. Gel electrophoresis of standard DNA with increasing amounts of DNA. Lane, 0.5 µg, lane 2, 1 µg, lane 3, 2 µg lane 3, 4 µg, lane 5, 4 µg. Lane 6 has 0.5 µg of DNA but shows the effect of a well that has punctured bottom.

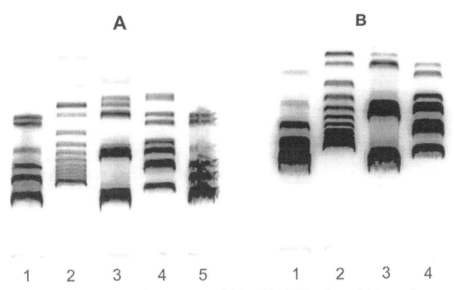

Fig. 4. Gel electrophoresis of DNA at normal (A) and high (B) voltage. Gel A was electrophoresed at a field strength of 4V/cm and gel B at 26V/cm. Lanes 1A and 1B lambda DNA standard restricted with *Hind* III. Lanes 2A and 2B 1 kb DNA ladder standards. Lanes 3A and 3B Boehringer Mannheim DNA standard III and lanes 4A and 4B are lambda DNA restricted with *Bst* EII. Lane 5A is the same as lane 1A with a poorly formed wel.l

Fig. 5. Electrophoresis of DNA using a gel prepared in water. Lane 1 lambda DNA restricted with *Hind* III. Lane 2 DNA 1 kb ladder standard. Lane 3 Boehringer Mannheim DNA standard III and lanes 4 lambda DNA restricted with *Bst* EII.

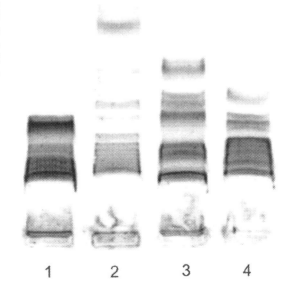

Silica powder method

Poor recovery is associated with:

- Inadequate solubilization of the agarose resulting from incorrect estimation of the volume of the agarose slice.
  Addition of more binding solution or longer incubation in binding buffer usually helps to overcome this problem.

- Irreversible binding of the fragments to the silica particles.
  Small DNA fragments of 200-500 bp frequently bind very strongly to the matrix. Raising the temperature during the binding step from room temperature to 55 °C often facilitates better recovery. Also, some large fragments bind irreversibly to the silica. However, the higher binding temperature does not improve recovery of the large fragments. The best remedy to overcome the irreversible-binding problem with both small and large fragments is to use the GELase recovery method.

- Mechanical shearing of DNA fragments.
  Large DNA fragments (>5000 bp) can be sheared while re-dissolving the pellet or by binding to the surface of more than one particle. Use of wide-bore pipette tips during all manipulations and careful, slow pipetting of the silica slurry will minimize DNA destruction. However, one can expect to recover only 50-60% of the large fragments when using the silica particle method. For better recovery of large fragments, the GELase method should be used.

- Poor elution of fragments from the slurry in the last step (step 13) of the procedure.
  This usually results from inadequately lowering the salt concentration with salt reduction buffer in step 11 or too low temperature in the elution step. Repeating the salt-lowering step or raising the elution temperature to 60 °C can help remedy this problem.

GELase method

Inadequate recovery using this method results from incomplete agarose digestion due to:

- Too high temperature during enzymatic treatment.
  The enzyme is inactivated very quickly at temperatures above 45 °C. Too short temperature equilibration at step 5 is the most common cause of such inactivation.

- Incomplete melting of the agarose slice.
  Increasing the time of incubation at step 3 and careful inspection of the melted slurry will remedy this problem.

- Re-gelling of melted agarose during incubation with enzyme. GELase can digest only melted agarose. The tube must be fully immersed in the water bath during treatment for successful recovery of the fragments.

- Inhibition of enzyme activity by electrophoresis buffer. This problem sometimes occurs when very large agarose slices are used or when TBE buffer is used. The manufacturer of the enzyme recommends incubating the agarose slices in GELase reaction buffer for several hours at room temperature to remove electrophoresis buffer from the slice before melting the agarose.

References

Adamson A W (1976) Physical chemistry of surfaces. John Willey & Sons Pub New York.

Adkins S, Burmeister M (1996) Visualization of DNA in agarose gels as migrating colored bands: Application for preparative gels and educational demonstrations. Analytical Biochem 240:17-23.

Baldwin RL (1995) A review of mini submarine electrophoresis gel boxes. BioConsumer Rev. 2:4-10.

Brunk CF, Simpson L (1977) Comparison of various ultraviolet sources for fluorescent detection of ethidium bromide-DNA complexes in polyacrylamide gels. Anal Biochem 82:455-462.

Burmeister M, Lehrach H (1989) Isolation of a large DNA fragments from agarose gels using agarase. Trends Genet 5:41.

Cariello NF, Keohavong P, Sanderson BJS, Thilly WG (1988) DNA damage produced by ethidium bromide staining and exposure to ultraviolet light. Nucleic Acids Res 16:4157.

Davis RW, Botstein D, Roth JR (1980) A manual for genetic engineering. Advanced Bacterial Genetics. Cold Spring Harbor Laboratory Pub Cold Spring Harbor, New York.

Dretzen G, Bellard M, Sassone Corsi P, Chambon P (1981) A reliable method for the recovery of DNA fragments from agarose and acrylamide gels. Anal Biochem 112:295-298.

Girvitz SC, Bacchetti S, Rainbow AJ, Graham FL (1980) A rapid and efficient procedure for purification of DNA from agarose gels. Annal Biochem 106:492-496.

Gründemann D, Schömig E (1996) Protection of DNA during preparative agarose gel electrophoresis against damage induced by ultraviolet light. BioTechniques 21:898-903.

Hartman PS (1991) Transillumination can profoundly reduce transformation frequencies. BioTechniques 11:747-748.

Helling RB, Goodman HM, Boyer HW (1974) Analysis of R.EcoRI Fragments of DNA from lambdoid bacteriophages and other viruses by agarose gel electrophoresis. J Virology 14:1235-1244

Himenz PC (1977) Principles of colloid and surface chemistry. Marcel Decker Pub.

Mäueler WA, Kyas F, Bröcker, Epplen JT (1994) Altered electrophoretic behavior of DNA due to short-time UV-B irradiation. Electrophoresis 15:1499-1505.

Nikogosyan DN (1990) Two-quantum UV photochemistry of nucleic acids: comparison with conventional low intensity UV photochemistry and radiation chemistry. Int J Radiation Biol 57:233-299.

Perbal B (1988) A practical guide to molecular cloning. 2nd edition. John Willey & Sons Publishers.

Rand KN (19960 Crystal violet can be used to visualize DNA bands during gel electrophoresis and to improve cloning efficiency. Elsevier Trends Journals Technical Tips on Line. Tip # T40022.

Sharp PA, Sugden B, Sambrook J (1973) Detection of two restriction endonuclease activities in Haemophilius parainfluenzae using analytical agarose-ethidium bromide electrophoresis. Biochemistry 12:3055-3060.

Scott TM, Dace GL, Atschuler M (1996) Low-cost agarose gel documentation system. BioTechniques 21:68-72.

Sealey PG, Southern EM (1982) Gel Electrophoresis of DNA. In: Gel electrophoresis of nucleic acids; a practical approach. Rickwood D and Hames BD Editors IRL Press.

Smith LS, Lewis TL, Mtsui SM (1995) Increased Yield of Small DNA Fragments Purified by Silica Binding. BioTechniques 18:970-975.

Vogelstein B, Gillespie D (1979) Preparative and analytical purification of DNA from agarose. Proc Natl Acad Sci USA 76:615-619.

Weinand U, Schwarz Z, Felix G (1987) Preparative and analytical elution of nucleic acids from gels adopted for subsequent biological tests: Application for analysis of mRNA from maize endosperm. FEBS Lett 98:319-323.

Weislander L (1979) A simple method to recover intact high molecular weight RNA and DNA after electrophoretic separation in low gelling temperature agarose gels. Annal Biochem 98:305-309.

Suppliers

Suppliers of silica-based elution kits

Amersham Pharmacia Biotech. (www.apbiotech.com)
This kit uses pre-packed glass fiber spin columns for recovery of DNA from TAE or TBE buffered agarose gels. Size range is 100 to 10 000 bp with recovery claimed to be about 60 to 80%.

Ambion, Inc. (www.ambion.com)
The GeniePrep(kit uses silica fiber in convenient spin cartridges. DNA fragments larger than 45 bp can be purified. The upper limit of DNA size is not specified.

Bio 101, Inc. (www.bio101.com)
This company offers three kits based on a proprietary silica matrix GLASS-MILK®. Size range of DNA fragments that can be purified with these matrices is between 200 bp to 20 000 bp. For DNA fragments below 200 bp, the company offers a specially developed matrix, GLASSFOG®.

Bio-Rad Laboratories (www.bio-rad.com)
Two kits are offered. Both kits use diatomaceous earth rather than a glass-based matrix. The matrix has a higher banding capacity than glass-based matrices and a size range 200 to 50 000 bp. DNA of 200 to 400 bp binds with varying efficiency that depends on the nature of the fragment.

Clontech Laboratories, Inc. (www.clontech.com)
Kit contains a uniform-size silica matrix that binds DNA fragments of 200 to 20 000 bp. Recovery up to 90% is claimed. Kit can be used with TAE and TBE buffered agarose gels.

Eppendorf 5′ (www.eppendorf.com)
This kit uses silica-based matrix. DNA fragments larger than 200 bp can be purified with this kit.

Life Technologies BRL (www.lifetech.com)
Kit uses a spin cartridge with silica-based membrane to bind DNA. Up to 50 μg of DNA can be purified by a single cartridge. DNA size range is 200 to 50 000 bp.

Qiagen, Inc. (www.qiagen.com)
This kit uses a silica-gel membrane in a spin column to bind DNA. Buffers supplied have pH indicators to optimize DNA recovery. DNA in the size range of 70 to 10 000 bp is recovered with 70 to 80 % efficiency. Kit can be used with TAE or TBE buffered agarose gels.

Roche Molecular Biochemicals, formerly Boehringer Mannheim.
(www.biochem.roche.com)
This kit uses a silica matrix capable of recovering DNA fragments in the range of 400 to 9500 bp with 80% efficiency.

Schleicher & Schuell (www.s-and-s.com/index.htm)
Sized, rod-shaped, glass particles that reduce DNA shearing are used to bind DNA. Isolation of very large DNA fragments is possible (up to 2 000 000 bp). Recovery of small (200 bp) DNA fragments is clamed to be 70 to 90%.

Sigma Chemical Co. (www.sigma.sial.com)
Kit uses an aqueous, glass particle suspension. The effective size range is 200 to 20 0000 bp with 70 to 90% recovery.

Preparation of Probes for Hybridization

STEFAN SURZYCKI

Introduction

Preparation of probes for hybridization involves in vitro incorporation of reporter molecules into nucleic acids. These reporters can be incorporated at one or both ends of nucleic acid molecules, giving specific, low density labeled probes. High density labeling is usually achieved by incorporating the reporters uniformly throughout entire length of nucleic acid molecules. For hybridization work internally labeled probes are preferred since they provide the strongest hybridization signal.

Two types of reporter molecules are presently in use, radioactive reporters and nonradioactive reporters. Radioactive reporters are usually tagged with ^{32}P or ^{35}S isotopes and can be directly detected. Nonradioactive reporters can be fluorescent tags, permitting their direct detection, or specific ligands such as biotin, haptens and hapten-like molecules, the detection of which is indirect.

Direct conjugation of reporter enzyme to the probe is used to detect the probe. Probes obtained by these methods have lower levels of background, thus allowing longer exposure time and consequently better sensitivity. The presence of enzyme (conjugated to the probe) during hybridization requires modification of the standard hybridization protocol to protect the enzyme from inactivation. This makes direct labeling procedures much less convenient and therefore less popular.

Indirect detection is carried out using enzymatic reactions catalyzed by either horseradish peroxidase or alkaline phosphatase. These enzymes can be conjugated directly to a secondary molecule that has a very high affinity for a specific ligand (for example biotin-avidin complexes) or they can be conjugated to antibody against a hapten tag. Since enzymes are only used for detection of labeled hybrids and are not present during hybridization, standard hybridization protocols can be used. This accounts for the wider use of indirect methods over direct detection procedures.

Two types of substrates are used to detect enzyme activity at the site of hybridization. Colorimetric detection uses a soluble, colorless substrate that is converted into an insoluble, colored product precipitated directly on the membrane. Chemiluminescent detection uses a chemiluminescent substrate that is converted by enzyme into a light-emitting substance easily detected by standard photographic film. The sensitivity of the chemiluminescent method is much greater than the colorimetric method and even exceeds the sensitivity of radioactive labeling (Kessler et al., 1990; Höltke et al., 1990). Chemiluminescent substrates now in use allow the detection of 0.03 pg or less DNA on the membrane in about 1 to 2 hours. The same amount of DNA would require 17 to 20 hours of exposure to detect a ^{32}P labeled probe.

Preparation of uniformly labeled probes, independent of the nature of the reporter, is carried out using four basic procedures: nick translation, random priming, RNA probe synthesis and PCR. Table 1 summarizes the major properties of these four methods.

Nick translation

E. coli DNA polymerase I using reporter tagged substrate (dNTP) synthesizes newly labeled DNA. Single-stranded nicks are introduced at random in the double-stranded DNA template with DNA exonuclease, DNase I. DNA polymerase I initiates DNA synthesis at the 3' end of the nick, whereas at the 5' end of the nick, the 5' to 3' exonuclease activity of this enzyme excises nucleotides. The position of the nick is "translated" down stream, with labeled nucleotides replacing non-labeled nucleotides between the original site of the nick and the new position of the nick. The resulting probe is labeled on both strands with labeled (newly synthesized) DNA interspersed with unlabeled (original template) DNA. The method works best with linear DNA fragments that are larger than 500 bp. The specific activity of these probes is somewhat lower, as compared to other methods, but nick-translated probes generally give very strong hybridization signals (Davis et al., 1994).

Random priming

Double-stranded DNA is denatured and DNA polymerase synthesizes a newly labeled DNA by template-dependent extensions of random hexamer primers (Feinberg and Vogelstein, 1983). Polymerases lacking 5' to 3' exo-

Table 1. Comparison of features of uniformly labeled probes prepared by different methods

	Nick Translation	Random Primer	RNA Synthesis	PCR
Template	ds DNA	ds DNA or ss DNA	ds DNA	ds DNA
Template optimal form	Linear or circular	Linear	Linear or circular	Linear or circular
Nature of the probe	dsDNA	dsDNA	RNA	dsDNA
Self-hybridization of the probe	Yes	Yes	No	Yes (No)
Amount of probe synthesized	up to 1 μg	20-500 ng	~300 ng	Unlimited
Probe length	~500 bp	100-200 bp	Defined	Defined
Insert specific	No	No	Yes	Yes
Requirement for subcloning	No	No	Yes	No or Yes
Density of nonradioactive label	High	Very high	Very high	Very high
Radioactive label (dpm/μg)	$\sim 2 \times 10^9$	$5\text{-}8 \times 10^9$	$5\text{-}8 \times 10^9$	not used
Labeling time	~2 hours	2-16 hours	~1 hour	1-2 hours
Sensitivity in Southern blots	++	+++	++	+++
Sensitivity in Northern blots	+	++	+++	+++
Colony/plaque hybridization	++	+++	not used	+++
Dot blots	++	+++	not used	+++

nuclease activity, such as Klenow fragment of *E. coli* DNA polymerase I, are used in this reaction. The template strand remains unlabeled whereas the newly synthesized strand is completely labeled. The enzyme can initiate and synthesize several new strands from every template, resulting in net synthesis of large amounts of labeled product, many times exceeding the amount of template input. For this reason the specific activity of probe produced by random priming is very high. Consequently, the radioactive probes, prepared this way, can be used only for a limited time because they disintegrate very quickly due to isotopic decay. The method is relatively insensitive to the purity of the DNA template or its size, and creates probes from 100 to 600 bp long. Random priming is particularly well suited for preparation of nonradioactive probes because their high specific activity does not result in decay with time.

PCR probes

Two alternative techniques are used to prepare probes using PCR. In the first procedure, uniformly labeled probes are generated by incorporation of a tagged nucleotide during PCR (Kessler 1992; McCreery and Helentijaris, 1994: Yamaguchi at al., 1994). In the second approach, a large number of specific DNA target molecules are synthesized by PCR that are subsequently used as a template to prepare random-primer or nick-translated probes (Rost, 1995). PCR labeling can be done either using genomic template without cloning the DNA fragment in question, or from a DNA fragment cloned into plasmid. Specific primers are required for the former, but not for the later PCR procedure. Advantages of generating DNA probes by PCR are numerous. First, large amounts of probe with high label density can be synthesized from very little DNA. Second, probes can be prepared using either purified or partially purified DNA as the source of the template. Third, preparation of the probe is highly flexible and does not depend on restriction enzyme site location. Fourth, it is possible to prepare specific, single-stranded probes using a single primer. The main disadvantage, is the difficulty in amplifying most sequences larger than 3000 bp in length.

RNA probes

This method generates single-stranded RNA probes using in vitro transcription by bacteriophage RNA polymerase. The probe sequences are cloned in a plasmid next to a sequence specific promoter using SP6, T7 or T3 RNA polymerases. After transcribing a single-stranded RNA fragment, the double-stranded DNA template is degraded with DNase giving strand-specific RNA probe that does not self-hybridize. The RNA probes can be used in Southern and Northern hybridization, as well as in ribonuclease protection assays. Large amounts of highly specific probes of very high specific activity can be prepared with this method. Frequently,

RNA based probes show higher sensitivity than equivalent nick-translated or random-primer probes. This is probably due to the comparatively high stability of DNA:RNA hybrids, over DNA:DNA hybrids and that reannealing of the probe to itself does not occur during hybridization. Disadvantages of using RNA probes are (a) the possibility of their degradation by RNase; (b) the necessity to re-clone target DNA sequences into plasmids with RNA polymerase promoters, and (c) the inability to reuse the same hybridization mixture due to the destruction of the probe during hybridization. RNA probes also generate very high background in filter hybridization when labeled with radioactive tags. This problem is not encountered when using nonradioactive probes.

Experimental procedures

All three procedures mentioned above are presented in this chapter. Labeling protocols for radioactive, as well as for nonradioactive, digoxigenin-labeled DNA and RNA are described. Nonradioactive probes have numerous advantages over radioactively labeled probes. Non-radioactive probes deliver equivalent sensitivity in a much shorter exposure time. They are stable, and once prepared, can be stored indefinitely. Preparation and use of the probes are safer and avoid the regulatory and cost issues associated with radioactivity. The stability of these probes permits the reuse of the hybridization mixture saving time and expense. Boehringer Mannheim introduced labeling probes with digoxigenin. This label has some additional advantages over use of other nonradioactive labeling systems. First, in our hands and others, DIG-labeled probes cause much less nonspecific background (McCreery and Helentijaris, 1994). Second, after synthesis of the probe, it is not necessary to remove unincorporated nucleotides. Third, standard hybridization conditions, identical to those of radioisotope-labeled probes, can be used. Incorporation of other nonradioactive labels, such as biotin, requires altering standard hybridization conditions (Leary et al., 1983). Fourth, digoxigenin is a very stable molecule and is not sensitive to most extreme conditions of storage and hybridization. This is not the case with other frequently used hapten-based labeling tags such as fluorescein, which is sensitive to prolonged light exposure. Fifth, a quick and simple assay exists to measure digoxigenin incorporation, permitting quantification of labeled probe. The procedures described here essentially follow those recommended by the manufacturer with some minor modifications. These protocols can be used for preparing probes labeled, not only with digoxigenin, but also with other reporters, including radioactive isotopes.

A brief description of other commercially available, nonradioactive labeling kits are given at the end of this chapter.

Subprotocol 1
Random Priming

This protocol labels 10 ng to 3 µg of DNA in a single reaction. The yield depends on the purity of the template and the time of labeling. Satisfactory labeling is usually achieved with 50 to 300 ng of template in 2 hours. Extending the incubation for 20 hours (overnight) will increase the yield three-fold to fourfold. To detect a single gene in a large genome, using a standard hybridization procedure with 10 ml of hybridization mixture requires 150 to 250 ng of labeled probe (15-25 ng/ml). The data presented in Table 2 can be used to determine the amount of template and time necessary to synthesize the desired amount of probe.

A schematic outline of random priming procedures is shown in Figure 1.

Materials

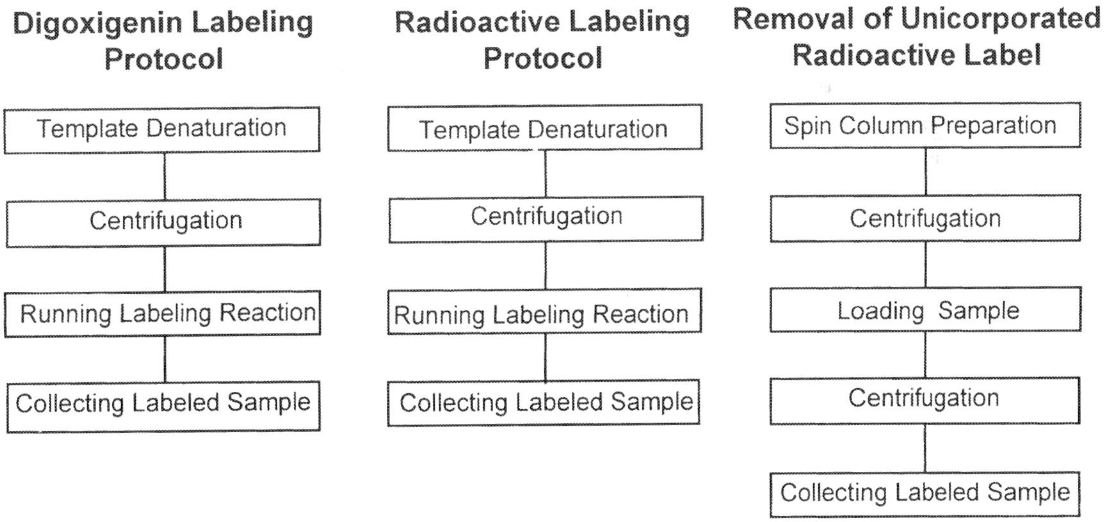

| Digoxigenin Labeling Protocol | Radioactive Labeling Protocol | Removal of Unicorporated Radioactive Label |

Digoxigenin Labeling Protocol

Template Denaturation

Centrifugation

Running Labeling Reaction

Collecting Labeled Sample

Radioactive Labeling Protocol

Template Denaturation

Centrifugation

Running Labeling Reaction

Collecting Labeled Sample

Removal of Unicorporated Radioactive Label

Spin Column Preparation

Centrifugation

Loading Sample

Centrifugation

Collecting Labeled Sample

Fig. 1. Schematic outline of the labeling procedures

Table 2. Amount of synthesized digoxigenin-labeled DNA using random a hexanucleotide priming method

Time of synthesis	Amount of template DNA used in labeling reaction					
	10 ng	30 ng	100 ng	300 ng	1000 ng	3000 ng
1 hour	15 ng	30 ng	60 ng	120 ng	260 ng	530 ng
20 hours	50 ng	120 ng	260 ng	500 ng	780 ng	890 ng

- Klenow fragment of DNA polymerase I (2 u/μl)
- Select-D (RF) columns (Eppendorf 5' # 2-750616)
- 10 x Digooxigenin labeling mix (Roche Molecular Biochemicals # 1277 065)
- dATP, dCTP, dGTP, dTTP; 100 mM stock solutions set (Roche Molecular Biochemicals # 1277 049)
- DNA to be labeled
- [α^{32}P] dCTP, 3000 Ci/mM (e.g., Amersham Life Sci. Inc. # AA0005) for radioactive labeling only.

Solutions
- **10 x Hexanucleotide Mixture**
  - 500 mM Tris-HCl, pH 7.5
  - 100 mM MgCl$_2$
  - 1 mM DTT
  - 2 mg/ml BSA
  - 62.5 A$_{260}$ U/ml random hexanucleotides
  The solution is stored at -20 °C. It can be obtained from Roche Molecular Biochemicals # 1277 081.
- **10 x Digoxigenin Labeling Mixture**
  - 1 mM of each dATP, dCTP and dGTP
  - 0.65 mM dTTP
  - 0.35 mM digoxigenin-11-dUTP (Roche Molecular Biochemicals # 1573 152)
  The solution is adjusted to pH 7.5 and stored at -20 °C. This reagent can be obtained from Roche Molecular Biochemicals # 1277 065.

Note: All the solutions for digoxigenin random-primer labeling of DNA probes are available in kit form from Roche Molecular Biochemicals # 1093 657 and # 1175 033

- **TE Buffer**
  - 10 mM Tris-HCl, pH 7.5 or 8.0

- 1 mM Na$_2$EDTA, pH 8.5
Sterilize by autoclaving and store at 4 °C.
- **0.5 M EDTA Stock Solution, pH 8.5**
Weigh an appropriate amount of EDTA (disodium ethylenediamine-tetraacetic acid, dihydrate) and add it to distilled water while stirring. EDTA will not go into solution completely until the pH is greater than 7.0. Adjust the pH to 8.5 using a concentrated NaOH solution or by adding a small amount of pellets while mixing and monitoring the pH. Bring the solution to a final volume with water and sterilize by filtration. The EDTA stock solution can be stored indefinitely at room temperature.

Procedure

Digoxigenin labeling

1. Place a sterilized 1.5 ml sterilized microfuge tube on ice and add DNA (see Table 2). Bring the volume up to 10 µl with sterilized water.

2. Denature the template DNA by placing the tube in a boiling water bath for 3 minutes. Quickly cool the tube on ice. Centrifuge for 5 seconds to recover all condensation accumulated on the side of the tube.

3. Place the tube back on ice and add 2 µl of the hexanucleotide mixture and 2 µl of digoxigenin labeling mixture.

4. Add water to bring the total volume to 19 µl.

5. Mix the sample by vortexing briefly and centrifuge it for 30 seconds.

6. Start the labeling reaction by adding 1 µl of Klenow fragment enzyme. Mix well by pipetting up and down. Avoid vigorous mixing that may result in loss of enzyme activity.

7. Incubate the reaction for at least 2 hours at 37 °C. Consult Table 2 for other incubation times. Best results are obtained by synthesizing the probe overnight.

8. Stop the labeling reaction by adding 1 µl of 0.5 M EDTA, pH 8.5. Store the probe at -20 °C. Probe can be stored at this temperature for several years.

Note: Removing unincorporated digoxigenin labeled nucleotide is not necessary. Measure the amount of digoxigenin incorporation using the protocol for estimation of DIG-dUTP concentration (see page 212).

Radioactive labeling

1. Place a 1.5 ml sterilized microfuge tube on ice and add DNA (see Table 2). Bring the volume up to 10 µl with sterilized water.

2. Denature the template DNA by placing it in a boiling water bath for 3 minutes. Quickly cool the tube on ice. Centrifuge for 5 seconds to recover all condensation accumulated on the side of the tube.

3. Select the labeled nucleotide to be used. Use of α^{32}P labeled dCTP is recommended over other α ^{32}P labeled dNTPs because of its high stability, particularly when probes of high specific activity are prepared. The specific activity of dCTP should be 3000 Ci/mM at a concentration of 10 mCi/ml.

4. Place the DNA on ice and add 2 µl of hexanucleotide mixture. Mix by pipetting up and down.

5. Add 3 µl of cold nucleotide mixture. This mixture contains three unlabeled dNTPs at a concentration of 1.5 mM each.

6. Add 5 µl of radioactive dCTP to the tube.
   Safety Note: Use a radioactive shield for this step and work in gloves.

7. Vortex the tube briefly and centrifuge it for 30 seconds in a microfuge.

8. Start the reaction by the addition of 1 µl of Klenow enzyme. Mix by pipetting up and down and centrifuge for 30 seconds in a microfuge.

9. Incubate the reaction for 30 to 60 minutes at 37 °C.

10. Stop the reaction by adding 1 µl of 0.5 M EDTA.

Removing unincorporated label

Removing unincorporated label is required for radioactive probes to lower background in Southern or Northern hybridization.

1. Separate unincorporated labeled nucleotides from labeled DNA using a Select-D (RF) column. Remove the column from the bag and invert it several times to resuspend the gel.

2. Remove top and bottom closures from the column. Remove the top closure first and the bottom closure second. Removing the closures in this order prevents a vacuum from forming in the column that will make draining the column buffer difficult.

3. Place the column in a vertical position and let any excess buffer drain from the column for 3 to 4 minutes. Usually 2 to 3 drops will flow from the column.

4. Place the column in one of the collection tubes provided. Place the whole assembly into a swinging bucket, tabletop centrifuge.

5. Centrifuge at 1100 x g for exactly 2 minutes. Begin timing from the moment the centrifuge reaches the desired speed.

6. Remove the column assembly from the centrifuge rubber adapter. Forceps may be necessary to assist you in removing the column and collection tube. Try to keep the column in a vertical position during this operation. Discard the collection tube and collected buffer.

7. Place the column in a vertical stand. Add 50 µl of the diluted labeling reaction to the top of the gel bed. Add the liquid to the center of the bed, avoid applying any sample to the side of the column.

8. Place this column in the second collection tube provided. Carefully place column assembly into the centrifuge and centrifuge for 4 minutes at 1100 x g.

Note: The column is now highly radioactive, work in gloves.

9. Withdraw the column and collection tube from the rubber centrifuge adapter using forceps. Discard the column into radioactive waste. The collection tube contains purified sample. Cap the tube and store at -20 °C. The best hybridization results are obtained when the probe is used immediately. Using ^{32}P-labeled probes more than 10 days old is not recommended.

Troubleshooting

Low yields of nick translated probe with a low level of DIG-label result from using impure template. Re-purifying the template usually helps in this case. Purification can be performed by ammonium acetate-ethanol precipitation. This procedure also helps to remove unincorporated label (McCreery and Helentjaris, 1994).

High background, when radioactive probes are used, results from incomplete removal of unincorporated radioactive nucleotide. Purifying the probe remedies this problem. Low specific activity usually results when old $[\alpha^{32}P]$-dNTP is used (more than 20 days). Similarly, using old probe will result in poor hybridization signals and higher background.

Subprotocol 2
PCR Labeling Protocol

Digoxigenin-labeled deoxynucleotides, like digoxigenin 11-dUTP (DIG-dUTP) can be used as a substrate by Taq polymerase in PCR. Sensitivity of PCR-generated probes is very high for all types of applications and the yield is higher than for other labeling methods (see Table 1). The efficiency of PCR, however, is decreased proportionally to the concentration of DIG-labeled substrate. When using this method, one has to consider the ratio of DIG-dUTP to dTTP in the labeling reaction because this ratio will affect the yield and sensitivity of the synthesized probe. When more DIG-dUTP is used, probe yield is less but sensitivity of the probe is higher. The relationship between yield and the DIG-dUTP:dTTP ratio is described by the equation: %probe yield = 100 x (DIGd-UTP:dTTP ratio) x 144. For probe with a high density of DIG dUTP, the ratio DIG dUTP:dTTP should be between 0.5 (1:2) to 0.16 (1:6). To prepare low sensitivity probe for applications that do not require maximum sensitivity (for example, Southern blots of plasmid DNA, library screening or colony hybridization) the ratio of DIG-dUTP:dTTP should be between 0.16 (1:6) to 0.11 (1:9).

A schematic outline of the PCR labeling procedure is shown in Figure 2.

Materials

- Taq DNA polymerase (1 to 5 u/μl)
- DNA template (about 100 ng/1 μl for eukaryotic DNA; about 10 ng/μl for plasmid DNA)
- Primers (1-10 μM)
- dATP, dCTP, dGTP and dTTP 100 mM stock solutions set (Roche Molecular Biochemicals # 1277 049).

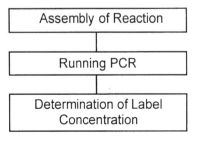

Solutions

Fig. 2. Schematic outline of the PCR labeling procedure

- **10 x PCR Buffer**
  - 100 mM Tris-HCl; pH 8.3
  - 500 mM KCl
  - 15 mM MgCl$_2$

  Buffer should be prepared using a 1 M stock solution of Tris-HCl adjusted to pH 8.3 at room temperature, sterilized by autoclaving for 20 minutes and stored at 4 °C. This buffer can be bought from Roche Molecular Biochemicals # 1271 318, or is supplied with Taq polymerase (Roche Molecular Biochemicals # 1146 165)

- **50 x dNTP (- dTTP) Mixture**
  - 10 mM dATP
  - 10 mM dGTP
  - 10 mM dCTP

  Use 100 mM stocks of dATP, dCTP and dGTP to prepare dNTP (-dTTP) mixture. Add 10 µl of each dNTP to 70 µl of sterilized water. Store at -20 °C.

- **5 x DIG-dUTP Mixture**

  Dilute 100 mM dTTP stock solution to concentration of 1 mM (100 fold dilution of a stock solution). Dilute DIG-dUTP solution to concentration of 1 mM. Use Table 3 to prepare DIG-dUTP mixture at the desired DIG-dUTP:dTTP ratio. Store the mixture at -20 °C.

Table 3. Preparation of DIG-dUTP mixture

DIG dUTP:dTTP ratio	Volume of 1 mM DIG-dUTP (µl)	Volume of 1 mM dTTP (µl)
0.5 (1:2)	33	67
0.33 (1:3)	25	75
0.25 (1:4)	20	80
0.16 (1:6)	14	86
0.11 (1:9)	10	90

▨ ▨ Procedure

Digoxigenin PCR labeling

1. Place a sterilized 0.5 ml microfuge tube on ice and add the ingredients following outline presented below. First, calculate the amount of water necessary for the desired volume and add it to the tube. Then add buffer and the remaining components. Add enzyme last and mix by pipetting up and down several times. Never mix by vortexing. Taq polymerase is **very sensitive to vortexing.**

   PCR Reaction Mixtures

Reagent	50 µl reaction	100 µl reaction	Final concentration
10 x PCR buffer	5 µl	10 µl	1 x
50 x dNTP (-)	1 µl	10 µl	200 µM each
5 x DIG-dUTP dTTP	10 µl	20 µl	200 µM total
Upstream primer	0.3-2.5 µl	0.5-5.0 µl	0.1- 1.0 µM
Downstream primer	0.3-2.5 µl	0.5-5.0 µl	0.1- 1.0 µM
Eukaryotic DNA (100 ng/µl)	1.0 µl	1.0 µl	100 ng
Prokaryotic DNA (10 ng/µl)	1.0 µl	1.0 µl	10 ng
Taq polymerase (1 u/µl)	1.0 µl	2.0 µl	2 u/100 µl
Water	to 50 µl	to 100 µl	-

2. Add 25 µl of mineral oil to each tube and centrifuge the tube for 5 seconds in a microfuge. Addition of oil is easier if you cut off the end of the yellow tip.

3. Amplify DNA using standard conditions. Typically these are: initial denaturation 95 °C for 3 minutes followed by 30 to 35 cycles consisting of denaturation (D) at 94 °C for 30 seconds, annealing (A) at 60 °C for 30 seconds, elongation (E) at 72 °C for 5 minutes. The reaction is finished with an extension step at 72 °C for 5 minutes.

4. Determine the concentration of DIG-dUTP in the probe using the protocol for estimation of DIG-dUTP concentration (see page 212). Store the probe at -20 °C. It is not necessary to remove unincorporated DIG-UTP.

Troubleshooting

Using the procedure described above, the incorporation of DIG-dUTP label is usually 20 to 50 ng/μl. Low incorporation (less than 20 ng/μl) can result from:

a. using a low ratio of DIG-dUTP:dTTP particularly for GC rich templates,

b. incorrect template or primer concentration.

Failure to obtain any products may result from:

a. incorrect template or primer concentrations,

b. a low DIG-dUTP:dTTP ratio, particularly for AT rich templates,

c. too high concentration of template, and

d. incorrect annealing temperature.

Subprotocol 3
Preparation of RNA Probes

This protocol describes the preparation of DIG-labeled RNA probes. The probes are labeled using SP6, T7 or T3 RNA polymerases. All three polymerases can use digoxigenin-labeled substrate with the same efficiency. DNA to be used for preparation of the probe should be first subcloned into a multiple cloning site next to the RNA polymerase promoter site. The plasmid should be linearized at the restriction site downstream from the cloned insert. This creates "run off" transcripts of uniform length. To avoid transcription of sequences other than the insert a restriction enzyme that leaves a 5' overhang or blunt end should be used. The linearized DNA should be purified by phenol/chloroform extraction. Normally about 1-10 μg of RNA probe is obtained using 1 μg of template.

Compared with DNA probes, preparation of RNA is more laborious because it requires more care to prevent RNase contamination. RNase-free disposable tubes and pipette tips should be used, and gloves worn at all times. A schematic outline of the preparation of RNA probes is shown in Figure 3.

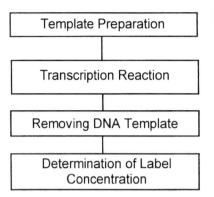

Fig. 3. Schematic outline of the preparation of RNA probes

Materials

- SP6 RNA polymerase, 20 u/µl (Roche Molecular Biochemicals # 810 266)
  T7 RNA polymerase, 20 u/µl (Roche Molecular Biochemicals # 881 767)
  T3 RNA polymerase, 20 u/µl (Roche Molecular Biochemicals # 1031 163)
- DNase, RNase free, 10 u/µl (Roche Molecular Biochemicals # 776785)
- RNase inhibitor, 20 u/µl (Roche Molecular Biochemicals # 7999 017)
- Diethyl pyrocarbonate (DEPC) Sigma Co. # D 5758
- DNA template (about 5 µg/µl)
- Phase Lock Gel (PLG I) microfuge tubes (Eppendorf 5' # pI-183182). PLG tubes contain a proprietary compound that, when centrifuged, migrates to form a tight barrier between organic and aqueous phases (Murphy and Hellwig, 1996). The interphase material is trapped in and below this barrier allowing the complete and easy collection of the entire aqueous phase free from contamination with organic solvents. The PLG barrier also offers increased protection from exposure to organic solvents used in deproteinization procedures. The tubes are not expensive and are strongly recommend to use when phenol extraction is carried out in microfuge tubes.
- Phenol:CIA mixture (PCI) (Eppendorf 5' # I-737642).

Solutions
- **DEPC-treated Water**
  - 0.05% Diethyl pyrocarbonate
    Mix DEPC with water and incubate at room temperature overnight. Autoclave for 30 minutes to remove residual DEPC. Store at 4 °C. Use this water to prepare all solutions.

Note: DEPC treated water can be bought from Ambion # 9920.

Safety Note: Diethyl pyrocarbonate (DEPC) is a suspected carcinogen and should be used in a fume hood.

- **10 x Transcription Buffer**
  - 400 mM Tris-HCl; pH 8.0
  - 100 mM NaCl
  - 60 mM $MgCl_2$
  - 100 mM Dithioerythritol (DTE)
  - 20 mM Spermidine
  - 1u/ml RNase inhibitor

  Buffer should be prepared using 1 M stock solution of Tris-HCl adjusted to pH 8.0 (20 °C) and sterilized by filtration. Buffer can be bought from Roche Molecular Biochemicals as a part of RNA labeling kit (Roche Molecular Biochemicals # 1175 025).

- **10 x NTP Mixture**
  - Tris-HCl 1 mM pH 7.5
  - 10 mM ATP
  - 10 mM CTP
  - 10 mM GTP
  - 6.5 mM UTP
  - 3.5 mM DIG-UTP

  Use 100 mM stocks of ATP, CTP and GTP to prepare NTP (-) mixture. Add 10 µl of each NTP to 70 µl of sterilized water. Store at -20 °C. Mixture can also be bought from Roche Molecular Biochemicals Co. # 1277073)

Note: All components necessary for synthesis of DIG-labeled RNA probes can be bought as a kit from Roche Molecular Biochemicals (# 1175 025).

- **0.5 M EDTA Stock Solution**

  Weigh out an appropriate amount of EDTA (disodium ethylendiaminetetraacetic acid, dihydrate) and stir into distilled water. EDTA will not go into solution completely until the pH is greater than 7.0. Adjust the pH to 8.5 using concentrated NaOH solution or by adding a small amount of pellets. Bring the solution to a final volume with water and sterilize by filtration. EDTA stock solution can be stored indefinitely at room temperature.

▓▓ Procedure

Digoxigenin RNA labeling

Preparation of template protocol

1. Linearize the template for transcription. Place a 1.5 ml microfuge tube on ice and add ingredients in the order indicated below. For probes cloned into the *Eco* RV site of the plasmid, pBluescript II KS+, use *Hind* III restriction enzyme.

Reagents	Volume (µl)	Final concentration
DEPC-treated water	to 20 µl	-
10 x Restriction buffer	2	1x
DNA template	as needed	5 µg
Restriction Enzyme (10 u/µl)	1	10 unit

Note: Mix the enzyme by pipetting up and down. Never vortex the reaction.

2. Incubate the reaction at 37 °C for 1-2 hours. Stop the reaction by adding 1 µl 0.5 M EDTA and 70 µl TE. Mix well by pipetting up and down.

3. Centrifuge the unopened PLG I tube in a microfuge for 30 seconds at 10 000 rpm to pellet the gel. Orient the tube in the centrifuge rotor with the lid connector pointing away from the center of rotation.

Note: Measure the time of centrifugation from the moment of **starting the microfuge**.

4. Transfer l00 µl of the reaction mixture into the PLG I tube.

5. Add 100 µl of PCI solution. Mix by inverting several times to form an emulsion.

6. Centrifuge the PLG I tube at 10 000 rpm for exactly 30 seconds. Be sure to orient the tube in the centrifuge rotor the same way as before.

Note: After centrifugation, the organic phase at the bottom of the tube will be separated from the aqueous phase at the top of the tube by the PLG barrier.

7. Add 100 µl of CIA solution. Mix by repeated inversion to form an emulsion. Do not vortex or allow the bottom organic phase to mix with the upper aqueous phase.

8. Centrifuge for 30 seconds to separate phases. Be very careful to place tube into the centrifuge in the same orientation as before.

9. Transfer the aqueous phase, collected from above the PLG barrier, to a fresh tube. Record the volume of the aqueous phase.

10. Add 50 µl of 7.5 M ammonium acetate (half the volume). Mix well by inverting the tube 4 to 5 times.

11. Add 300 µl of 95% ethanol (2 x the total volume). Mix well by inverting the tube 4 to 5 times.

12. Place the tube in the centrifuge and orient the attached end of the lid pointing away from the center of rotation. Centrifuge the tube at maximum speed for 10 minutes at room temperature.

13. Remove the tube from the centrifuge and open the lid. Holding the tube by the lid, touch the tube edge to the edge of an Erlenmeyer flask and gently lift the end to drain ethanol. You do not need to remove all the ethanol from the tube. Place the tubes back into the centrifuge, orienting the tube as before.

Note: When pouring off ethanol do not invert the tube more than once because this can loosen the pellet.

14. Wash the pellet with 700 µl of cold 70% ethanol. Holding a P1000 Pipetman vertically, slowly deliver the ethanol to the side of the tube opposite the pellet, that is, the side facing the center of the rotor. Hold the Pipetman as shown in Figure 3, Chapter 2. **Do not start the centrifuge,** in this step the centrifuge rotor is used as a "tube holder" that keeps the tubes at an angle, convenient for ethanol washing. Remove the tube from the centrifuge by holding the tube by the lid. Pour off the ethanol.

15. Place the tube back into the centrifuge and repeat the 70% ethanol wash one more time.

Note: This procedure makes it possible to quickly wash a large number of pellets without centrifugation and vortexing. Vortexing and centrifuging are time consuming and frequently lead to substantial loss of the material.

16. After the last wash, place the tube into the centrifuge, making sure that the tube orientation is the same as before. Without closing the tube lids, start the centrifuge for 2-3 seconds and collect the remaining ethanol at the bottom of the tube. Remove all ethanol with a P200 Pipetman outfitted with a capillary tip.

Note: Never **dry the DNA pellet** in a vacuum. This will make dissolving the DNA pellet very difficult if not impossible.

17. Resuspend the pelleted DNA in 20 to 30 µl of TE. Store at - 20 °C.

Labeling protocol

1. Place a 1.5 ml microfuge tube on ice and assemble the transcription reaction as follows:

Reagents	Volume (µl)	Final concentration
DEPC-treated water	to 20 µl	-
10 x Transcription buffer	2	1x
10 x NTP labeling mixture	2	1x
DNA template	as needed	1 µg
RNA polymerase (SP6, T7, T3)	2	10 units

Note: Mix the enzyme by pipetting up and down.

2. Incubate the reaction at 37 °C for at least 2 hours.

3. Add 2 µl of DNase and incubate for 15 minutes at 37 °C to remove template DNA (optional).

Note: This step is usually unnecessary because DIG-labeled DNA greatly exceeds the amount of DNA template. Remaining template DNA does not interfere with hybridization.

4. Stop the reaction by adding 1 µl 0.5 M EDTA. Determine labeling efficiency of RNA using the protocol for estimation of DIG-dUTP concentration (see page 212). Store the probe at -20 °C.

Note: Removing unincorporated DIG-labeled substrate is not necessary.

Troubleshooting

Poor incorporation of DIG and high background during hybridization usually results from incomplete restriction enzyme digestion during template preparation. To confirm complete linearization of the plasmid, run agarose gel electrophoresis with a small amount of restricted probe with suitable size markers.

Very high level of unlabeled nucleotide in the probe can result in a "spotty" background. These impurities can be removed by ethanol precipitation in the presence of ammonium acetate as described below.

1. Dilute the DIG-labeled probe to a final volume of 100 µg with DEPC treated water.

2. Add 50 µl (half the volume) of 7.5 M ammonium acetate. Mix well by inverting the tube 4 to 5 times.

3. Add 300 µl of 95% ethanol (2 x of the total volume) and mix well by inverting the tube 4 to 5 times.

4. Place the tube in the centrifuge and orient the attached end of the lid pointing away from the center of rotation. Centrifuge at maximum speed for 10 minutes at room temperature.

5. Remove the tube from the centrifuge. Pour off the ethanol by holding the tube by the open lid and gently inverting it, touch the lip of the tube to the rim of an Erlenmeyer flask. Hold the tube to drain the ethanol. You do not need to remove all the ethanol from the tube. Place the tubes back into the centrifuge in the same orientation as before.

Note: When pouring off ethanol do not invert the tube more than once because this can dislodge the pellet.

6. Wash the pellet with 700 µl of cold 70% ethanol. Holding a P1000 Pipetman vertically slowly deliver the ethanol to the side of the tube opposite the pellet. Hold the Pipetman as shown in Figure 3, Chapter 2. **Do not start the centrifuge**, in this step the centrifuge rotor is used as a "tube holder" which keeps the tubes at an angle, convenient for ethanol washing. Remove the tube from the centrifuge by holding the tube by the lid. Pour off the ethanol as described in step 5. Place the tube back into the centrifuge and repeat the 70% ethanol wash one more time.

Note: This procedure makes it possible to quickly wash the pellet in a large number of the tubes without centrifugation and vortexing. Vortexing and centrifuging the pellet is time consuming and frequently leads to substantial loss of material and DNA shearing.

7. After the last wash, place the tube into the centrifuge, making sure that the tube side containing the pellet faces away from the center of rotation. Without closing the lid of the tubes start the centrifuge for 2-3 seconds and collect the remaining ethanol at the bottom of the tube. Remove all ethanol with a P200 Pipetman fitted with a capillary tip.

Note: Never **dry the DNA pellet** in a vacuum. This will make dissolving the DNA pellet very difficult, if not impossible.

8. Dissolve the pellet in DEPC-treated water by gently tapping the tube with your finger. Store at -20 °C.

Subprotocol 4
Estimation of DIG-dUTP Concentration

In this procedure DIG-labeled DNA is detected colorimetrically. The procedure takes about 20 minutes and can be performed in a Petri dish. A schematic outline of the procedure is shown in Figure 4.

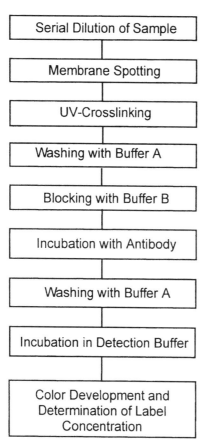

Fig. 4. Schematic outline of the determination of the label concentration

▣▣ Materials

- 15 ml and 50 ml sterilized polypropylene, conical centrifuge tubes (for example Corning # 25319-15; 25330-50).
- Anti-digoxigenin antibody conjugated with alkaline phosphatase (Anti-DIG-AP) from Roche Molecular Biochemicals (# 1093 274).
- Blocking reagent (Roche Molecular Biochemicals # 1096 176).
- Maleic acid (Sigma Co. # MD375).
- NBT solution (Roche Molecular Biochemicals # 1383 213).
- MagnaGraph nylon membrane (Micron Separation Inc. # NJTHY00010) MagnaGraph membrane is recommended since it gives lower background in detection of DIG-labeled DNA.
- Standard DNA labeled with digoxigenin (Roche Molecular Biochemicals # 1585 738).
- X-phosphate solution (Roche Molecular Biochemicals # 1383 221).

- **Washing Buffer (Buffer A)**   Solutions
  - 100 mM Maleic acid
  - 150 mM NaCl
  - 3% (v/v) Tween 20
  Sterilize by filtration. Store at 4 °C.
- **Blocking Buffer (Buffer B)**
  - 100 mM Maleic acid
  - 150 mM NaCl
  - 1% (w/v) blocking reagent
  Add first two ingredients, adjust pH to 7.5 with 10 N NaOH. Sterilize by autoclaving. After liquid cools, add 10 x blocking reagent. Store at 4 °C.
- **Detection Buffer (Buffer C)**
  - 100 mM Tris-HCl, pH 9.5
  - 100 mM NaCl
  Sterilize by autoclaving and store at 4 °C.
- **Blocking Regent Solution**
  - 10% (w/v) Blocking Reagent
  Add 10 g of blocking reagent to 100 ml Buffer A. Place on stir plate and heat to 60 °C for approximately one hour. Sterilize by filtration and store at 4 °C. Do not boil. The solution should have a uniformly milky appearance.
- **10 x SSC**
  - 1.5 M NaCl
  - 0.15 M sodium citrate pH 7.5

Adjust pH to pH 7.5 with concentrated NaOH. Sterilize by autoclaving. Store at 4 °C.

Note: All of the solutions can be bought as DIG Wash and Blocking Buffer Set from Roche Molecular Biochemicals (# 1585 762).

▓▓ Procedure

1. Add 9 μl of water to 5 microfuge tubes. Prepare a ten-fold serial dilution of the newly labeled DNA. Serially dilute as follows: 1:10, 1:100, 1:1 000, 1:10 000 and 1:100 000. Prepare the same a serial dilution for the DIG-labeled standard.

2. Cut a piece of nylon membrane 7 cm x 2 cm and place it at the bottom of the Petri dish. Spot 1 μl of each DIG-labeled dilution onto the membrane. Start from the largest dilution (1:100 000) and continue spotting toward the less dilute samples. Use the same yellow tip for all serial spotting. Repeat this procedure using standard DNA. Place each standard directly under the appropriate dilution of your labeled DNA. Arrange the spots on the membrane in the following way:

Sample DNA	1/100 000	1/10 000	1/1000	1/100	1/10
Standard DNA	1/100 000	1/10 000	1/1000	1/100	1/10

Note: One μl of 1:10 dilution of standard DNA contains 0.5 ng.

3. Let the spotted samples dry completely. Wet the membrane with 10 x SSC. Blot excess liquid with Whatman filter paper. Immobilize the DNA on the **damp** membrane by UV-crosslinking or baking at 120 °C for 30 minutes.

Note: The membrane must be damp to crosslink the DNA. Best results are obtained by application of a calibrated UV-light source such as Stratalinker UV-oven (Stratagen Co. # 400071). Crosslinking of DNA to nylon membrane requires 120 mJ/cm^2 of energy. The use of the "Autocrosslinking" function of an UV-oven is recommended.

4. Transfer the membrane to a Petri dish and add 15 ml of buffer A. Incubate on a rotary shaker at 10-20 rpm for 5 minutes. Discard buffer A.

5. Add 15 ml of buffer B and incubate on a rotary shaker as above for 5 minutes. Discard buffer B.

6. During incubation in buffer B, prepare anti-DIG-AP solution. Add 20 ml of buffer B and 4 µl of anti-DIG-AP conjugate antibody to a sterile 50-ml plastic tube. Mix well by inverting tube several times.

7. Add anti-DIG-AP solution to the membrane and incubate with gentle rotation for 5 minutes at room temperature. Discard the antibody solution.

8. Add 15 ml of buffer A to the membrane and wash with gentle rotation for 5 minutes. Discard buffer A and repeat the wash once more.

9. Add 15 ml of buffer C and incubate on the rotary shaker for 2 minutes. Discard buffer C.

10. Prepare color development solution as follows: Add 45 µl of NBT and 35 µl of X-phosphate to 10 ml of buffer C. Protect the solution from direct light.

11. Add color development solution to the membrane. Make sure the solution covers the membrane. Cover the dish with aluminum foil to protect it from light. Incubate **without shaking** for 30 to 60 minutes checking occasionally for color development.

12. Compare the spot intensities of probe with control DNA and estimate the concentration of DIG-labeled probe.

Subprotocol 5
Determination of Specific Activity of Radioactive Probe

This protocol describes a procedure to calculate the specific activity of probe. A schematic outline of the determination of specific activity of radioactive probe is shown in Figure 5.

Materials

– Glass Microfibre Filters 2.4 cm (Whatman # GF/C 1820-024 or equivalent)
– BioSafe Scintillation fluid

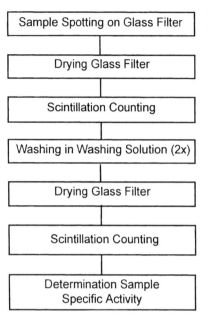

Fig. 5. Schematic outline of the determination of specific activity of radioactive probe

Solutions   Washing Solution

HCl (concentrated)	200 ml
Sodium pyrophosphate	20 g
Water	to 2000 ml
Store at 4 °C.	

▩▩ Procedure

1. Dilute the reaction mixture to a final volume of 50 μl using TE buffer. Withdraw 1 μl and spot onto a glass filter.

2. Dry the spotted filter under a heat lamp. Place the filter in a vial with scintillation fluid and count in a scintillation counter. Record your results as CPM before washing.

3. Retrieve the filter from the scintillation vial and dry the filter under a heat lamp. Place the filter into a 100 ml beaker and add 50-70 ml of cold washing solution. Place the beaker on ice and swirl it carefully for 1 to 2 minutes. Pour the washing solution into a radioactive waste bottle. Repeat the washing procedure twice. Drain all the washing solution well.

4. Add 50 ml of cold 95% ethanol to the beaker and wash by swirling for 2 to 3 minutes. Pour ethanol into the radioactive waste bottle. Drain ethanol well and place the filters under a heat lamp to dry. Put the filter into the same scintillation vial used before and count again. Record your results as CPM after washing.

5. Calculate the percent of label incorporated using the formula:
%Inc = (CPM after washing/CPM before washing) x 100.

6. Calculate the specific activity of the probe. First, calculate the amount of newly synthesized DNA using equation 1 and then, calculate specific activity using equation 2

$$\text{nDNA (ng)} = \frac{\mu\text{Ci dCTP added x \%inc x 13.2}}{\text{Sp. activity dCTP (Ci/mM)}}$$

$$\text{Sp. activity (CPM/}\mu\text{g)} = \frac{\mu\text{Ci dCTP added 2.2 x \%inc x } 10^7}{\text{Template DNA (ng)} + \text{nDNA (ng)}}$$

References

Davis L, Kuehl M, Battey J (1994) Basic methods in molecular biology. 2nd edition. Paramount Publishing Business and Professional Group. Appelton & Lange. Norwalk, Connecticut.

Feinberg AP, Vogelstein B (1983) A technique for radiolabeling DNA restriction endonuclease fragments to high specific activity. Anal Biochem 132:6-13.

Höltke HJ, Seibl R, Burg J, Mühlegger J, Kessler C (1990) Nonradioactive labeling and detection of nucleic acids: II. Optimization of the digoxigenin system. Mol Gen Hoppe-Seyler. 371:929-938.

Kessler C (1992) Nonradioactive labeling and detection of biomolecules. Springer, Berlin, Heidelberg, New York, pp 206-211.

Kessler C, Höltke HJ., Seibl R, Burg J, Mühlegger J (1990) Nonradioactive labeling and detection of nucleic acids. I. A novel DNA labeling and detection system based on digoxigenin:anti-digoxigenin ELISA principle. Mol Gen Hoppe-Seyler. 371:917-927.

Leary JJ, Brigati DJ, Ward DC (1983) Rapid and sensitive colorimetric method for visualizing biotin DNA probes hybridized to DNA or RNA immobilized on nitrocellulose: Bioblots. Proc Natl Acad Sci USA 80:4045-4049.

McCreery T, Helentijaris T (1994) Production of DNA hybridization probes with digoxigenin-modified nucleotides by random hexanucleotide priming. In: Protocol for nucleic acid analysis by nonradioactive probes. Methods in Molecular Biology Vol 28:73-76. Isaac PG. Editor. Humana Press. Totowa, New Jersey.

McCreery T, Helentijaris T (1994) Production of hybridization probes by the PCR utilizing digoxigenin-modified nucleotides. In: Protocols for nucleic acid analysis by nonradioactive probes. Methods in Molecular Biology. Vol 28:67-71. Isaac PG Editor. Humana Press. Totowa, New Jersey.

Murphy NR, Hellwig RJ (1996) Improved nucleic acid organic extraction through use of a ubique gel barrier material. BioTechniques 21:934-939.

Rost A-K (1995) Nonradioactive Northern blot hybridization with DIG-labeled DNA probes. In: Quantitation of mRNA by Polymerase Chain Reaction. Nonradioactive PCR Methods. Ed. Köhler TH, Laßner D, Rost A-K, Thammm B, Pustowoit B and Remke R Springer, Berlin, Heidelberg Springer Lab Manual.

Yamaguchi K, Zhang D, Byrn RA (1994) Modified nonradioactive method for Northern blot analysis. Anal Biochem 218:343-346.

Suppliers

Chemiluminescent DNA labeling kits

Ambion, Inc. (www.ambion.com)
Bright-Star kit uses psoralen to conjugate biotin to nucleotides. Labeling does not require an enzymatic reaction since the psoralen-biotin conjugate is covalently attached to nucleic acids by treatment with long wave UV light. Secondary conjugate is streptavidin-alkaline phosphates (streptavidin-AP). CDP-Star is used as the substrate for AP. Probes are stable for at least a year at -70 °C.

Amersham Pharmacia Biotech. (www.apbiotech.com)
Several random primer labeling kits use fluorescein tagged dUTP to label DNA. Detection uses AP or HRP (horseradish peroxidase) enzymes and CDP-Star or ECL as substrate, respectively. The HRP/ECL system is less sensitive compared to the AP system and more difficult to use. Fluorescein labeling has limitations during hybridization that are discussed in this chapter. There is also a kit that labels DNA by crosslinking AP or HRP to the DNA directly. Secondary conjugate is not needed with this system but sensitivity of the method is low and the method requires application of special hybridization conditions.

Cruachem Inc. (www.cruachem.com)
This random primer labeling kit uses biotinylated nucleotides. Secondary conjugate is streptavidin conjugated to HRP. Chemiluminescent substrate is LumiGLO. As in all HRP based kits, its weakness is high background and low sensitivity.

ICN Pharmaceuticals Inc. (www.icnpharm.com)
These random primer kits use biotinylated nucleotides and AP conjugated to streptavidin. Chemiluminescent substrate is CSPD. Kits have moderate sensitivity but are very inexpensive.

Life Technologies BRL (www.lifetech.com)
Company offers number of DNA labeling kits all based on biotin labeled nucleotides and streptavidin conjugated with AP. Chemiluminescent substrates are enhanced Lumi-Phos 530 or CDP-Star.

NEN Life Science Products (www.nenlifesci.com)
There are several kits based on fluorescein or biotin labeled nucleotides. Secondary conjugates are either anti-fluorescein conjugated to HRP or AP or streptavidin linked with AP or HRP. The chemiluminescent substrate offered is CDP-Star.

New England Biolabs (www.neb.com)
This random primer labeling kit uses biotin-dATP labeling and streptavidin-AP detection system. CDP-Star is used as AP substrate. This kit provides sensitive labeling with low background presumably because dATP is the labeling nucleotide.

PanVera (www.panvera.com)
This random primer labeling kit uses digoxin labeled nucleotides with anti-DIG HRP secondary conjugate. The kit is inexpensive but suffers from the limitation of a HRP detection system. This company also offers DNA and RNA labeling kits that attach digoxin directly to DNA or RNA.

Roche Molecular Biochemicals (biochem.roche.com)
The company offers a number of labeling kits all based on digoxygenin-labeled nucleotides. The secondary detection system is anti-DIG conjugated to AP and CDP-Star. This is a very sensitive labeling system with exceptionally low background. The methods described in this book are based on these products.

Schleicher & Schuell (www.s-and-s.com)
Nonenzymatic labeling of DNA by photocrosslinking psoralen conjugated with biotin. Biotin is detected by a secondary conjugate, streptavidin-AP. CDP-Star is used as chemiluminescent detection substrate. Labeling DNA with this kit is efficient. The streptavidin-biotin system can give high background with some membranes.

Sigma Chemical Co. (www.sigma.com)
DNA labeling kits use biotin-streptavidin labeling and a detection system with CSPD or CDP-Star substrates.

Stratagene, Inc. (www.stratagen.com)
This random primer labeling kit uses fluorescein labeled nucleotides and anti-fluorescein antibody conjugated to AP. Nonstandard hybridization conditions are required when DNA is labeled with fluorescein.

Tropix, Inc. (www.tropix.com)
This company manufactures the chemiluminescent substrates used with most kits. This company also offers several detection systems. Detection kits use anti-fluorescein antibody conjugated to AP for fluorescein labeled DNA or streptavidin conjugated to AP for biotin labeled DNA. Tropix is an inexpensive source of chemiluminescent substrates.

Nucleic Acid Hybridization. A Theoretical Consideration

STEFAN SURZYCKI

1 Introduction

DNA-DNA and RNA-DNA hybridization reactions are the basis of many assays in DNA analysis and are presently some of the most frequently used techniques in molecular biology. The hybridization reaction is the formation of partial or complete double-stranded nucleic acid molecules by sequence-specific interaction of two complementary single-stranded nucleic acids. The hybridization reaction, using labeled probes, is the only practical way to detect the presence of specific nucleic acid sequences in a complex nucleic acid mixture. The most frequently used hybridization technique is the membrane hybridization technique. Denatured DNA or RNA is immobilized on an inert support in a way that self-annealing is prevented but bound sequences are available for hybridization with labeled single or double stranded probes. Extensive washing of the membrane to remove unbound probe and poorly matched hybrids follows the hybridization reaction. Membrane hybridization is used in many different applications such as Southern and Northern blot hybridization, dot blot hybridization and phage plaque or bacterial colony hybridization. This chapter will briefly consider theoretical aspects of membrane hybridization.

2 Hybridization Reaction

T_m temperature

Hybrid formation between complementary strands is commonly called reassociation, renaturation or reannealing reaction. The reverse reaction is called strand separation, dissociation or melting of the DNA. To determine the temperature of hybridization, the melting temperature (T_m), defined as the temperature at which the DNA (or RNA) is 50% denatured, must be known. Melting of DNA (or DNA:RNA duplex) is an intramolecular, first order reaction, and therefore, is independent of substrate concentration. It depends only on the base composition of the duplex and the composition of the solvent. Determination of T_m for DNA:DNA duplexes was first empirically established by Marmur and Doty (Marmur and Doty, 1962) for molecules shorter than 500 bp (equation 1).

$$T_m = 81.5 + 0.41 \, (\%GC) + 16.6 \log [M^+] \tag{1}$$

Where $[M^+]$ is the molar concentration of monovalent cations, Na^+ or K^+, in a range of 0.01 to 0.4 M, and %GC is the percent of GC bases in a range of 30% to 70%.

Equation 1 was modified to incorporate DNA:RNA and RNA:RNA hybrids and extends the monovalent cation concentration range from 0.01 to 4.0 M, the concentrations at which most hybridization reactions are carried out (Wetmur, 1991):

$$T_m = A + 16.6 \log \frac{[M^+]}{1 + 0.7 \, [M^+]} + B \, (\%GC) - \frac{500}{L} \tag{2}$$

Where, A is 81.5 °C, 67 °C or 78 °C for DNA:DNA, DNA:RNA and RNA:RNA hybrids, respectively. B is 0.41 °C, 0.8 °C or 0.78 °C for DNA:DNA, DNA:RNA and RNA:RNA hybrids, respectively. L is the length of hybrid in base pairs, and %GC is percent of GC bases in range of 30% to 70%.

The probes used in membrane hybridization are usually prepared by random primer or nick translation methods and are shorter than 500 bp. When these probes are used, it is possible to drop the last term of equation 2 since it does not significantly change the T_m. When divalent cations are present, term $[M^+]$ should be changed to $[salt] = \{[M^+] + 4 \, [M^{+2}]\}^{0.5}$. However, most hybridization solutions used in membrane blot hybridization do not contain divalent cations.

Kinetics of hybridization

The hybridization reaction proceeds in two steps: the nucleation reaction and "zippering" reaction. Nucleation is the formation of short hybrids, a few bases long, between reacting strands. The nucleation reaction, with some approximation, is a diffusion-limited reaction defined by the Smoluchowski and Deby equation. Therefore, the reaction rate depends on solvent viscosity, temperature and ionic strength of the medium (Wetmur and Davidson, 1968; Chang et al, 1974).

Many nucleation events will take place, until by chance, the correct base pair is formed. A rapid zippering process follows this. The zippering reaction is an extension of the hybrid from the nucleation site throughout the entire molecule. This reaction is very fast and largely independent of the factors mentioned above. Thus, the limiting step in a hybridization reaction is the nucleation reaction and not the zippering process. The overall hybridization rate is dependent on a nucleation rate constant (k_n,), probe length and target complexity as described by following equations:

$$k_2 = \frac{k_n \sqrt{L}}{N} \tag{3}$$

or when N = L

$$k_2 = \frac{k_n}{\sqrt{L}} \tag{4}$$

Where: k_n is the nucleation rate constant, N is the number of nucleotide pairs of the target (complexity), and L, the probe length in bases. The nucleation rate, k_n, for monovalent cation concentrations commonly used in filter hybridization (0.2 - 4.0 M) and solvent viscosity comparable to 1.0 M NaCl, can be evaluated using the following equation (Orosz and Wetmur, 1974):

$$k_n = \{4.35 \log [M^+] + 3.5\}10^5 \tag{4a}$$

Hybridization temperature

Since the rate of the nucleation reaction equals zero at T_m temperature, hybridization should be carried out at temperatures below the T_m temperature calculated from equation 1 or 2. The difference between T_m temperature and hybridization temperature is defined as "criterion of hybridization" (Britten et. al, 1974). The overall rate of the hybridization reaction k_2, is strongly affected by temperature. This dependence has a bell-shaped curve and increases as the criterion is increased, reaching a broad maximum between 20 °C to 25 °C below T_m for the DNA:DNA hybrid and 15 °C to 20 °C below T_m for a DNA:RNA hybrid. As the temperature of hybridization falls farther below the T_m, the hybridization rate decreases very fast due to intramolecular base pair formation that decreases the availability of nucleation sites. Therefore, the optimal temperature for hybrid formation (T_h) for a DNA: DNA hybrid is:

$$T_h = T_m - (20°C \text{ to } 25°C) \tag{5}$$

and for a DNA:RNA hybrid:

$$T_h = T_m - (15°C \text{ to } 20°C) \tag{6}$$

At T_h temperature, not only does hybrid formation occur at maximum speed, but is most reliable. This is because at T_m, the formation of perfectly matched hybrids by the zippering reaction is faster than the formation of mismatched hybrids (10% or more mismatch). Moreover, the T_m of imperfectly matched hybrids is lower than perfectly matched hybrids (approximately 1 °C for each percent of mismatch). Consequently, the maximum hybridization rate of mismatched hybrids occurs much below the hybridization temperature of a perfect hybrid. The formation of poorly matched hybrids shows a similar bell-shaped dependence on temperature but maximum rate (k_2) are several orders slower than for a well-matched hybrid and the entire curve is displaced toward lower temperatures.

Hybridization time

Membrane blot hybridization is usually performed with a large excess of probe DNA. For example, about 5-10 μg of eukaryotic genome is used in Southern blot analysis. Assuming that a single gene size is 2000 bp, genome size is 3×10^9 bp (e.g., human genome) and only 2% of DNA bound to the nylon membrane is open for hybridization (Vernier et al., 1996), the amount of a single gene present on a Southern blot is about $0.06 \ 10^{-6}$ μg $\{(2 \ 10^3 \times 5 \ 10^{-6}/3 \ 10^9) \ 0.02\}$. The concentration of probe is usually 20 to 25 ng/ml. If the probe is the same size as the genomic target and 10 ml of hybridization solution is used, the amount of probe present is 0.2 to 0.25 μg or nearly 3×10^6 fold excess of probe DNA over target DNA. Using these hybridization conditions, the reaction is pseudo-first order and its half time is:

$$t_{1/2} = 2/k_2 \ C_0 \tag{7}$$

where k_2 is calculated from equation 4 or 4a and C_0 is the initial concentration of probe in moles of nucleotides per liter (Wetmur, 1991). Since most hybridization reactions are carried out at 1 M Na^+, the k_n for this reaction is equal to $3.5 \ 10^5 \ M^{-1} sec^{-1}$ (Wetmur, 1995). The concentration of the probe used in Southern hybridization is usually equal to $C_0 = 6.1 \ 10^{-8} M$ (20 ng/ml) and k_2, for a single gene probe, as calculated from equation 4a is $7.8 \ 10^3$ $M^{-1}sec^{-1}$. Using equation 7, t of the reaction can be calculated to be approximately 1 hour.

If an excess of double-stranded probe is hybridized to an immobilized target, self-annealing of the probe limits the time of effective hybridization to approximately 2 to 3 times t calculated for single-stranded probe. Thus, the above reaction will reach completion in about 2 to 3 hours, due to self-annealing of the probe. The half-time for pseudo-first order hybridization can be approximated (in hours) when double-stranded probe is used with standard conditions of hybridization (i.e., 1 M salt at T_h equal to 20 °C to 25 °C below T_m) from the following equation (Sambrook et al, 1989).

$$t_{1/2} = \frac{1}{X} \times \frac{Y}{5} \times \frac{Z}{10} \times 2 \tag{8}$$

Where X is μg of probe added, Y is probe complexity in kilobases (length of probe) and Z is the volume of hybridization in ml. For example, for the DNA hybridization reaction described above, that is, hybridization with a 2000 bp probe at a concentration of 0.02 μg/ml in 10 ml, t, calculated from equation 8 is 2 hours (1/0.2 x 2/5 x 10/10 x 2), a value close to that calculated from equation 7.

Hybridization reactions, however, are usually carried for 13 to 17 hours to increase the signal strength. Probes prepared by random primer labeling can form extended networks in which the single-stranded tails from one duplex hybridize to a complementary single-stranded tail of another duplex. This network formation occurs 5 to 6 times slower than the annealing of a single-stranded probe, and its formation will increase the hybridization signal intensity (Geoffrey et al, 1987). Hybridization times longer than 20 hours are not recommended because this would increase background, and most importantly, degrade target and probe, making it impossible to reuse membrane blots and nonradioactive probes in subsequent hybridization reactions.

The hybridization kinetics described above only apply strictly to liquid hybridization. When nucleic acid is immobilized on a membrane, the hybridization rate is decreased because the rate of access of the probe to the target is decreased significantly. Consequently a nucleation rate (k_n) is 2 to 4 times lower than the nucleation rate in liquid hybridization but the effect of T_m and ionic strength are not changed. For a more detailed analysis of filter hybridization, see Anderson and Young (1985).

Washing reaction

The hybridization reaction is followed by a washing reaction that removes any unhybridized probe and melts mismatched hybrids. This reaction, unlike the hybridization reaction, is a first order reaction and depends on the thermal stability of the hybrid (T_m). The T_m of the hybrid is lowered by approximately 1 °C for each percent of mismatch (Bonner et al, 1973). To obtain 95% faithful hybrids, the washing reaction is carried out at a washing temperature expressed by equation:

$$T_w = T_m - 5°C \tag{9}$$

Where T_m is melting temperature calculated from equation 2. Reactions carried out at this temperature are called high stringency washes. Reactions done at temperatures lower than this are usually referred to as "low stringency" washes.

3 Stringency of Hybridization

Conditions of hybridization and washing that favor the formation and maintenance of high-fidelity hybrids are called **high stringency hybridizations** and are used for the detection of closely related sequences. Such hybridization is best achieved by applying a stringent criterion of hybridization followed by stringent washing condition. Thus, to achieve high-stringency hybridization for DNA, hybridization reactions should be carried out at $T_h = T_m$ - (20 to 25)°C and washing reactions at $T_w = T_m$ - 5 °C.

Conditions of hybridization that allow the formation of hybrids with many mismatched bases are called **low stringency hybridizations**. Low stringency hybridization is used for detecting distantly related sequences and is frequently referred to as heterologous hybridization. Heterologous hybridization is best achieved by using a low stringency hybridization reaction followed by a low stringency wash. Reactions should be performed at temperatures much below T_m - (20 °C to 25 °C), and a washing temperature equal to the hybridization temperature or lower.

4 Hybridization Solutions

Hybridization reaction solutions

The rate limiting step of hybridization is the nucleation reaction. Because the rate of nucleation is a very strong function of salt concentration, the hybridization reaction should not be carried out at low salt concentration. To achieve fast hybridization rates, a monovalent cation concentration between 0.75 to 1.0 M should be used. Because DNA used in most hybridization reactions has approximately 50% GC, the T_h is usually between 72 °C to 78 °C, as calculated from equations 3 and 4. However, single-stranded DNA is particularly prone to depurination at temperatures higher than 50 °C. Prolonged hybridization at high temperature will result in degradation of probe and target DNAs (Blake, 1995). RNA probes and targets are even more sensitive to high-temperature degradation than single-stranded DNA. To lower hybridization temperature, the hybridization reactions are usually carried out in the presence of denaturing solvents while maintaining high ionic strength (Hutton, 1977). The most commonly used solvent in membrane hybridization is formamide. On average this solvent lowers the T_m of DNA by 0.70 °C per 1% formamide. The effect of formamide is greater on AT nucleotide pairs than on GC pairs (Anderson and Young, 1985). Formamide has no apparent effect on the rate of hybridization reaction at concentrations between 30 and 50% for membrane hybridization, making it an ideal solvent for lowering the incubation temperature. The concentration of formamide most frequently used is 50%. This lowers the temperature required for hybridization reactions to below 50 °C without substantially lowering the rate of hybridization (72 °C - 50 {0.72} °C = 36 °C or 78 °C - 50 {0.72} °C = 42 °C).

Formamide can also be used to alter the stringency of a hybridization reaction. By holding temperature of incubation constant (e.g., 42 °C) and varying the concentration of formamide below 50%, the temperature of hybridization can be lowered to a temperature that favors formation of mismatched hybrids over well-matched hybrids. For example for a DNA probe of 50% GC the T_m temperature at 1 M of monovalent cation concentration in the presence of 30% formamide is equal to 77.7 °C (81.5 + 0.41 x 50 + 16.6 log 0.588 - 0.72 x 30). Thus at 42 °C the hybridization temperature is 36 °C below the T_m of the DNA used, a condition that favors the formation of poorly matched hybrids (see the above discussion of hybridization stringency).

Four hybridization solutions are presently in use: standard hybridization solution, high SDS (Church's) solution, Denhardt solution and DIG Easy Hyb solution (Denhardt, 1966; Church and Gilbert, 1984). The first three solutions usually incorporate formamide as a solvent to lower the T_m of the DNA, whereas the DIG Easy Hyb solution uses a nontoxic, non-ionic detergent at a concentration equivalent to 50% formamide. Table 1 lists standard buffers used in hybridization, as well as, ingredients of standard and high SDS hybridization solutions. Boehringer Mannheim Co. formulated the DIG Easy Hyb ingredients which are proprietary. All of these solutions contain blocking agents to prevent nonspecific binding of the probe to the membrane. Denhardt's solution contains a mixture of sonicated, single-stranded salmon sperm DNA, BSA, Ficoll and polyvinyl pyrrolidine (PVP) as blocking agents. Standard and high SDS solutions use

0.5% casein (e.g., dry powdered milk) as a blocking agent. Because Denhardt's solution is expensive and difficult to prepare and sterilize, it is no longer widely used. Moreover, Denhardt's solution is a very poor blocking agent for nonradioactive probes, resulting in high background. Casein-containing blocking solutions are more effective than Denhardt's solution with nonradioactive probes and are, by far, much cheaper and easier to prepare.

Table 1. Hybridization buffers and hybridization solutions presently in use

Salt Solutions			Hybridization Solutions		
1 x SSC	1 x SSPE	Denhardt's blocking (10 x)	Standard SSC	High SDS	DIG Easy
0.15 M NaCl	0.15 M NaCl	5-6 x SSC or SSPE	5-6 x SSC or SSPE	5-6 x SSC or SSPE	
0.015 Na citrate	0.01 M NaH$_2$PO$_4$	50% Formamide	50% Formamide	50% Formamide	
	1 mM EDTA	0.5 % SDS	0.1% Sarcosyl	0.1% Sarcosyl	
		2.0% Ficoll 400	0.02% SDS	7% SDS	
		0.2 % PVP	2-5% casein	2-5% casein	
		0.2% BSA Fraction V		50 mM phosphate buffer pH 7.0	
		100 µg/ml salmon sperm DNA			
Recommended use					
All hybridization and washing	Prolonged hybridization with formamide	Hybridization with radioactive probes	Hybridization with radioactive and non radioactive probes	DNA fingerprinting with radioactive and non radioactive probes	All hybridization

Washing reaction solutions

The washing reaction is used to remove unhybridized probe and poorly matched hybrids. Membranes are treated at temperatures that dissociate nonspecific hybrids. To keep washing reaction temperature low, low concentration of salts in reaction are used. The solution most frequently used is 0.1 to 0.2 SSC (SSEP) supplemented with 0.1% SDS to prevent enzymatic degradation of DNA. Stringent washes are done at 62 °C to 68 °C with 0.1 SSC 0.1% SDS solutions. For example, DNA with 50% GC content at 0.015 M salt concentration (0.1 x SSC, see Table 1) has a T_m about 72 °C. To achieve stringent washing conditions, the temperature should be 5 °C to 10 °C below the T_m temperature of the DNA, i.e., a temperature between 62 °C to 67 °C (72 - 10 = 62 or 72 - 5 = 67). For moderate stringency, 0.2 x SSC + 0.1% SDS is used at temperature of 42 °C (about 35 °C below T_m). Low stringency washes are carried out with 0.2 to 0.5 x SSC + 0.1% SDS solution at room temperature.

Since the DNA melting reaction is practically instantaneous, it is not necessary to wash for a long time. DNA melting and washing of the unbound probe can be achieved in 10 to 15 minutes. This is particularly important for stringent and moderately stringent washing reactions that are carried out above 50 °C, a temperature that could degrade unhybridized target and probe DNA and prevent their subsequent use.

5 Hybridization Membranes

Different types of support membranes are commonly used for nucleic acid immobilization. These are nitrocellulose membrane, uncharged nylon membrane and positively charged nylon membrane. All types of commercially available nylon membranes were recently evaluated and reviewed (Brush, 1995).

Nitrocellulose

Nitrocellulose membrane is the oldest in use and was first introduced in the early 1960s. However, it has many limitations. Nucleic acid hydrophobic attachment to the membrane is weak, resulting in a slow release of nucleic acids during hybridization and washing. Nitrocellulose membrane is fragile and cannot withstand multiple probing. Blots cannot easily be stored for very long because the membrane disintegrates in air. Nitrocellulose membrane cannot be used for chemiluminescent detection. Attachment of nucleic acids to the membrane requires drying the membrane at 80 °C in a vacuum oven due to the explosive nature of the membrane in an oxygen environment, and finally, nitrocellulose membrane is highly hydrophobic and difficult to wet with transfer solutions. These limitations led to the development of nylon as a support media.

Uncharged nylon

Uncharged nylon membrane was introduced in the early 1980s originally for its outstanding durability. The surface of the uncharged nylon consists of amine and carboxyl groups, providing a net neutral charge on the membrane. This membrane also has a number of advantages over nitrocellulose membrane. Nylon membrane is much stronger and can withstand exposure to a broad range of solvents and a number of sequential hybridizations. Nylon covalently binds nucleic acids when crosslinked by UV irradiation or baking, preventing release of nucleic acid during repeated hybridizations and washings (Nerzwicki-Bauer et al, 1990). Nylon membrane has a higher binding capacity than nitrocellulose and can efficiently bind fragments as short as 25 to 50 bases, and nylon blots can be stored indefinitely at room temperature. There are some disadvantages of uncharged nylon membrane. The background is relatively high with some radioactive probes as compared to nitrocellulose. Uncharged nylon gives a high background with biotinylated and chemiluminescent detection systems, and hybridization is less sensitive because the bound nucleic acids are less available for hybridization. It has been determined that the amount of available target for hybridization is only 2% of the total bound nucleic acids, as opposed to 4% for nitrocellulose (Vermier et al, 1996). These shortcomings led to the development of positively charged, nylon membrane.

Positively charged nylon

Positively charged nylon membrane, introduced in the early 1990s, retained nylon as a support, but its surface charges have been modified by attachment of high-density, quaternary ammonium groups. This makes the membrane strongly cationic and positive charges are retained over a pH range of 3 to 10. Positively charged membrane retains all of the advantages of nylon membrane and gains some new characteristics. The membrane has a very high electrostatic attraction for nucleic acids, permitting transfer of DNA under alkaline conditions. It has a much greater capacity for nucleic acids resulting in lower background and stronger hybridization signals. Also this membrane has very low background when chemiluminescent or biotinylated detection systems are used. Some producers have developed chemically optimized, positively charged nylon membrane for chemiluminescent detection systems. The positively charged nylon membrane is presently recommended for use in all hybridization applications. Table 2 lists some manufacturers along with important characteristics of commercially available, positively charged nylon membrane.

Table 2. Manufactures and properties of nylon membranes

Company	Membrane	Binding Capacity[a]	Application	Chemilumine-scent bkg[b]
Amersham	Hybond N+	480-600	DNA, RNA hybridization	High
Roche Molecular Biochemicals	Nylon membrane		DNA, RNA hybridization and DNA fingerprinting	Low
Bio-Rad	Zeta-Probe and Zeta-Probe GT	400-600	DNA, RNA hybridization	Moderate
	C/P lift	150	Colony and plaque lift	Unknown
	Immun-lite	150	Protein blotting	Low for proteins
DuPont	GeneScreen Plus	400-500	DNA hybridization	High
GibcoBRL	PhotoGene Nylon	450	DNA, RNA hybridization	Moderate
ICN Biochemical	BioTrans(+)	400-500	DNA fingerprinting and DNA, RNA hybridization	Low
Millipore	Immobilon-S	-	DNA, RNA hybridization	Very low
MSI	Magna Charge	450	DNA, RNA hybridization	Very low
	Magna Graph	400	DNA, RNA hybridization	Exceptionally low
National Labnet	HyBlot Plus	100	DNA, RNA hybridization	Unknown
Pall	Biodyn B	600	DNA, RNA hybridization	Very low
	Biodyn Plus	›400	DNA, RNA hybridization	Very low
Schleicher & Schuell	Maximum Strength	›400	DNA, RNA hybridization	Moderate
Tropix	Tropilon+	-	DNA, RNA hybridization	Very low

[a] Binding of DNA in $\mu g /cm^2$

[b] Data from author's laboratory experience with designation for background when using single copy human gene detection DIG labeled probe, DIG Easy Hyb solution and CPD chemiluminescent substrate.

High - exposure for more than 10 minutes leads to totally "black" background; Moderate- exposure possible for 10 to 20 minutes with dark gray background; Low- exposure for 30 minutes possible with slightly gray background

References

Aderson MLM, Young. BD (1985) Quantitive filter hybridization. In: Nucleic acid hybridization a practical approach. Ed. Hames BD. and Higgins SJ IRL Press 1985. pp. 73-111.

Blake RD (1995) Denaturation of DNA. In: Molecular Biology and Biotechnology. A Comprehensive Desk Reference. Ed. Meyers RA VCH Publishers, Inc pp 207-210.

Bonner TI, Brenner DJ, Neufeld BR, Britten RJ (!973) Reduction in the rate of DNA reassociation by sequence divergence. J Mol Biol 81:123-135

Britten RJ Graham DE, Neufeld BR (1974) Analysis of repeating DNA sequences by reassociation. Method Enzymol 29E:363-420.

Brush M (1995) A run on nylons: A survey of nylon blotting membranes. BioConsumer 2:14-22. Chang C-T, Hain TC, Hutton JR, Wetmur JG (1974) The effects of microscopic viscosity on the rate of renaturation of DNA. Biopolymers 13:1847-1855.

Church GM, Gilbert W (1984) Genomic sequencing. Proc Natl Acad Sci USA 81:1991-1995.

Denhardt DT (1966) A membrane-filter technique for the detection of complementary DNA. Biochem Biophys Res Commun 23:641-646

Geoffrey M, Wahl L, Shelby C, Berger, Kimme AR (1987) Molecular hybridization of immobilized nucleic acids: Theoretical concepts and practical consideration. Method Enzymol 152:399-407.

Hutton JR (1977) Renaturation kinetics and thermal stability of DNA in aqueous solutions of formamide and urea. Nucl Acid Res 4:3537-3555.

Marmur J, Doty P (1962) Determination of the base composition of deoxyribonucleic acid from its thermal melting temperature. J Mol Biol 5:109-118.

Nierzwicki-Bauer SA, Gebhardt JS, Linkkila L and Walsh K (1990) A comparison of UV crosslinking and vacuum baking for nucleic acids immobilization and retention. BioTechniques 9:472-478.

Orosz JM, Wetmur J (1977) DNA melting temperatures and renaturation rates in concentrated alkylammonium salt solutions. Biopolymers 16:1183-1190.

Sambrook J, Fritsch EF, Maniatis T (1989) Molecular Cloning. A laboratory Manual. Cold Spring Harbor Laboratory Press. 1998. pp. 9.48

Vernier P, Mastrippolito R, Helin C, Bendali M, Mallet J, Tricoire H (1996) Radioimager quantification of oligonucleotide hybridization with DNA immobilized on transfer membrane: Application to the identification of related sequences. Anal Biochem 235:11-19.

Wetmur JG (1991) DNA probes: Application of the principles of nucleic acid hybridization. Critical Rev Biochem Molec Biol 26:227-259.

Wetmur JG (1995) Nucleic acid hybrids, formation and structure of. In: Molecular Biology and Biotechnology. A Comprehensive Desk Reference. Ed. Meyers RA, VCH Publishers, Inc pp 605-608.

Wetmur JG, Davidson N (1968) Kinetics of renaturation of DNA. J Mol Biol 3:349-370.

DNA Transfer and Hybridization

STEFAN SURZYCKI

Introduction

Nyagaard and Hall, and Southern (Nyagaard and Hall, 1964; Southern, 1975) introduced immobilizing target DNA to membrane for hybridization studies. The technique permits hybridization of various probes to immobilized target DNA under controlled conditions, coupled with fast detection of the hybrid. Because of the sensitivity, speed, and convenience of this "solid state" hybridization procedure, it has been widely applied in applied and basic research. The Southern hybridization procedure described in this chapter is separated into two general protocols: transfer of the DNA to the membrane and hybridization of immobilized DNA to DNA or RNA probes. A description of agarose gel electrophoresis of DNA is given in Chapter 8.

Three types of DNA transfer are described: Southern blot, dot blot and colony or plaque transfer. Although the condition of the transfer and immobilization of DNA to the membrane vary somewhat in the three types of transfer described, the conditions of hybridization do not. The theoretical background of the hybridization process and parameters for optimization of hybridization are described in Chapter 10. The Southern blot hybridization protocol described here can be used as a hybridization protocol for dot blot, and colony and plaque immobilized DNA.

Subprotocol 1
Southern Blot Transfer

The Southern blot method, originally described by Southern (1975), combines the high resolving power of agarose gel electrophoresis in the separation of DNA fragments with the specificity of DNA-DNA or DNA-RNA hybridization reactions. The basic principle of the technique is that DNA fragments, separated by agarose gel electrophoresis, are transferred and immobilized to a solid support, such as a nylon or nitrocellulose membrane. Once immobilized, the DNA is available for hybridization with labeled DNA or RNA probes. This technique is applicable to the analysis of small, cloned DNA fragments, as well as to the analysis of genomic DNA. DNA transfer to solid support is generally accomplished by capillary methods but electroblotting, positive pressure, and vacuum transfer procedures can also be used (Peferoen et al, 1982; Smith et al, 1984). These other methods, in general, are faster than capillary transfer but are less efficient and require expensive equipment.

There are two capillary transfer methods, upward capillary transfer and downward capillary transfer. Upward capillary or "standard" transfer results in very efficient transfer of DNA or RNA of all sizes but requires overnight exposure. Downward capillary transfer is just as efficient as upward transfer and requires much shorter transfer time (3-4 hr). It uses a Schleicher & Schuell TurboBlotter transfer apparatus and therefore is more expensive. Capillary transfer can be carried out with neutral or alkaline transfer solutions (Chomczynski, 1992). Protocols for both neutral and alkaline transfers are given below.

A schematic outline of the transfer procedure is shown in Figure 1.

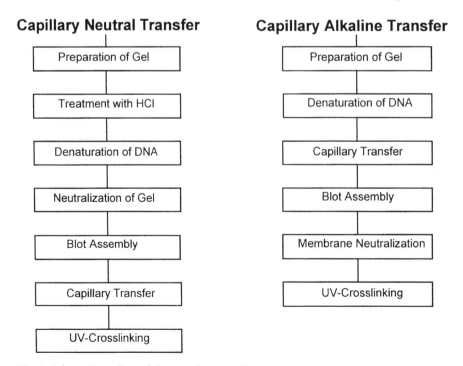

Fig. 1. Schematic outline of the transfer procedures

▧▧ Materials

–   MagnaGraph nylon membrane 0.22 µ pore size (Micron Separation, Inc. # NJTHY00010). See Chapter 10 for a description of equivalent nylon membranes from other manufacturers. If fragments larger than 300 bp are to be transferred, it is not necessary to use membrane with a 0.22 µm pore size, membranes with a pore size of 0.45 µm can be used.
–   Whatman 3MM chromatography paper (Whatman Co. # 3030917)
–   Pyrex glass dishes
–   Stratalinker® UV oven (Stratagen Co. # 400071 or equivalent)
    To efficiently crosslink DNA to nylon membrane the UV source should be capable of delivering 120 mJ/cm². Excessive crosslinking will decrease hybridization efficiency.

Solutions

–   **10 x SSC Solution**
    –   1.5 M NaCl
    –   0.15 M sodium citrate
    Dissolve 87.5 g NaCl and 44.1 g of sodium citrate in 850.0 ml of distilled or deionized water. Adjust pH to 7.5 with 10 N NaOH, and add water to 1000 ml. Store at 4°C.
–   **Denaturation Solution**
    –   0.5 N NaOH
    –   1.5 M NaCl
    Prepare the solution using 10 N NaOH. Solution can be stored at room temperature for a few months. If a white precipitate forms, the solution should be discarded.
–   **Neutralization Solution**
    –   0.5 M Trisma Base
    –   1.5 M NaCl
    Add 60.5 g of Trisma Base and 87.45 g of NaCl to 850 ml of deionized water. Dissolve salts and adjust pH to 7.5 with concentrated HCl. Fill with water to 1000 ml. Store at 4°C.
–   **Alkaline Denaturation Solution**
    –   0.4 M NaOH
    –   3 M NaCl
    Prepare solution using 10 N NaOH and store at room temperature.
–   **Alkaline Transfer Solution**
    –   8 mM NaOH
    –   3 M NaCl
    –   0.5 % Sarcosyl

Prepare the solution using 10 N NaOH. Addition of Sarcosyl is optional but its use can decrease background. Store solution at room temperature. Discard if a precipitate forms.

Procedure

Capillary neutral transfer

1. After electrophoresis, transfer the gel to a Pyrex dish and trim away any unused areas of the gel with a razor blade. Cut the lower corner of the gel at the bottom of the lane with size standards. This will provide a mark to orient the hybridization bands on the membrane with the bands in the gel. Transfer the gel to a transilluminator and draw the outline of the gel on an acetate sheet with a felt marker pen. Mark the positions of the wells, the positions of standard DNA bands and the position of the cut corner. Cut away and remove the gel above the wells.

Note: Because the gel is thinner in the well area, the transfer solution may pass preferentially through this part of the gel causing uneven DNA transfer. Safety Note: Wear gloves and UV-protective glasses during this operation. Use only powder free gloves when using a chemiluminescent detection system. The presence of talcum powder will result in the formation of a "spotted" background.

2. Transfer the gel back to the Pyrex dish and add enough 0.25 N HCl to allow the gel to move freely in the solution. This will take about 150 to 200 ml of solution for a standard gel size.

Note: This procedure breaks large DNA molecules by depurination and is not necessary if you are transferring small DNA fragments. However, it is important not to let the hydrolysis reaction proceed too far; otherwise, the DNA is cleaved into fragments that are too short to bind efficiently to the membrane (less then 200 bp).

3. Place the dish on an orbital shaker and incubate for 15 minutes rotating at 10-20 rpm. Decant the acid carefully holding the gel with the palm of your gloved hand.

4. Rinse the gel for 10-20 seconds in 200 ml of distilled water. Discard the water and proceed immediately to the next step.

5. Add 200 ml of denaturation solution to the dish and gently agitate it for 15 to 20 minutes on a rotary shaker.

6. Decant the denaturation solution as described above and repeat step 5 one more time.

7. Add 200 ml of water and rinse the gel for 10 to 20 seconds to remove most of the denaturation solution trapped on the surface of the gel. Decant the water, holding the gel with palm of your gloved hand. Be very careful during this procedure because the denaturation solution contains NaOH making the gel very slippery.

8. Add 100-200 ml of the neutralization solution to the dish and treat the gel for 20 minutes with gentle agitation.

9. Discard neutralization solution and repeat step 8 one more time.

10. While the gel is being treated, prepare the nylon membrane for transfer. Cut the nylon membrane to the size of the gel. Use the outline of the gel drawn on the acetate sheet as a guide. Use gloves and only touch the edges of the nylon membrane. Place the membrane in a separate Pyrex dish filled with distilled water. Leave the membrane in water for 1 to 2 minutes. Decant the water and immerse the membrane in 10 x SSC. Cut three sheets of Whatman 3MM paper to the size of the nylon membrane. Prepare a long strip of Whatman 3MM paper to use as a wick. The wick should be approximately 30 cm long and 10 cm wide.

11. Assemble the blot sandwich. Refer to Figure 2 for details. Add 400 - 600 ml of 10 x SSC to a large Pyrex dish. Place a glass platform across the center of the dish and cover it with the wick. Make sure that both ends of the wick are immersed in SSC solution. Wet the wick with 10 to 20 ml of transfer solution and remove trapped air bubbles by rolling a 10 ml glass pipette over it. Carefully lift the gel from the Pyrex dish and place it in the center of the wick with the **sample wells down**. Smooth the gel and remove trapped air bubbles by gently rolling a glass pipette over the surface.

Note: The gel is now upside-down with the well openings facing the wick. This is necessary to obtain the best results during transfer (sharper resolution due to less diffusion during the transfer), and to maintain the left-to-right sample orientation on the membrane.

12. Cover the entire dish, including the surface of the gel, with saran wrap. With a razor blade "cut away" the saran wrap covering the gel itself. This will leave an opening over the gel while the remaining area of the wick will be covered by saran wrap.

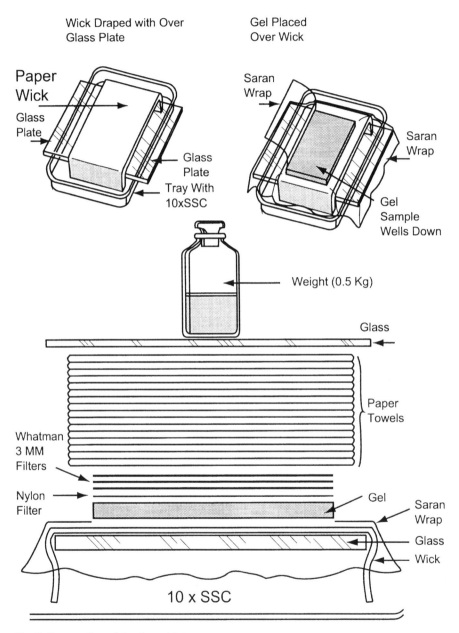

Fig. 2. Preparation of Southern blot

13. Place the nylon membrane on top of the gel. Add 5 ml of 10 x SSC to the top of the membrane and remove air bubbles by rolling a pipette over it. Cut the left bottom corner of the membrane to coincide with the cut made in the gel. Place three sheets of dry Whatman 3MM paper, prepared in step 10, on the top of the membrane. Place several inches of paper towels on top of the 3MM paper. Because the wick area is protected by saran wrap, it is not necessary to cut the paper towels to the size of the gel. Place a glass plate on top of the paper towel stack and weigh it down with a one-liter Erlenmeyer flask filled with 500 ml of water. Allow minimum of 17 hours for the transfer.

Note: To prevent the gel from collapsing, the weight placed on the top of the stack should never exceed 500 g (i.e., about 500 ml of water).

14. Dissemble the blot. Remove the weight, glass plate and paper towels. Using forceps, remove the membrane and place it "DNA side" up (the side that was in contact with the gel) on a clean sheet of Whatman 3MM paper. Mark the DNA side with pencil on the corner of the membrane.

15. Place the membrane on a sheet of dry Whatman 3MM paper. Do not allow the membrane to dry at any time. Place the membrane into an UV oven.

16. UV irradiate the damp membrane to crosslink DNA to the membrane using the automatic setting of the UV oven. Irradiate **both sides** of the membrane. Alternatively, you can wrap the membrane in aluminum foil and bake it in an oven at 80 °C for 1 hour. The baking step immobilizes DNA on the membrane. Membranes can be stored at room temperature practically indefinitely.

Capillary alkaline transfer

Alkaline transfer of DNA is possible only with positively charged nylon membrane. Capillary alkaline transfer is faster because the neutralization step is omitted from the procedure and the time required for transfer is shorter. Alkaline transfer solution can cause some depurination of the DNA and frequently results in weaker hybridization signals than the neutral transfer. Alkaline blotting is not recommended when reprobing is desired.

1. After electrophoresis, transfer the gel to a glass Pyrex dish and trim away any unused areas of the gel with a razor blade. Cut the lower corner

of the gel at the bottom of the lane with size standards. This will provide a mark to orient the hybridization bands on the membrane with the bands in the gel. Transfer the gel to a transilluminator and draw the outline of the gel on an acetate sheet with a felt marker pen. Mark the positions of the wells, the positions of standard DNA bands and the position of the cut corner. Cut away and remove the gel above the wells.

Note: Because the gel is thinner in the well area, the transfer solution may pass preferentially through this part of the gel causing uneven DNA transfer. **Safety Note:** Wear gloves and UV-protective glasses during this operation. Use only powder free gloves when using a chemiluminescent detection system. The presence of talcum powder will result in the formation of a "spotted" background.

2. Transfer the gel back to the Pyrex dish and add enough 0.25 N HCl to allow the gel to move freely in the solution. This will take about 150 to 200 ml of solution for a standard gel size.

Note: This procedure breaks large DNA molecules by depurination and is not necessary if you are transferring small DNA fragments. However, it is important not to let the hydrolysis reaction proceed too far; otherwise, the DNA is cleaved into fragments that are too short to bind efficiently to the membrane (less then 200 bp).

3. Place the dish on an orbital shaker and incubate for 15 minutes rotating at 10-20 rpm. Decant the acid carefully holding the gel with the palm of your gloved hand.

4. Rinse the gel for 10-20 seconds in 200 ml of distilled water. Discard the water and proceed immediately to the next step.

5. Transfer the gel to 200 ml (4:1 ratio of buffer to gel) of alkaline denaturation solution and denature for 30 minutes with gentle agitation on an orbital shaker.

6. Decant the alkaline denaturation solution carefully, holding the gel with the palm of your gloved hand and repeat steps 2 and 3 one more time.

7. Add 200 ml of alkaline transfer solution and incubate the gel for 15 minutes.

8. Transfer the DNA from the gel to the membrane as described in steps 10 to 13 of the neutral capillary transfer procedure. At least 3 hours is required for adequate transfer. The transfer time can be extended to overnight.

Note: An extended time of alkaline transfer may contribute to high background and should be used only when very large DNA fragments are to be transferred. If a high background is encountered, it can be lowered by incubating the membrane in 2 x SSC at 65 °C for 30 minutes after immobilizing the DNA.

9. Disassemble the blot. Remove the weight, glass plate and paper towels. Using forceps, remove the membrane and place it "DNA side" up (the side that was in contact with the gel) on a clean sheet of Whatman 3MM paper. Mark the DNA side with pencil on the corner of the membrane.

10. Incubate the membrane in 100 ml of alkaline neutralization solution for 10 minutes.

11. Place the membrane on a sheet of dry Whatman 3MM paper. Do not allow the membrane to dry at any time. Place the membrane into a UV oven.

12. UV irradiate the damp membrane to crosslink DNA to the membrane using the automatic setting of the UV oven. Irradiate **both sides** of the membrane. Alternatively, you can wrap the membrane in aluminum foil and bake it in an oven at 80 °C for 1 hour. The baking step immobilizes DNA on the membrane. Membranes can be stored at room temperature for an indefinite time.

Subprotocol 2
Dot blot

Dot blot is a rapid method for quantitative screening of purified DNA, RNA, cell lysate and PCR-amplified DNA. A schematic outline of the dot blot procedure is shown in Figure 3.

Materials

– Manifold Dot-Blot or Slot-Blot apparatus (e.g., Bio-Rad # 170-6545 or Schleiher & Schull # SRC 96D).
– Vacuum unit
– MagnaGraph nylon membrane 0.22 μm pore size (Micron Separation, Inc. # NJ2HYA0010). See Chapter 10 for a description of equivalent nylon membranes from other manufacturers. If fragments larger than

Dot Blot Protocol

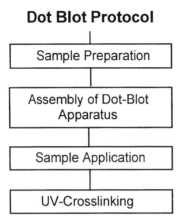

Fig. 3. Schematic outline of the dot blot procedure

300 bp are going to be transferred, it is not necessary to use 0.22 μm pore size membrane and membrane of 0.45 μm can be used.
– Whatman 3MM chromatography paper (Whatman Co. # 3030917)
– Stratalinker® UV oven (Stratagen Co. # 400071 or equivalent)
   To efficiently crosslink DNA to nylon membrane the UV source should be capable of delivering 120 mJ/cm². Excessive crosslinking will decrease hybridization efficiency.
– Pyrex glass dishes

Solutions – **TE Buffer**
   – 10 mM Tris HCl, pH 7.5
   – 1 mM Na_2EDTA, pH 8.5.
   Sterilize by autoclaving for 20 minutes and store at 4 °C.
– **2 M Ammonium Acetate Solution**
   Weigh the appropriate amount of ammonium acetate and dissolve it in distilled or deionized water. Sterilize by filtration and store in a tightly capped container at 4 °C.
– **20 x SSC Solution**
   – 3.0 M NaCl
   – 0.3 M Sodium citrate
   Dissolve 175 g NaCl and 88.2 g of sodium citrate in 850.0 ml of distilled or deionized water. Adjust pH to 7.5 with 10 N NaOH, add water to 1000 ml. Store at 4 °C.
– **6 x SSC Solution**
   Prepare from stock solution of 20 x SSC using deionized or distilled water. Store at 4 °C.

Procedure

1. Prepare a serial dilution of samples in TE buffer. Use 1:2, 1:10 and 1:100 dilution series for each sample. The final volume of diluted samples should be 50 μl.

2. Denature DNA with the addition of 5 μl (0.1 volume) of freshly prepared 3 M NaOH. Incubate at room temperature for 30 minutes.

3. Cool samples on ice and neutralize by the addition of 50 μl (1 volume) of 2 M ammonium acetate. Mix gently by pipetting up and down. If desired, 6 x SSC can be used for neutralization instead of ammonium acetate. Keep samples on ice until loading onto membrane.

4. Cut two pieces of Whatman 3MM paper to the size of the manifold plate and soak them in 6 x SSC.

5. Cut the membrane to the size of the manifold and wet it in distilled water. Transfer the membrane to 6 x SSC and soak it for 5 to 10 minutes. Alternatively 1 M ammonium acetate can be used instead of 6 x SSC.

6. Place the two pieces of wet filter paper on the filter support plate. Make sure that one corner is cut off for orientation of the transfer.

7. Place the membrane on the filter paper matching the cut-off corners. Place the sample well plate on the top and clamp the manifold together.

8. Apply a low vacuum, and wash the wells with 500 μl of 6 x SSC.

9. Apply the chilled samples as quickly as possible. Continue suction of the fluid for about 5 minutes until all of the samples have been loaded onto the membrane.

10. Remove the membrane from the manifold and place it on sheet of Whatman 3MM paper. Mark the DNA-side of the membrane with a pencil. Do not let membrane dry.

Note: If the sample cannot be filtered in 5 to 10 minutes, the concentration of DNA, RNA or lysate is too high. This could result in a weak or non-specific hybridization signal due to high background to signal ratio. Prepare a new dot blot. Determine which dilution of sample filtered first and start a new dilution from this initial concentration.

11. Crosslink DNA to the membrane in a UV oven using the automatic setting. Irradiate **both sides** of the membrane. Alternatively, you can wrap the membrane in aluminum foil and bake it in an oven at 80 °C for

1 hour. The baking step immobilizes DNA on the membrane. Membranes can be stored at room temperature practically indefinitely.

12. Continue the hybridization and signal detection as described in the Southern hybridization procedure.

Subprotocol 3
Plaque and Colony Transfer

To transfer, the membrane is applied to the surface of a plate containing bacteriophage plaques, resulting in direct contact between the plaques and the membrane (Benton and Davis, 1977). Unpacked bacteriophage DNA, present in the plaque, binds to the membrane and can be hybridized with the desired probe. A variation of this technique for use with single-stranded phage also has been described (Wei and Surzycki, 1989). Colony transfer is accomplished by transferring bacteria from a master plate to a membrane (Grunstein and Hogness, 1975; Hanahan and Meselson, 1980). Colonies grown on the membrane are lysed and the liberated DNA is bound to the membrane. A schematic outline of the transfer procedures is given in Figure 4.

Materials

- MagnaLift nylon circle membranes, 82 mm diameter (Micron Separation, Inc. # NL4HY08250).
- See Chapter 10 for a description of equivalent nylon membranes from other manufacturers. Membrane circles of this size can be used with regular size Petri plates.
- Whatman 3MM chromatography paper (Whatman Co. # 3030 917).
- Pyrex glass dishes.
- Stratalinker® UV oven (Stratagen Co. # 40071 or equivalent).
  Efficient crosslinking DNA to nylon membrane requires a UV source capable of delivering 120 mJ/cm^2. Excessive crosslinking will decrease hybridization efficiency.

Solutions
- **10 x SSC Solution**
  - 1.5 M NaCl
  - 0.15 M sodium citrate

Plaque Transfer

Colony Transfer

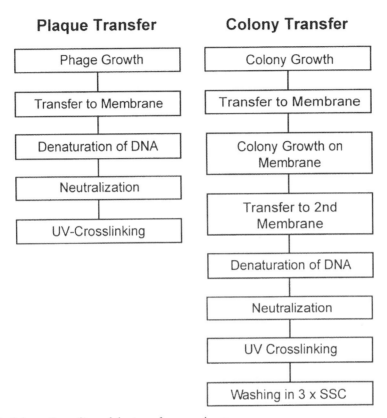

Fig. 4. Schematic outline of the transfer procedures

Dissolve 87.5 g NaCl and 44.1 g of sodium citrate in 850.0 ml of distilled or deionized water. Adjust pH to 7.5 with 10 N NaOH, and add water to 1000 ml. Store at 4 °C.

– **2 x SSC Solution**
Use 10 x SSC stock solution and deionized or distilled water to prepare this solution. Store at 4 °C.

– **Denaturation Solution**
  – 0.5 N NaOH
  – 1.5 M NaCl

Prepare solution using 10 N NaOH. Solution can be stored at room temperature for a few months. If a white precipitate forms the solution should be discarded.

– **Neutralization Solution**
  – 0.5 M Trisma Base
  – 1.5 M NaCl

Add 60.5 g of Trisma Base and 87.45 g of NaCl to 850 ml of deionized water. Dissolve salts and adjust the pH to 7.5 with concentrated HCl. Fill with water to 1000 ml. Store at 4 °C.

Procedure

Plaque transfer protocol

1. Cells should be plated with phage in soft agar using any standard plating technique. Incubated at 37 °C until plaques are about 1 to 2 mm in diameter. For screening individual plaques, pick a plate with about 100 to 200 plaques per plate. For primary screening of a library, use a plate with about 10 000 plaques.

2. Cool agar plates at 4 °C for 15 minutes to solidify soft agar.

3. To place a membrane on the soft agar surface hold the membrane by its edges and bend it slightly into a U shape. Contact the middle of the filter with the center of the plate. Release both edges slowly onto the plate, being careful not to trap air bubbles and to wet the membrane evenly. This operation should take about 30 seconds.

Note: Wear gloves during this procedure. Finger oils prevent wetting of the membrane and interfere with DNA transfer.

4. Mark the orientation of the membrane on the plate by stabbing three holes through membrane and the agar with a sterile 19 g needle. Position the holes asymmetrically. Mark the approximate positions of the holes on the bottom of the plate. Allow phage to transfer for 5 to 10 minutes.

Note: The holes will be used to determine the position of any plaques giving a positive hybridization signal.

5. With flat-tip forceps pull the membrane off the agar surface. Place filter on a Whatman 3MM paper DNA side up and mark the DNA side with a pencil.

Note: Store the plates upside down at 4 °C until the results of hybridization become available. Pick plaques that give a positive hybridization signal.

6. Fill a Pyrex dish with denaturation solution and with DNA side up, float the membrane on the surface. After a few seconds immerse the membrane and soak it for 30 seconds. This step denatures phage DNA and binds the DNA to the filter.

7. Place the membrane, DNA side up, on Whatman 3MM paper and blot the excess of denaturation solution.

8. Transfer the membrane to a Pyrex dish with neutralization solution and soak it, DNA side up, for 30 seconds.

9. Transfer the membrane to 2 x SSC and soak it for 30 seconds. Place the membrane onto dry Whatman 3MM paper with the DNA side up and blot the excess SSC. Do not let membrane dry.

10. Transfer the Whatman paper and membrane into a UV oven and cross-link DNA to the membrane using the automatic setting of the UV oven. Irradiate **both sides** of the membrane. Alternatively, you can wrap the membrane in aluminum foil and bake it in an oven at 80 °C for 1 hour. The baking step immobilizes DNA on the membrane. Membranes can be stored at room temperature practically indefinitely.

11. Store the crosslinked membrane in a plastic bag at room temperature until ready to hybridize with probe. Hybridization and signal detection should be carried out as described for the Southern blot hybridization procedure.

Note: The presence of cellular debris on the membrane can lead to high background. To remove the debris, incubate the membrane in 3 x SSC containing 0.1% SDS for 1 hour at 68 °C, rotating at slow speed in a hybridization oven.

Colony transfer protocol

1. Sterilize the membrane circles by autoclaving for 20 minutes. Label the membrane with a pencil and using flat-tip forceps, place it, labeled side down, on the surface of a plate with medium containing the appropriate antibiotic. When the membrane becomes wet, pull it off with forceps and place the labeled side up on another agar plate with the same medium and antibiotic as above (e.g., LB agar supplemented with ampicillin). Spread 0.1 ml of the appropriate bacterial dilution over the surface of the membrane with a sterile bent glass rod.

2. Let the plates stand at room temperature until all of the liquid has been absorbed. Invert the plates and incubate at 37 °C until small colonies (about 0.1 cm diameter) appear. This will take about 8 to 10 hours.

3. With forceps, lift the membrane from the surface of the plate and place it colony side up, on two sheets of Whatman 3MM paper. Wet a new membrane as described in step 1. Carefully place a fresh wet membrane over the one with the colonies, cover with a glass plate. Apply gentle pressure to transfer the colonies to the new membrane. While the membranes are

sandwiched together, use a sharp needle to make three asymmetrical holes to mark the position of colonies.

4. Gently peel the membranes apart and return the second membrane to its agar plate, colony side up. Incubate the plate for several hours and store at 4°C, seal with parafilm and store until hybridization results become available.

5. Fill a Pyrex dish with denaturation solution and with DNA side up, float the membrane on the surface. After a few seconds immerse the membrane and soak it for 30 seconds. This step denatures DNA and binds the DNA to the filter.

6. Place the membrane, DNA side up, on Whatman 3MM paper and blot the excess of denaturation solution.

7. Transfer the membrane to a Pyrex dish with neutralization solution and soak it, DNA side up, for 30 seconds.

8. Transfer the membrane to 2 x SSC and soak it for 30 seconds. Place the membrane onto dry Whatman 3MM paper with the DNA side up and blot the excess SSC. Do not let membrane dry.

9. Transfer the Whatman paper and membrane into a UV oven and cross-link DNA to the membrane using the automatic setting of the UV oven. Irradiate **both sides** of the membrane. Alternatively, you can wrap the membrane in aluminum foil and bake it in an oven at 80 °C for 1 hour. The baking step immobilizes DNA on the membrane. Membranes can be stored at room temperature practically indefinitely.

10. Incubate the membrane in 3 x SSC containing 0.1% SDS for 1 hour at 68 °C, rotating at slow speed in a hybridization oven. This step will remove cellular debris that can result in high background. Perform hybridization and signal detection using a hybridization oven or plastic bags as described for Southern blot hybridization.

Subprotocol 4
Southern Blot Hybridization

This procedure describes the hybridization procedure using a hybridization oven. The hybridization oven considerably shortens the hybridization procedure, decreases the amount of reagents needed and results in much lower background. The hybridization procedure, using plastic bags, is essentially the same as this procedure and will also be described because not all laboratories have a hybridization oven. The hybridization procedure described below uses stringent conditions for hybridization and washing, assuming 50% GC content of target DNA (see equation 2, 5 and 9 in Chapter 10). For less stringent conditions, see a discussion of this topic in Chapter 10. A schematic outline of the hybridization and signal detection procedures is shown in Figure 5.

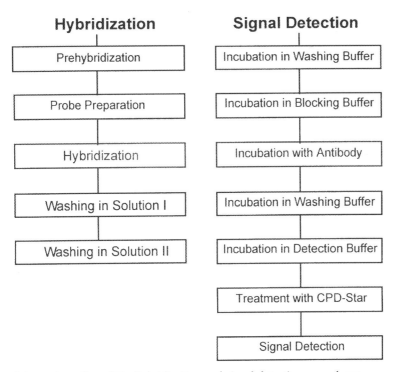

Fig. 5. Schematic outline of the hybridization and signal detection procedures

Materials

- Hybridization oven (e.g., HyBaid Co. # H9320 or equivalent)
  Oven should be capable of rotation at variable speeds.
- Hybridization bottles 150 x 35 mm (e.g., HyBaid Co. # H9084 or equivalent)
- Plastic bags (e.g., Kapak Co. # 402 or Roche Molecular Biochemicals # 1666 649)
- Rotary platform shaker

    – Rocking platform shaker
    – Plastic bag sealer (Fisher Scientific # 01-812-13 or equivalent)
    – Plastic containers with air-tight lids
    – CDP-Star solution (Tropix Co. # MS100R).
      Store solution in the dark at 4 °C. CDP-Star is easily destroyed by ubiquitous alkaline phosphatase. Wear gloves and use sterilized tips when handling CDP-Star solution.
    – Dig Easy Hyb solution (Roche Molecular Biochemicals # 1603 558)
    – Blocking reagent for nucleic acid hybridization (Roche Molecular Biochemicals # 1096 176)
    – Anti-DIG-alkaline phosphatase 750 u/ml (Roche Molecular Biochemicals # 1093 274)
      Stock solution should not be frozen. Store at 4°C.
    – Antifoam C (Sigma Co. # A 8001)
    – Maleic acid (Sigma Co. # MD 375)
    – Ion exchange resin AG 501-X8 (Bio-Rad # 142-6424)
    – Formamide (Fluka # 47671)

Solutions – **20 x SSC Solution**
    – 3.0 M NaCl
    – 0.3 M Sodium citrate
    Dissolve 175 g NaCl and 88.2 g of sodium citrate in 900.0 ml of distilled or deionized water. Adjust pH to 7.5 with 10 N NaOH, and add water to 1000 ml. Store at 4°C.

    – **Sarcosyl Stock Solution, 20% (w/v)**
    Dissolve 40 g of Sarcosyl in 100 ml of double distilled or deionized water. Fill up to 200 ml with water. Sterilize by filtration through a 0.45 µm filter. Store at room temperature.

    – **SDS Stock Solution, 10% (w/v)**
    Add 10 g of powder to 70 ml of distilled water and dissolve by slow stirring. Add water to a final volume of 100 ml and sterilize by filtration through a 0.45 µm filter. Store at room temperature.
    Safety Note: SDS powder is a nasal and lung irritant. Weigh the powder carefully and wear a face mask.

    – **Formamide (deionized)**
    – 100% Formamide
    Add 500 ml of formamide to 50 g of ion exchange resin AG 501-X8 and stir slowly for 30 minutes. Remove resin by filtration through a Whatman 1MM filter. Store at - 20 °C.

    – **Hybridization Solution**
    – 5 x SSC

- 2.5% blocking-reagent
- 0.001% Antifoam A
- 0.02% (w/v) SDS
- 0.1% (w/v) Sarcosyl
- 50% formamide, deionized

Prepare 2 x hybridization mixture as follows: Add 5 g of blocking reagent powder to 40 ml of 10 x SSC Add 2 µl of antifoam A, 0.2 ml of 10% SDS and 0.5 ml of 20% Sarcosyl. Dissolve by stirring for 2 hours at 50 °C to 60 °C. Fully dissolved blocking solutions will have uniformly "milky" appearance. Fill up with water to 50 ml. Store at -20 °C. Immediately before use, add an equal volume of 100% deionized formamide.

Note: Instead of this solution, Dig Easy Hyb solution can be used. The solution is nontoxic and gives much lower background when DIG-labeled probes are used.

- **Washing Solution I**
  - 2 x SSC
  - 0.1% SDS
- **Washing Solution II (high stringency)**
  - 0.1 x SSC
  - 0.2% SDS
- **10 x Maleic Acid Solution**
  - 1.0 M Maleic acid
  - 1.5 M NaCl

Adjust pH to 7.5 with concentrated NaOH. Sterilize by autoclaving for 20 minutes. Store at 4 °C. This solution is available in the Wash and Block Buffer Set from Roche Molecular Biochemicals (# 1585 762).

- **10 x Blocking Solution**
  - 10% (w/v) Blocking reagent
  - 1 x Maleic acid solution

Add 10 ml of maleic acid solution to 90 ml sterilized water. Add 10 g of blocking reagent powder. Dissolve by stirring slowly at 60 °C. Do not boil because it will cause the reagent to coagulate. Autoclave for 20 minutes. Blocking reagent must be completely in solution before autoclaving. Store at 4 °C. This solution is available in the Wash and Block Buffer set from Roche Molecular Biochemicals (# 1585 762)

- **Washing Buffer (Buffer A)**
  - 1 x Maleic acid solution
  - 0.3% (w/v) Tween 20

Dilute 10 x maleic acid solution in sterilized water and Tween 20. Do not autoclave. Store at 4 °C. This solution is available in the Wash and Block Buffer set from Roche Molecular Biochemicals (# 1585 762)

- **Blocking Buffer (Buffer B)**
  - 1 x Maleic acid solution
  - 1 x Blocking solution

Dilute 10 x maleic acid solution and 10 x blocking solution 1:10 in sterilized water. Prepare only the amount necessary for use. The buffer can be prepared from components of the Wash and Block Buffer Kit (Roche Molecular Biochemicals # 1585 762).

- **Detection Buffer (Buffer C)**
  - 0.1 M Tris HCl, pH 9.5
  - 0.1 M NaCl
  - 0.05 M $MgCl_2$

Prepare this buffer from stock solutions. Adjust the pH before the addition of $MgCl_2$ to avoid a precipitate. Sterilize by filtration. Store at 4 °C.

Procedure

Protocol using a hybridization oven

Hybridization reaction protocol

The experimental procedure described uses Dig Easy Hyb solution. Hybridization solution can be used in place of Dig Easy Hyb.

1. Place the dry membrane with DNA crosslinked to it into a glass hybridization bottle. Make sure that the side of the membrane with DNA is facing away from the glass.

Note: Several membranes can be placed in a single bottle. Some membrane overlapping does not affect hybridization or increase background.

2. Pour 10 ml of Dig Easy Hyb solution into a 15 ml plastic centrifuge tube and add it to the hybridization bottle. Save the centrifuge tube for further use. The minimum amount of liquid per hybridization bottle is 5 to 6 ml.

3. Close the hybridization bottle tightly and place it in the hybridization oven. Incubate for 1-3 hours at 41 °C rotating at a **slow speed** (3 to 5 rpm). This is the prehybridization step that lowers background.

4. Ten minutes before the end of prehybridization, begin to prepare the probe for hybridization. Pour 10 ml of Dig Easy Hyb solution into

the previously used 15 ml conical centrifuge tube. Add probe to a final concentration of 20 to 25 ng/ml. Close the tube tightly and place it into an 80 °C water bath. Incubate for 10 minutes to denature the DNA.

Note: Leave the probe in the water bath until you are ready to start the hybridization procedure. However, do not incubate the tube longer than 20 minutes at 80 °C. This can lead to the degradation of the probe.

5. After prehybridization has been completed, retrieve your hybridization bottle from the hybridization incubator and pour off the prehybridization solution.

Note: Prehybridization solution can be stored and used again.

6. Add the Dig Easy Hyb containing the denatured probe. Return the bottle to the oven and allow it to rotate slowly at 41 °C overnight.

7. Pour off the hybridization solution into a 15 ml centrifuge tube. The probe can be stored at -20 °C, remelted and reused 3 to 4 times. **Washing reaction protocol**

8. Add 20 ml of washing solution I to the bottle. Place it into the hybridization oven and rotate at maximum speed for 15 minutes at room temperature.

9. Pour off solution I and discard it. Drain the liquid well by placing the bottle on end on a paper towel for 1 minute. Repeat the wash one more time and drain the bottle.

10. Add 20 ml of washing solution II prewarmed to 62 °C. Place the bottle into the hybridization oven preheated to 62 °C. Rotate it at a slow speed for 20 minutes.

11. Pour off solution II and discard it. Drain remaining liquid well by placing the bottle on end on a paper towel for 1 minute. Repeat the washing with solution II one more time.

12. Pour off washing solution II. Drain the liquid well by placing the bottle on end on a paper towel for 1 minute. **Detection protocol**

13. Add 20 ml of washing buffer (buffer A). Cool the oven to room temperature and rotate the bottle at maximum speed for 2-5 minutes.

14. Pour off and discard washing buffer (buffer A). Add 10 ml of blocking buffer (buffer B). Incubate for 1-3 hours **rotating slowly** at room temperature.

15. Pour off buffer B and discard it. Invert the bottle over a paper towel and let it drain well.

16. Add 10 ml of buffer B to a plastic 15 ml centrifuge tube and add 2 μl of anti-DIG-alkaline phosphatase stock solution. Mix well and add the solution to the bottle with the membrane. Incubate for 30 minutes rotating slowly at room temperature.

Note: The working antibody solution is stable for about 12 hours at 4 °C. Do not prolong incubation with antibody over 30 minutes this will result in high background.

17. Pour off antibody solution and drain the bottle well. Discard antibody solution.

18. Add 20 ml of buffer A and wash the membrane for 15 minutes at room temperature rotating at maximum speed.

19. Discard buffer A and drain the bottle well. Add 50 ml of detection buffer (buffer C) and equilibrate the membrane in it for 20 minutes rotating at maximum speed. Repeat this procedure one more time.

20. Move the membrane in the bottle toward the open end by shaking the bottle. Open the bottle and pour half of buffer C into a plastic bag. Wearing powder-free gloves, transfer the membrane from the bottle to the bag. This can be easily accomplished by keeping the wet membrane immersed in a pool of buffer C and moving it with your fingers. Move the membrane to the end of the bag and leave as little space as possible between the membrane and the end of the bag. This will limit the amount of expensive chemiluminescent substrate necessary to fill the bag.

21. Poor off buffer C from the bag. Place the bag on a Whatman 3MM paper sheet and remove the remaining liquid from the bag by gently pressing it out with Kimwipe tissue.

Note: Do not press strongly on the membrane because this will increase the background. Most of the liquid should be removed from the bag leaving the membrane slightly wet. At this point a very small amount of liquid will be visible at the edge of the membrane.

22. Open the end of the bag slightly, leaving the membrane side that does not contain DNA attached to the side of the bag. Add 0.9-1 ml of CDP-Star solution directing the stream toward side of the bag. Do not add solution directly onto the membrane. Place the bag on a sheet of Whatman 3MM paper with the DNA-side up and distribute the liquid over the surface of the membrane by gently moving the liquid around with a Kimwipe paper towel. Make sure that the entire membrane is evenly covered. Do not press on the membrane because this will cause "press marks" on the film. Gently remove excess CDP-Star from the bag by guiding excess solution toward the open end of the bag and onto the Whatman paper with a Kimwipe towel. Make sure that the membrane remains damp. At this point, small liquid droplets are visible on the edge of the membrane but there is no liquid present on its surface. Seal the bag with a heat sealer.

23. Place the bag in an X-ray film cassette leave it at room temperature for 30 minutes. In a darkroom, place X-ray film over the membrane. If the film is two-sided, place the less shiny, emulsion side against the membrane. Exposure time is 10-15 minutes at room temperature for single copy gene detection. This time can be increased up to 12 hours. See Chapter 10 for a description of membrane types that can be used and recommended times of exposure for each type of membrane. Open the cassette and develop the film using standard procedures for film development.

Note: Maximum light emission for CDP-Star is reached in 20 to 30 minutes, the light emission remains constant for approximately 24 hours and the blot can be exposed to film a number of times during this period.

Procedure using hybridization bags

The experimental procedure described uses Dig Easy Hyb solution. 1x Hybridization Solution can be used in place of Dig Easy Hyb.

1. Place the bag on a clean sheet of Whatman 3MM paper. Place the dry membrane on the sheet of Whatman 3MM paper cut to a size slightly larger than the nylon membrane. Insert the nylon membrane and Whatman paper carefully into the plastic bag. Gently withdraw the Whatman paper from the bag leaving the nylon membrane inside.

Hybridization reaction protocol

2. Add 10 ml of Dig Easy Hyb (or 1 x prehybridization solution) to the bag. Place the bag on the lab bench and carefully eliminate all bubbles.

Note: Use at least 2.5 ml hybridization solution per 100 cm^2 of membrane.

3. Seal the top of the bag leaving 4 to 8 cm between the top seal and the top of the membrane. It is very important to leave this large space between the membrane and the sealed edge.

4. Put a platform shaker into a 41 °C incubator and place the bag on the platform. Incubate for 2-4 hours with constant rocking.

5. Ten minutes before the end of prehybridization begin to prepare the probe for hybridization. Pour 10 ml of Dig Easy Hyb solution into the 15 ml conical centrifuge tube. Add probe to a final concentration of 20 to 25 ng/ml. Close the tube tightly and place it into an 80 °C water bath. Incubate for 10 minutes to denature the DNA.

Note: Leave the probe in the water bath until you are ready to start the hybridization procedure. However, do not incubate the tube longer than 20 minutes at 80 °C. This can lead to the degradation of the probe.

6. Open the bag by cutting it with scissors along the line of the top seal, as far away from the edge of the membrane as possible. Pour off the prehybridization mixture.

Note: Prehybridization mixture can be stored and reused several times.

7. Add the Dig Easy Hyb with the probe to the bag. Remove air bubbles and seal the bag. Return the sealed bag to the 41 °C incubator and rock it overnight.

Washing reaction protocol

8. Carefully cut open the top of the bag. Pour the hybridization mixture into a 15 ml plastic centrifuge tube. The probe can be stored at -20 °C and reused several times.

9. Cut open the bag completely, open it like a book and remove the membrane with forceps. Place the membrane in a plastic box fitted with an airtight cover. Work over bench paper and wear gloves.

10. Add 100-150 ml of washing solution I to the membrane. Wash at room temperature on a rotary shaker set at slow speed for 15 minutes. Drain the solution well.

11. Repeat the washing one more time and discard washing solution I.

12. Preheat 100 ml of washing solution II to 62 °C and pour onto the membrane. Close the lid of the plastic container tightly and place it into a 62 °C water bath. Wash for 20 minutes with occasional shaking. Keep the plastic container fully immersed in water by placing a weight on the container. Discard washing solution and repeat the washing procedure one more time. Discard washing solution II.

Detection protocol

13. Add 50 ml of washing buffer (buffer A) and equilibrate the membrane for 1 minute at room temperature.

14. Remove the membrane from the plastic container and allow excess liquid to drain off. Do not allow the membrane to dry. Add 15 ml of blocking buffer (buffer B) to new plastic bag and transfer the membrane to it. This can be accomplished easily by keeping the wet membrane immersed in a pool of buffer B in the bag and moving it with your fingers. Move the membrane to the end of the bag and leave as little as possible space between the membrane edges and the edges of the bag. **Wear powderless gloves during this procedure.** Place the bag on Whatman 3MM paper and remove all bubbles. Seal the top of the bag leaving 4 to 8 cm between top of the bag and the top of the membrane. Place the bag on a rocking platform shaker and incubate with gentle rocking for 1-3 hours at room temperature.

15. Open the bag and remove buffer B from the bag. Place the bag on a sheet of Whatman 3MM paper and remove the remaining buffer by gently rolling a 10 ml pipette over the bag. Prepare 10 ml of anti-DIG -alkaline phosphatase solution. Add 10 ml of Buffer B to a plastic 15 ml centrifuge tube and add 2 μl of anti-DIG-alkaline phosphatase stock solution. Mix well and add the solution to the bag. Remove all bubbles and close it. Incubate for 30 minutes at room temperature with gentle rocking.

Note: Prolonging incubation with antibody over 30 minutes may result in high background.

16. Open the bag and discard the antibody solution. Add 20 ml of buffer A to the bag, close it and incubate for 15 minutes by rocking at room temperature. Repeat the washing procedure one more time.

17. Add 40 ml of buffer C to the bag and incubate for 10 minutes at room temperature. Repeat this procedure one more time.

18. Poor off buffer C from the bag. Place the bag on a Whatman 3MM paper sheet and remove the remaining liquid from the bag by gently pressing it out with a Kimwipe tissue.

Note: Do not press strongly on the membrane because this will increase the background. Most of the liquid should be removed from the bag leaving the membrane slightly wet. At this point a very small amount of liquid will be visible at the edge of the membrane.

19. Open the end of the bag slightly, leaving the membrane side that does not contain DNA attached to the side of the bag. Add 0.9-1 ml of CDP-Star solution directing the stream toward side of the bag. Do not add solution directly onto the membrane. Place the bag on a sheet of What-

man 3MM paper with the DNA-side up and distribute the liquid over the surface of the membrane by gently moving the liquid around with a Kimwipe paper towel. Make sure that the entire membrane is evenly covered. Do not press on the membrane because this will cause "press marks" on the film. Gently remove excess CDP-Star from the bag by guiding excess solution toward the open end of the bag and onto the Whatman paper with a Kimwipe towel. Make sure that the membrane remains damp. At this point, small liquid droplets are visible on the edge of the membrane but there is no liquid present on its surface. Seal the bag with a heat sealer.

20. Place the bag in an X-ray film cassette and leave it at room temperature for 30 minutes. In a darkroom, place X-ray film over the membrane. If the film is two-sided, place the less shiny, emulsion side against the membrane. Exposure time is 10-15 minutes at room temperature for single copy gene detection. This time can be increased up to 12 hours. See Chapter 10 for a description of membrane types that can be used and recommended times of exposure for each type of membrane. Open the cassette and develop the film using standard procedures for film development.

Note: Maximum light emission for CDP-Star is reached in 20 to 30 minutes, the light emission remains constant for approximately 24 hours and the blot can be exposed to film a number of times during this period.

Stripping the membrane for reprobing

The nylon membrane can be stripped and reprobed at least 4 times. When reprobing of the membrane is required, make sure that the membrane does not become dry at any time. Use only alkali-labile DIG-labeled probes for hybridization. Store membrane before stripping in 2 x SSC sealed in a plastic bag at 4 °C. The procedure can be carried out in a hybridization oven or a Pyrex dish.

Stripping protocol

1. Place the membrane into a hybridization bottle and wash in sterile water for 1 minute. Discard water.

2. Prepare fresh alkaline probe-stripping solution containing 0.4 N NaOH and 0.1% SDS.

3. Add 20 ml of alkaline probe-stripping solution and incubate for 10 minutes at 37 °C at maximum rotation speed. Discard alkaline probe-stripping solution.

4. Add 20 ml 2 x SSC and wash at room temperature at maximum speed for 20 minutes. Discard the solution and repeat washing procedure one more time.

5. Proceed with reprobing, starting from the prehybridization step of the hybridization procedure.

Results

Figure 6 presents the result of single gene probe (labeled with DIG) hybridization with genomic DNA of *Chlamydomonas*. *Chlamydomonas* genomic DNA was double-digested with restriction enzymes *Hind* III/*Ava* I, *Hind* III/*Bam* HI and *Hind* III/*Cla* I. Three different time exposures are shown to illustrate low background obtained with MagnaGraph membrane.

Troubleshooting

There are two basic problems that are encountered in hybridization procedures, problems with low sensitivity of detection and problems with high background (Genius System User's Guide, 1995).

Low sensitivity

Low sensitivity of detection can result from any of the reasons listed below:

- Incomplete DNA transfer to the membrane.
  This can occur because:
  - Collapsing of the gel during transfer. Apply less weight during the transfer step. The gel should be at least 4 mm thick.
  - Transfer time is not long enough or agarose gel is too thick. Allow at least 17 hours for capillary transfer. The gel should not be thicker than 5 mm.
  - Inadequate denaturation of DNA before transfer. This usually results from using old denaturation solution containing precipitate.
  - Presence of bubbles between membrane and the gel during transfer or during hybridization.

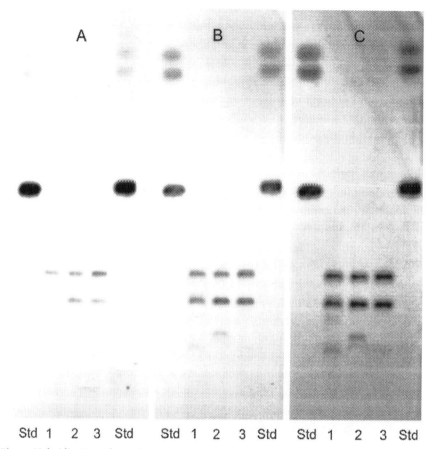

Fig. 6. Hybridization of a single gene probe with genomic DNA of *Chlamydomonas*. The same hybridization membrane was exposed for 5 minutes (A), 20 minutes (B) and 2 hours (C). Lane 1, DNA digested with *Hind* III and *Ava* I restriction enzymes. Lane 2, DNA digested with *Hind* III and *Bam* HI restriction enzymes. Lane 3, DNA digested with *Hind* III and *Cla* I restriction enzymes.

- Inadequate depurination step. This will usually result in bands larger than 10 kb missing from the blot. Increase the treatment time in 0.25 N HCl.
- DNA bands "blow-through" the membrane. This usually occurs with fragments smaller than 300 bp. For transfer of small fragments, do not use the depurination step, and use membrane with a pore size of 0.22 μm. Use 20 x SSC for the transfer when unsure of size of the DNA being transferred.

- Incorrect estimation of the labeling efficiency of the probe. Use a new probe or correctly estimate probe labeling efficiency using the protocol for estimation of DIG-dUTP concentration (Chapter 9).
- Incorrect estimation of GC content of the target or probe. Lower temperature of hybridization and washing reactions. See Chapter 10 for a detailed discussion of the conditions of hybridization and washing reactions
- Using inactivated antibody. Repeated freezing and thawing cycles can easily destroy antibody. Always store anti-DIG alkaline phosphatase stock solution at 4°C. Test suspected antibody using the procedure described for testing efficiency of labeling of the probe using labeled standard DNA (Chapter 9).
- Exposure of membrane to film too short. Increase exposure time up to 24 hours.

High background

High background can result from:

- Inadequate blocking of the membrane.
  Increase blocking reagent concentration in hybridization solution to 5% or use Dig Easy Hyb solutions.

- Presence of precipitate in the antibody solution.

- Using gloves containing powder.

- Hybridization solution with incompletely dissolved blocking reagent can cause spotty or uneven background.

- Probe concentration too high.
  Perform hybridization at lower probe concentration and wash for a longer time at 68 °C. Using a higher concentration of probe is unlikely to significantly improve sensitivity and can cause an increase in background.

- Non-uniform distribution of chemiluminescent substrate during the detection procedure.

- Drying the membrane during any hybridization steps. The background will not be uniform and will appear as irregular smears.

- Uneven contact of the bag with X-ray film. The background will appear as irregular smears. Make sure that the surface of the bag is smooth and does not contain wrinkles caused by incorrect sealing of the bag.

- Dirt on the surface of the bag. This will result in dark areas on the film caused by electrostatic charges outside of the bag. Clean the surface of the bag with 70% ethanol and remove all fingerprints.

- Uneven distribution of the probe due to delivering the probe directly onto the membrane. This will result in "clouds" of background on an otherwise clear autoradiograph.

- Inadequate washing with buffer C. Repeat the washing procedure, and add fresh CPD-store solution.

References

Benton WD, Davis RW (1977) Screening λgt recombinant clones by hybridization to single plaques in situ. Science 196:180-182.

Chomczynski P (1992) One-hour downward alkaline capillary transfer for blotting of DNA and RNA. 201:134-139.

Grunstein M, Hogness D (1975) Colony hybridization: A method for isolation of cloned DNAs that contain a specific gene. Proc Natl Acad Sci USA 72:3961-3965.

Genius system user's guide for membrane hybridization. Version 3. (1995) Boehringer Mannheim Biochemicals. 9115. Hague Road. P.O. Box 50414. Indianapolis, IN 46250.

Hanahan D, Meselson M (1980) A protocol for high density screening of plasmids in 1776. Gene 10:63- 68.

Nyagaard AP, Hall BD (1964) J Mol Biol 9:125-130.

Peferoen M, Huybrecht R, De Loof A 1982 Vacuum-blotting: A new simple and efficient transfer of proteins from sodium dodecyl sulfate-polyacrylamide gels to nitrocellulose. FEBS Lett 145:369-372.

Smith MR, Devine CS, Cohn SM, Lieberman MW 1984 Quantitative electrophoretic transfer of DNA from polyacrylamide or agarose gels to nitrocellulose. Anal Biochem 137:120-124.

Southern EM (1975) Detection of specific sequences among DNA fragments separated by gel electrophoresis. J Mol Biol 98:502-517.

Wei Y-G, Surzycki SJ (1986) Screening recombinant clones containing sequences homologous to Escherichia coli genes using single-stranded bacteriophage vector. Gene 48:251-256

Northern Transfer and Hybridization

STEFAN SURZYCKI

Introduction

RNA gel blots, or Northern hybridization analysis, was introduced shortly after the DNA blotting technique (Southern, 1975) and was humorously named Northern blot (Alwine et al, 1977). The technique is primarily used for studying gene expression, quantification of transcription, and analysis of RNA processing. It essentially consists of RNA:DNA or sense RNA and antisense RNA blot hybridization. RNA molecules are fractionated by size, using denaturing agarose gel electrophoresis. The fractionated RNA is transferred to a solid support, such as nylon or nitrocellulose membrane. The membrane is hybridized with a specific, labeled probe and results are visualized by autoradiography. The method is sensitive enough to detect mRNA present in the cell in 5 to 10 copies when analyzing 10 μg of total RNA. As little as one copy of mRNA per cell can be detected when 1 to 2 μg of poly-A$^+$ mRNA is analyzed (Davis et all, 1994).

Northern blot analysis is presented here as three separate protocols: a protocol for electrophoresis of an RNA on formaldehyde or glyoxal gels, a protocol for transfer of RNA from the gel to a nylon membrane and a protocol for hybridization analysis of the transferred RNA using DIG-labeled probes.

Northern blot analysis is more difficult to perform than Southern transfer and hybridization largely, because RNA is sensitive to degradation by RNases. The most difficult task is to inactivate RNase, the presence of which is universal. Most RNases are exceptionally stable enzymes that can withstand autoclaving or phenolic treatment and do not require any cofactors for their activity.

To prevent contamination of equipment and solutions with RNase, the following precautions should be taken:

- Gloves should always be worn. Because gloves can be easily contaminated with RNase, they should be changed frequently.

- Whenever possible disposable, certified RNase-free, plasticware should be used.

- RNase-free micropipette tips and microfuge tubes should purchased whenever possible. Regular microfuge tubes and tips usually are not contaminated with RNase, and they do not require special treatment if they are used from unopened bags. Gloves should be worn when preparing supplies for sterilization.

- All glassware should be treated with 0.1% DEPC aqueous solution and autoclaved to remove DEPC. It also is possible to inactivate RNase by baking glassware at 180 °C for at least 2 hours or overnight. Alternatively, RNase can be easily and efficiently eliminated from glassware, countertops, pipettors and plastic surfaces using a commercially available, non-toxic RNaseZap® solution. We strongly recommend its use.

- Gel apparatus, combs and gel casting trays, etc., should be treated with 0.2 N NaOH for 15 minutes and rinsed with RNase-free water before use. They should be reserved, if possible, for exclusive use with RNA gels. RNaseZap® solution can be used instead of alkaline treatment.

- All solutions should be made with DEPC-treated water or sterilized MilliQ water.

Subprotocol 1
RNA Agarose Gel Electrophoresis

Gel electrophoresis of RNA molecules requires techniques different from that used for DNA. To separate RNA molecules according to their size, it is necessary to maintain their complete denaturation before and during electrophoresis. Nondenatured RNA can form secondary structures such as "hairpins" that profoundly influence their electrophoretic mobility and their transfer to solid support media. A number of denaturants have been used. Among these are glyoxal with DMSO (McMaster and Carmicheal, 1977), formaldehyde (Lehrbach et al, 1977; Rave et al., 1979) and methylmercuric hydroxide (Bailey and Davidson, 1977; Thomas, 1980). Presently formaldehyde and glyoxal-DMSO are used more often than the highly toxic methylmercuric hydroxide. The buffers used for RNA electrophoresis differ from those used for DNA. These buffers are of very low ionic strength resulting frequently in the creation of a pH gradient along the length of the gel that causes overheating of the gel and distortion of

RNA bands. To prevent this, RNA gels should be run at low field strength (<5 V/cm) using a large volume of buffer and constant stirring to prevent gradient formation.

Procedures for both formaldehyde and glyoxal-DMSO agarose gels are described here as well as a method for running native RNA agarose gels. Native gels do not include toxic denaturants in agarose, but this does not affect electrophoretic separation of RNA (Liu and Chou, 1990). Use of native or formaldehyde gel is recommended because they are simple and fast. However, precaution should be taken when building and running formaldehyde gels because formaldehyde is toxic. To lower exposure to formaldehyde, its concentration has been decreased from 2.2 M originally described to 0.66 M. Usually formaldehyde is removed from the gel before transfer of RNA to the membrane.

The glyoxal-DMSO method does not use toxic chemicals, but it is more difficult to use than the formaldehyde method. This method requires very careful control of pH during electrophoresis to a pH below 8. This is because glyoxal denatures RNA by binding covalently to the guanine residue, form-

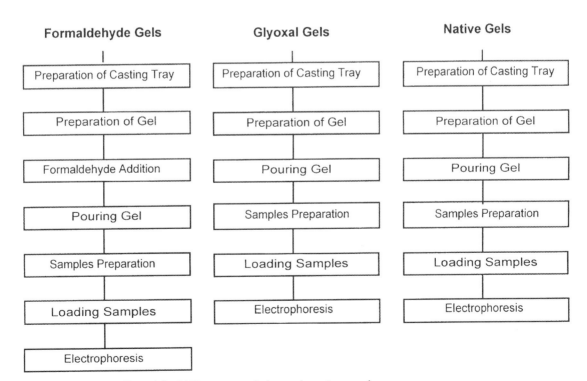

Fig. 1. Schematic outline of the RNA agarose gel electrophoresis procedures

ing products that are stable only at a pH below 8. At a pH above 8, glyoxal dissociates from RNA. Submarine gels require continuous recirculation and mixing of electrophoresis buffer to maintain the pH within an acceptable limit. Also, commercially available glyoxal must be purified before use to remove glyoxylic acid that is readily formed by oxidation and degrades RNA. Electrophoresis time is longer than for formaldehyde gels. However, glyoxal-DMSO gels can be used for blotting without removing the denaturant. A schematic outline of RNA agarose gel electrophoresis procedures is given in Figure 1.

Materials

- Agarose powder SeaKem® LE (FMC BioProducts # 5 0001 or equivalent)
- Ethidium bromide (Sigma Co. # E8751)
- Formaldehyde (Sigma co. # F1268 or equivalent)
  Commercially available solutions are 37% or 12 M. If the pH of the solution is ›4.0 and there is a yellow precipitate at the bottom of the bottle, use a fresh solution.
- Formamide 100% (Fluka # 47671 or equivalent)
- Diethyl pyrocarbonate (DEPC) (Sigma Co. # D5758)
- MOPS (3-(n-morpholino)propanesulfonic acid (Sigma Co. # M1254)
- Glyoxal (40% v/v) (Sigma Co. # G3140)
- AG 501-X8 mixed bed resin (Bio-Rad # 142-6424)
- DMSO (Sigma Co. # D5879 or D8779)
- Gel electrophoresis apparatus with power supply (minimum 13 x 20 cm gel size, e.g., Owl Scientific # A1)
- RNaseZap® Solution (Ambion # 9780)
- RNA size standards (for example, Ambion # 7140 or equivalent)

Solutions – **DEPC Treated Water**
Add 0.1 ml DEPC to 100 ml of deionized water. Shake vigorously to dissolve DEPC in water. Incubate 2 hours to overnight at 37 °C. Autoclave to inactivate DEPC. DEPC must be completely inactivated before using any solution treated with it because it will carboxylate RNA. Carboxylated RNA will inhibit reverse transcriptase, DNA:RNA hybridization and *in vitro* translation. Use DEPC-treated water to prepare any solutions that cannot be treated with DEPC directly. Primarily these are solutions containing primary amines, such as Tris or ammonium acetate that will react with DEPC.

Safety Note: When using DEPC, wear gloves and use a fume hood because DEPC is highly flammable and a suspected carcinogen.

Note: Sterilized, deionized water from a MilliQ apparatus (Millipore Co. # ZD5111584) does not contain RNase and can be used directly in all applications instead of DEPC-treated water.

– **0.5 M EDTA Stock Solution, pH 8.5**

Weigh an appropriate amount of EDTA (disodium ethylenediamine-tetraacetic acid, dihydrate) and add it to distilled water while stirring. EDTA will not go into solution completely until the pH is greater than 7.0. Adjust the pH to 8.5 using a concentrated NaOH solution or by adding a small amount of pellets while mixing and monitoring the pH. Bring the solution to a final volume with water and sterilize by filtration. EDTA stock solution can be stored indefinitely at room temperature.

– **Deionized Glyoxal Solution (40%)**

Measure the pH of the glyoxal solution with pH paper. If the pH is below 4.5, glyoxal should be deionized. Put about 5 ml of AG 501-X8 resin into a small disposable plastic column and load it with 10 ml of glyoxal. Collect eluent in an RNase-free tube and measure the pH with pH paper. Pass the eluent over the column one more time or until the pH is 4.5 or higher. Place small aliquots (about 50 μl) into screw cap tubes and store at -70 °C. Before using, check the pH. Use each aliquot only once and discard.

– **Deionized Formamide Solution (100%)**

– 100% Formamide

Add 500 ml of formamide to 50 g of ion exchange resin, AG 501-X8, and stir slowly for 30 minutes. Remove resin by filtering through Whatman 1 MM filter. Store at - 20 °C.

– **Ethidium Bromide Stock Solution (5 mg/ml)**

– 50 mg ethidium bromide

Dissolve the powder in 10 ml sterilized MilliQ water by stirring under a chemical fume hood. Store at room temperature in a tightly closed, dark bottle.

Safety Note: Ethidium bromide is a powerful mutagen and suspected carcinogen. It is particularly toxic when inhaled in powder form. Take all possible precautions not to inhale it during preparation of the stock solution. Weigh powder under chemical fume hood and wear a face mask and gloves.

– **10 x MOPS Formaldehyde Gel Electrophoresis Buffer**

– 0.2 M MOPS

– 80 mM Sodium acetate

– 10 mM EDTA

Dissolve MOPS and sodium acetate in DEPC-treated water or MilliQ water, add EDTA from a 0.5 M stock solution and adjust pH to 7.0 with 10 N NaOH. Sterilize by filtration. Store at room temperature in the dark. During storage a yellow color can develop, but this does not interfere with gel electrophoresis.

Note: Tris based buffers cannot be used when formaldehyde is present because it interacts with amine groups in this buffer.

- **0.1 M Phosphate Solution pH 7.0**
  This solution is used as electrophoresis buffer for glyoxal gels. Dissolve 0.1 moles of Na_2HPO_4 in a minimum amount of water, adjust pH to 7.0 with concentrated phosphoric acid and fill up to 1 liter with RNase-free water. Sterilize by autoclaving and store at room temperature.
- **10 x TBE Electrophoresis Buffer**
  - 890 mM Tris-base
  - 890 mM boric acid
  - 20 mM EDTA
  This buffer is used for native gel electrophoresis. Dissolve Tris and boric acid in RNase-free water and add the appropriate amount of 0.5 M EDTA, pH 8.5. Store at room temperature.
- **5 x RNA Sample Buffer for Formaldehyde Gels**
  - 1 x MOPS formaldehyde gel electrophoresis buffer
  - 50% Deionized formamide
  - 5.6% Formaldehyde
  - 10% Glycerol
  - 0.1 % Bromophenol blue
  Prepare 5 ml of solution, aliquot in small portions into RNase free tubes and store at -20 °C.
- **Glyoxal Gel Loading Buffer**
  - 50% Glycerol
  - 10 mM sodium phosphate pH 7.0
  - 0.05% Bromophenol blue
  Prepare using MilliQ or DEPC-treated water and store in small aliquots at -20 °C.

▨▨ Procedure

Formaldehyde gels protocol

The quantities indicated in this protocol are for use with a gel size of 20 cm x 15 cm x 0.4 cm. The well length is 0.7 cm and well width is 2 mm, dimensions optimal for obtaining sharp RNA bands. The volume of the sample that can be loaded in the well is 25 μl. For other gel and comb sizes, the volumes should be adjusted accordingly. Samples are stained with ethidium bromide that is added directly to RNA at a concentration of 5 μg per sample. This concentration of ethidium bromide does not interfere with transfer or Northern hybridization (Gong, 1992).

1. Seal the opened ends of the gel casting tray with tape. Regular labeling tape or electrical insulation tape can be used. Check that the comb teeth are about 0.5 mm above the bottom of the tray. To adjust this height, it is most convenient to place a plastic charge card (e.g., MasterCard) under the comb and adjust the comb height to a position where the card is easily removed from under the comb teeth. Wipe the comb with RNaseZap immediately before use.

2. Prepare 1500 ml of 1x MOPS buffer by adding 150 ml of 10 x MOPS stock solution to final volume of 1500 ml of RNase-free water.

3. Place 142 ml of the buffer into a 500 ml conical flask and add the appropriate amount of agarose. Weigh 1.5 g of agarose for a 1.0% agarose gel. Melt the agarose by heating the solution in a microwave oven at full power for approximately 3 minutes or by boiling for 10 minutes in a boiling water bath. Swirl the agarose solution to ensure that the agarose is dissolved; that is, no shiny agarose particles are visible. If evaporation occurs during melting, adjust the volume to 142 ml with water.

Note: 1.0% agarose formaldehyde gels have an efficient range of separation for linear RNA molecules between 500 to 7000 bases. For other ranges consult Table 1.

4. Cool the agarose solution to approximately 50 °C and add 7.5 ml of formaldehyde (1:20 dilution of 37% stock). Slowly pour the agarose into the gel casting tray. Remove any air bubbles by trapping them in 10 ml pipette.

Note: Final formaldehyde concentration in the gel is approximately 1.8% or 0.66 M. This concentration should be increased to 4% (1.2 M) if the RNA of interest is very large (›10 000 b).

Safety Note: Formaldehyde vapor is toxic. Prepare the gel under a chemical fume hood or in well-ventilated area. Wear gloves at all times.

5. Position the comb approximately 1.5 cm from the edge of the gel. Let the agarose solidify for about 30 to 60 minutes in a chemical fume hood. After the agarose has solidified, remove the comb with a gentle back and forth motion, taking care not to tear the gel.

6. Remove the tape from the ends of the gel casting tray and place the tray on the central supporting platform of the gel box.
   Safety Note: For safety purposes, the electrophoresis apparatus always should be placed with the positive electrode (red) facing away from the investigator. Place the gel apparatus in chemical hood because formaldehyde vapor is toxic.

7. Add electrophoresis buffer to the buffer chamber until it reaches a level approximately 2 to 3 mm above the surface of the gel.

8. Prepare the sample as follows. Add 5 µl of 5 x RNA sample buffer, 1 µl of ethidium bromide and desired amount of RNA sample to a sterile microfuge tube. If the total volume is less than 25 µl fill it up with sterile RNase-free water. Incubate the RNA sample at 65 °C for 10 minutes then transfer the tube immediately to ice for 2 minutes. Centrifuge for 20 seconds to collect condensation from the sides of the tube and place it back on ice.

Note: Generally, 10 to 20 µg total RNA or 0.5 to 2 µg of poly(A)$^+$ RNA per lane can be loaded using the gel described in this procedure. Samples must be loaded in sequential sample wells. Do not "skip" wells when loading fewer samples than the number of wells. In this case, load samples in the wells at the center of the gel.

9. Load the samples into the wells using a yellow, RNase-free tip. Place the tip **under** the surface of the electrophoresis buffer and **above** the well. Expel the sample slowly, allowing it to sink to the bottom of the well. Take care not to spill the sample into a neighboring well. During sample loading, it is very important not to place the end of the tip into the sample well or touch the edge of the well with the tip. This can damage the well, resulting in uneven or smeared bands. The RNA size standard, if desired, should be loaded in the first well from the left. This is because many computer programs that are used for calculating fragment size require the size standards in this position.

10. Place the lid on the gel box and connect the electrodes. RNA will travel toward the positive (red) electrode. Turn on the power supply. Adjust the voltage to about 2 to 5 V/cm. For example, if the distance between electrodes (not the gel length) is 40 cm, to obtain field strength of 3 V/cm, voltage should be set to 120 V. Continue electrophoresis until the tracking dye moves 1/2 to 2/3 the length of the gel.

Note: It will take the tracking dye approximately 2 to 3 hours to reach this position on a gel 20 cm long. Consult Table 2 in this chapter for the size of

RNA molecules that co-migrates with bromophenol blue dye. Electrophoresis should not be carried out for longer than 3 hours because formaldehyde will elute from the gel causing the RNA to renature.

11. Turn the power supply off and first disconnect the positive (red), and next, the negative lead from the power supply. This order of disconnecting leads prevents the occurrence of accidental electrical shock. Remove the gel from the electrophoresis chamber. To transfer RNA to a membrane follow the procedure described in Subprotocol 2.

Running glyoxal gels requires recirculation of the electrophoresis buffer between anodal and catodal compartments because of its very low buffering capacity. Place a magnetic stirrer under both chambers and set up a peristaltic pump to recirculate buffer between the two tanks. The quantities indicated in this protocol are for use with a gel size of 20 cm x 15 cm x 0.4 cm. **Glyoxal gels protocol**

1. Seal the opened ends of the gel casting tray with tape. Regular labeling tape or electrical insulation tape can be used. Check that the comb teeth are about 0.5 mm above the bottom of the tray. To adjust this height, it is most convenient to place a plastic charge card (e.g., MasterCard) under the comb and adjust the comb height to a position where the card is easily removed from under the comb teeth. Wipe the comb with RNaseZap® immediately before use.

2. Prepare 1500 ml of 10 mM sodium phosphate buffer by adding 150 ml of a 0.1 M sodium phosphate stock solution to a final volume of 1500 ml of RNase-free water.

3. Place 150 ml of the buffer into a 500 ml conical flask and add the appropriate amount of agarose. Weight 1.5 g of agarose for a 1.0% agarose gel. Melt the agarose by heating the solution in a microwave oven at full power for approximately 3 minutes or by boiling for 10 minutes in a boiling water bath. Swirl the agarose solution to ensure that the agarose is dissolved; that is, no agarose particles are visible. If evaporation occurs during melting, adjust the volume to 150 ml with water.

Note: Because glyoxal reacts with ethidium bromide, the gels are poured and run without the dye.

4. Cool the agarose solution to approximately 70 °C and add solid sodium iodoacetate to final concentration of 10 mM to inactivate RNase. If agarose used is certified RNase-free, this step can be omitted. Cool agarose to 50 °C and slowly pour it into the gel casting tray. Remove any air bubbles by trapping them in a 10 ml pipette.

5. Position the comb approximately 1.5 cm from the edge of the gel. Let the agarose solidify for about 30 to 60 minutes. After the agarose has solidified, remove the comb with a gentle back and forth motion taking care not to tear the gel.

6. Remove the tape, and place the tray on the central supporting platform of the gel box.
Safety Note: For safety purposes, the electrophoresis apparatus should be oriented with the positive electrode (red) away from the investigator.

7. Add the electrophoresis buffer prepared in step 2 to the buffer chamber until it reaches a level approximately 2 to 3 mm above the surface of the gel.

8. Before loading the gel denature the RNA sample as follows. To sterilize a microfuge tube add:

Ingredients	Amount	Final Concentration
Deionized glyoxal (6 M)	4.0 µl	1 M
DMSO (100%)	12.5 µl	50 %
Na phosphate (0.1 M)	2.5 µl	10 mM
RNA	as required	up to 20 µg
Water	fill up to 25 µl	

Incubate the tube at 50 °C for 1 hour, cool on ice and add 5 µl of glyoxal gel loading buffer. Mix by pipetting up and down.

Note: Generally, 10 to 20 µg total RNA or 0.5 to 2 µg of poly(A)$^+$ RNA per lane can be loaded using a gel of this size. Samples must be loaded in sequential sample wells. Do not "skip" wells when loading fewer samples than the number of wells. In this case, load samples in the wells at the center of the gel.

9. Load the samples into the wells using a yellow, RNase-free tip. Place the tip **under** the surface of the electrophoresis buffer and **above** the well. Expel the sample slowly, allowing it to sink to the bottom of the well. Take care not to spill the sample into a neighboring well. During sample loading, it is very important not to place the end of the tip into the sample well or touch the edge of the well with the tip. This can damage the well, resulting in uneven or smeared bands. The RNA size standard, if desired, should be loaded in the first well from the left. This is because

many computer programs that are used for calculating fragment size require the size standards in this position.

10. Place the lid on the gel box and connect the electrodes. RNA will travel toward the positive (red) electrode. Turn on the power supply. Adjust the voltage to about 2 to 5 V/cm. For example, if the distance between electrodes (not the gel length) is 40 cm, to obtain a field strength of 3 V/cm, voltage should be set to 120 V. Continue electrophoresis until bromophenol blue moves at least 1/2 to 2/3 of the length of the gel. It will take the tracking dye approximately 3-5 hours to reach this position on a gel 20 cm long. Consult Table 2 of this chapter for the size of RNA molecules that co-migrate with bromophenol blue dye in glyoxal gels.

Note: Run the gel for 10 minutes before starting buffer recirculation. Start stirrers under buffer chambers and begin recirculation by starting the peristaltic pump.

11. Turn the power supply off and first disconnect the positive (red), and next, the negative lead from the power supply. This order of disconnecting leads prevents the occurrence of accidental electrical shock. Remove the gel from the electrophoresis chamber. To transfer RNA to a membrane follow the procedure described in Subprotocol 2.

Native gels can be used as an analytical tool to assess the efficiency of RNA purification. They are easy to prepare and do not use toxic chemicals. Northern blot transfer from native gel is not recommended.

Native gels protocol

1. Seal the opened ends of the gel casting tray with tape. Regular labeling tape or electrical insulation tape can be used. Check that the comb teeth are about 0.5 mm above the bottom of the tray. To adjust this height, it is most convenient to place a plastic charge card (e.g., MasterCard) under the comb and adjust the comb height to a position where the card is easily removed from under the comb teeth. Wipe the comb with RNaseZap® immediately before use.

2. Prepare 1500 ml of 1 x TBE using RNase-free water.

3. Place 150 ml of the buffer into a 500 ml conical flask and add the appropriate amount of agarose. Weigh 1.5 g of agarose for a 1.0% agarose gel. Melt the agarose by heating the solution in a microwave oven at full power for approximately 3 minutes or by boiling for 10 minutes in a boiling water bath. Swirl the agarose solution to ensure that the agarose is dissolved. If evaporation occurs during melting, adjust the volume to 150 ml with water.

Note: Ethidium bromide can be incorporated into the agarose gel at a concentration of 0.5 µg/ml. If the gel contains ethidium bromide, it is not necessary to add it to the sample.

4. Cool the agarose solution to approximately 60 °C and slowly pour the agarose into the gel casting tray. Remove any air bubbles by trapping them in a 10 ml pipette.

5. Position the comb approximately 1.5 cm from the edge of the gel. Let the agarose solidify for about 30 minutes. After the agarose has solidified, remove the comb with a gentle back and forth motion taking care not to tear the gel. Remove the tape, and place the tray on the central supporting platform of the gel box.

Note: Native agarose gels should be as thin as possible (3-4 mm) to shorten electrophoresis time. Electrophoresis time should not be longer than 3 hours to maintain RNA molecules in a denatured state.

6. Add electrophoresis buffer until it reaches a level approximately 2 to 3 mm above the surface of the gel.

7. Prepare the sample as follows. To a sterile microfuge tube, add 5 µl of 5 x RNA sample buffer, 1 µl of ethidium bromide and the desired amount of RNA sample. If the total volume is less than 25 µl, fill it up with sterile RNase-free water. Incubate at 65 °C for 10 minutes and transfer it immediately to ice for 2 minutes. Centrifuge the tube for 20 seconds to collect condensation and place the tube back on ice until ready to load onto the gel. **If ethidium bromide is present in the gel do not add it to the sample.**

8. Load the samples into the wells using a yellow, RNase-free tip. Place the tip **under** the surface of the electrophoresis buffer and **above** the well. Expel the sample slowly, allowing it to sink to the bottom of the well. Take care not to spill the sample into a neighboring well. During sample loading, it is very important not to place the end of the tip into the sample well or touch the edge of the well with the tip. This can damage the well, resulting in uneven or smeared bands. The RNA size standard, if desired, should be loaded in the first well from the left. This is because many computer programs that are used for calculating fragment size require the size standards in this position.

9. Place the lid on the gel box and connect the electrodes. RNA will travel toward the positive (red) electrode. Turn on the power supply. Adjust the voltage to about 2 to 5 V/cm. For example, if the distance between

electrodes (not the gel length) is 40 cm, to obtain a field strength of 3 V/cm, voltage should be set to 120 V. Continue electrophoresis until bromophenol blue moves at least 1/2 to 2/3 of the length of the gel. It will take the tracking dye approximately 2 hours to reach this position on a gel 20 cm long.

10. Turn the power supply off and first disconnect the positive (red), and next, the negative lead from the power supply. This order of disconnecting leads prevents the occurrence of accidental electrical shock. Remove the gel from the electrophoresis chamber. You can photograph the gel using Polaroid film 667 with a Polaroid camera at a speed of 1 and F8 or use a computer imaging system to record the results (see Chapter 8 for details).

Table 2. Suggested SeaKem® agarose concentrations for optimal resolutions of RNA molecules

Agarose Concentration (%)	Efficient Range of Separation
0.3	5 000 - 60 000
0.6	1 000 - 20 000
0.7	800 - 10 000
1.0	500 - 7 000
1.2	400 - 6 000
1.5	200 - 3 000
2.0	100 - 2 000

Table 3. Mobility of RNA molecules in relation to bromophenol blue dye in SeaKam® LE agarose gels

Concentration of Agarose (%)	Formaldehyde Gels[a]	Glyoxal Gels
0.6	900	650
1.0	320	740
1.5	140	370
2.0	60	220

[a] Mobility of RNA in native gels is identical to that in formaldehyde gels

Results

Figure 2 presents the result of electrophoresis of corn total RNA using a 1% agarose formaldehyde gel. Different concentrations of RNA were loaded and electrophoresis was continued until the bromophenol blue had moved 1/2 the length of the gel. Two sharp bands appear on the gel, 28S RNA (4.7 kb) and 18S RNA (1.9 kb). The 5S and 5.8S RNA bands are located on the leading edge of the gel, running together with tRNA. The smears between, above and below rRNA bands represent the mRNA.

Note: The integrity of the RNA samples and lack of RNA degradation can be easily judged from the appearance of rRNA bands. Degradation of the sample appears as diffused 28S and 18S bands or an incorrect ratio of stain between rRNA bands. This ratio should be approximately 2:1 for 28S and 18S, respectively.

Troubleshooting

The most common problems with gel electrophoresis of RNA are:
- Inadequate denaturation of the samples
  This will appear either as multiple rRNA bands or rRNA bands that appear to be smeared but at a correct ratio. This is seen more frequently with glyoxal gels than formaldehyde gels and is caused by inadequate recirculation of the buffer.

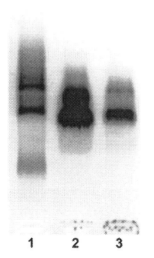

Fig. 2. Gel electrophoresis of total RNA isolated

- Overloading the gel with RNA sample
  This will result in very broad rRNA bands that run on the gel with excessive "smearing". Bands could have a U-shape appearance and their mobility might be faster than expected from their base number.

- Degradation of the sample before or during electrophoresis
  This will be indicated by an incorrect ratio between 28S and 18S rRNA, or, in more severe cases, the total disappearance of these bands. If degradation of RNA occurs during electrophoresis because RNase contamination of buffer or agarose, this can be recognized by the disappearance or "smearing" of the RNA size standards.

Subprotocol 2
Northern Blot Transfer

Blotting and hybridization conditions differ from those used for DNA as described in Chapter 11. The denaturation step before transfer is not necessary because RNA molecules are single stranded. Alkaline transfer is not recommended as alkaline conditions result in degradation of RNA. Small molecules can be completely hydrolyzed or broken down to short fragments that cannot be retained by the membrane. Transfer to the membrane requires a higher salt concentration than that used for DNA. The transfer solution is 20 x SSC rather than 10 x SSC used in DNA transfers. All other standard conditions of Northern transfer such as choice of membranes, time of transfer, architecture of the blot etc., are identical to those for DNA and are described in Chapter 11. Two transfer protocols are presented: transfer from formaldehyde gels and transfer from glyoxal gels. A schematic outline of the Northern blot transfer is shown in Figure 3.

Materials

- MagnaGraph nylon membrane 0.22 μm pore size (Micron Separation Inc. # NJTHY00010). See Chapter 10 for a description of equivalent nylon membranes from other manufacturers. If fragments larger than 300 bp are to be transferred, it is not necessary to use 0.2 μm pore size membrane and a membrane of 0.45 μm can be used.
- Whatman 3MM chromatography paper (Whatman Co. # 3030917 or equivalent)
- Pyrex glass dishes

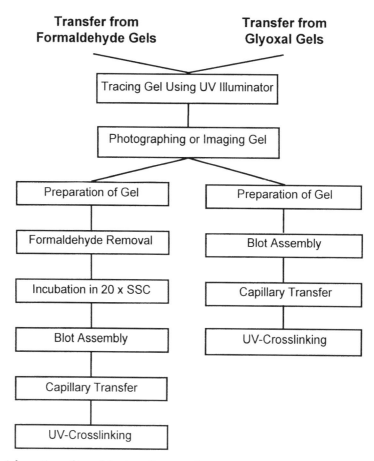

Fig. 3. Schematic outline of the transfer procedures

– Stratalinker® UV oven (Stratagen Co. # 400071 or equivalent)
  To efficiently crosslink RNA to nylon membrane, the UV source should
  be capable of delivering 120 mJ/cm². Excessive crosslinking will decrease
  hybridization efficiency.

Solutions   – **20 x SSC Solution**
  – 3.0 M NaCl
  – 0.3 M Sodium citrate
    Dissolve 175 g NaCl and 88.2 g of sodium citrate in 850.0 ml of DEPC-
    treated or MilliQ deionized water. Adjust pH to 7.5 with 10 N NaOH,
    and add water to 1000 ml. Store at 4 °C.

Procedure

1. After electrophoresis, transfer the gel to a Pyrex dish. Be very careful when transferring the gel because **formaldehyde makes the gel very brittle**. Trim away any unused areas of the gel with a razor blade. Cut the lower corner of the gel at the bottom of the lane with size standards. This will provide a mark to orient the hybridized bands on the membrane with the bands in the gel. Transfer the gel to a transilluminator and draw the outline of the gel on an acetate sheet with a felt marker pen. Mark the position of wells, the positions of the bands, the RNA standards, and the position of the cut corner. Cut away and remove the gel above the wells. You can photograph the gel using Polaroid film 667 with a Polaroid camera at a speed of 1 and F8 or use a computer imaging system to record the results (see Chapter 8 for details). It is not necessary to photograph the gel with a ruler because the location of hybridized bands can be directly determined from the schematic drawing.

Transfer from formaldehyde gel protocol

Note: Because the gel is thinner in the wells, if they are not removed, transfer solution may pass preferentially through this part of the gel causing uneven RNA transfer.
Safety Note: Wear gloves and UV-protective glasses during this operation. Use only powder free gloves if you intend to use a chemiluminescent detection system. The presence of talcum will result in a "spotted background".

2. Transfer the gel back to an RNase-free Pyrex dish filled with enough RNase-free water to cover the gel. This will take about 150 to 200 ml for a standard gel size.

Note: When using formaldehyde agarose gels containing less than 0.4 M (12%) formaldehyde, gel washing is not required before transfer. Go on to step 5.

3. Place the dish on an orbital shaker and incubate for 20 minutes rotating at 10 - 20 rpm. Decant the water carefully holding the gel with the palm of your gloved hand. Repeat this wash three more times.

Note: This procedure removes formaldehyde from the gel. The presence of formaldehyde will hinder RNA transfer to the membrane.

4. Add 20 x SSC to the gel. Use approximately the same volume as in step 2. Incubate for 45 minutes on an orbital shaker.

Note: This step is optional, but it may improve transfer efficiency of large RNA molecules.

5. While the gel is being treated, prepare the nylon membrane for transfer. Cut the nylon membrane to the size of the gel. Use the gel outline drawn on the acetate sheet as a guide. Wear gloves and only touch the edges of the filter. Place the filter in a separate baking dish filled with RNase-free water. Leave the filter in water for 1 to 2 minutes. Decant the water and immerse the filter in 20 x SSC. Cut three sheets of Whatman 3MM paper to the size of the nylon membrane. Prepare a long strip of Whatman 3MM paper to be used as a wick. The wick should be approximately 30 cm long and 10 cm wide.

6. Set up the blot as follows. Figure 4 shows details of the procedure. Add 400 to 600 ml of 20 x SSC to a large Pyrex dish. Place a glass platform across center of the dish and cover it with the wick. Make sure that both ends of the wick are immersed in SSC solution. Wet the wick with 10 to 20 ml of transfer solution and remove trapped air bubbles by rolling a 10 ml glass pipette over it. Carefully lift the gel from the baking dish and place it in the center of the wick **sample wells down**. Smooth the gel and remove trapped air bubbles by gently rolling a glass pipette over the surface.

Note: The gel is now upside-down with the wells facing the wick. This is necessary to obtain the best results during transfer (sharper resolution due to less diffusion during the transfer), and to maintain the left-to-right sample orientation on the blotted membrane.

7. Cover the entire dish, including the surface of the gel, with saran wrap. With a razor blade "cut away" the saran wrap covering the gel itself. This will leave an opening over the gel, whereas the remaining area of the wick will be covered by saran wrap.

8. Place the nylon membrane on top of the gel. Add 5 ml of 20 x SSC to the top of the filter and remove any air bubbles by rolling a pipette over it. Cut the left bottom corner of the filter to coincide with the cut made in the gel. Place three sheets of dry Whatman 3MM paper, prepared in step 5, on top of the membrane. Place 5 to 10 cm of paper towels on top of the 3MM paper. Because the wick area is protected by saran wrap it is not necessary to cut the paper towels to the size of the gel. Place a glass plate on top of the paper towel stack and weigh it down by placing a one-liter Erlenmeyer flask filled with 500 ml of water. Transfer overnight at 4 °C or 4 to 5 hours at room temperature.

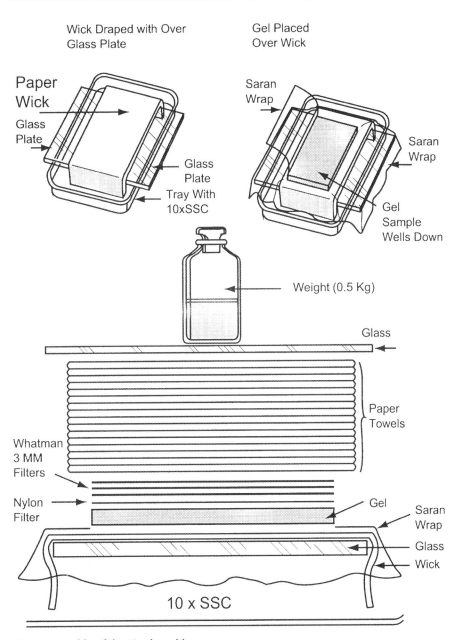

Fig. 4. Assembly of the Northern blot

Note: To prevent the gel from collapsing the weight should never exceed 500 g (that is, about 500 ml of water).

9. Dissemble the blot. Remove the weight, glass plate and paper towels. Using forceps, remove the membrane and place it "RNA side" up (the side that contacted the gel) on a clean sheet of Whatman paper. Mark the RNA side with a pencil on the corner of the membrane.

10. Place the filter on a sheet of dry Whatman 3MM paper. Do not allow the membrane to dry at any time. Place the membrane into a UV oven.

11. UV irradiate the damp membrane to crosslink RNA to the membrane using the automatic setting of the "UV oven". Irradiate **both sides** of the membrane. Alternatively, you can wrap the membrane in aluminum foil and bake it in an oven at 80 °C for 1 hour. The baking step immobilizes RNA on the membrane. The membrane can be stored at room temperature almost indefinitely.

Transfer from glyoxal gel protocol

1. After electrophoresis, transfer the gel to a Pyrex dish and stain it with ethidium bromide. Use between 200 to 300 ml of an aqueous solution of ethidium bromide (0.5 µg/ml). Place the dish on an orbital shaker and stain for 20 minutes rotating at 10-20 rpm.

Note: Glyoxal gels cannot be run in the presence of ethidium bromide because ethidium bromide reacts with glyoxal. The gel can be stained with ethidium bromide after electrophoresis and destaining is not necessary.

2. Transfer the gel to a new Pyrex dish and trim away any unused areas of gel with a razor blade. Cut the lower corner of the gel at the bottom of the lane with the size standards. This will provide a mark to orient the hybridization bands on the membrane with the bands in the gel. Transfer the gel to a transilluminator and draw the outline of the gel on an acetate sheet with a felt marker pen. Mark the position of wells, the position of standards, ribosomal RNA bands and the cut corner. Cut away and remove the gel above the wells. You can photograph the gel using Polaroid film 667 at a speed of 1 and F8 or use a computer imaging system to record the results (see Chapter 8 for details). It is not necessary to photograph the gel with a ruler because the location of hybridized bands can be directly determined from the schematic drawing.

Note: Because the gel is thinner in the wells, if the wells are not removed the transfer solution may pass preferentially through this part of the gel causing uneven RNA transfer.

Safety Note: Wear gloves and UV-protective glasses during this operation. Use only powder-free gloves when using a chemiluminescent detection system. The presence of talcum will result in a "spotted background".

3. Return the gel to the Pyrex dish and add 200 - 300 ml of 20 x SSC solution. Incubate for 45 minutes on an orbital shaker.

Note: This step is optional, but it may improve transfer efficiency of large RNA molecules.

4. While the gel is being treated, prepare the nylon membrane for transfer. Cut the nylon membrane to the size of the gel. Use the gel outline drawn on the acetate sheet as a guide. Wear gloves and only touch the edges of the filter. Place the filter in a separate baking dish filled with RNase-free water. Leave the filter in water for 1 to 2 minutes. Decant the water and immerse the filter in 20 x SSC. Cut three sheets of Whatman 3MM paper to the size of the nylon membrane. Prepare a long strip of Whatman 3MM paper to be used as a wick. The wick should be approximately 30 cm long and 10 cm wide.

5. Assemble the blot. Figure 4 shows details of the procedure. Add 400 to 600 ml of 20 x SSC to a large Pyrex dish. Place a glass platform across center of the dish and cover it with the wick. Make sure that both ends of the wick are immersed in SSC solution. Wet the wick with 10 to 20 ml of transfer solution and remove trapped air bubbles by rolling a 10 ml glass pipette over it. Carefully lift the gel from the baking dish and place it in the center of the wick **sample wells down.** Smooth the gel and remove trapped air bubbles by gently rolling a glass pipette over the surface.

Note: The gel is now upside-down with the wells facing the wick. This is necessary to obtain the best results during transfer (sharper resolution due to less diffusion during the transfer), and to maintain the left-to-right sample orientation on the blotted membrane.

6. Cover the entire dish, including the surface of the gel, with saran wrap. With a razor blade "cut away" the saran wrap covering the gel itself. This will leave an opening over the gel, whereas the remaining area of the wick will be covered by saran wrap.

7. Place the nylon membrane on top of the gel. Add 5 ml of 20 x SSC to the top of the filter and remove any air bubbles by rolling a pipette over it. Cut the left bottom corner of the filter to coincide with the cut made in the gel. Place three sheets of dry Whatman 3MM paper, prepared in step 5, on top of the membrane. Place 5 to 10 cm of paper towels on top of the

3MM paper. Because the wick area is protected by saran wrap it is not necessary to cut the paper towels to the size of the gel. Place a glass plate on top of the paper towel stack and weigh it down by placing a one-liter Erlenmeyer flask filled with 500 ml of water. Transfer overnight at 4 °C or 4 to 5 hours at room temperature.

Note: To prevent the gel from collapsing, the weight should never exceed 500 g (that is, about 500 ml of water).

8. Dissemble the blot. Remove the weight, glass plate and paper towels. Using forceps, remove the membrane and place it "RNA side" up (the side that contacted the gel) on a clean sheet of Whatman paper. Mark the RNA side with a pencil on the corner of the membrane.

9. Place the filter on a sheet of dry Whatman 3MM paper. Do not allow the membrane to dry at any time. Place the membrane into a UV oven.

10. UV irradiate the damp membrane to crosslink RNA to the membrane using the automatic setting of the "UV oven". Irradiate **both sides** of the membrane. Alternatively, you can wrap the membrane in aluminum foil and bake it in an oven at 80 °C for 1 hour. The baking step immobilizes RNA on the membrane. The membrane can be stored at room temperature almost indefinitely.

Troubleshooting

Efficiency of the transfer can be checked by staining the gel with ethidium bromide. Inefficient transfer will result in a low fluorescence of ribosomal RNA bands. Alternatively, if RNA samples were stained with 5 μg/sample of ethidium bromide, the staining pattern will be visible on nylon membrane. Low transfer efficiency could result from:

- Too thick gel.
  Gels that are thicker than 5 mm should not be used.

- Agarose concentration higher than 2%.
  If high concentration of agarose is used, transfer time should be increased.

- Collapsing of the gel during transfer.
  The top weight should never exceed 500 g for gels 4 mm thick.

- Too high a concentration of ethidium bromide in the sample.

Subprotocol 3
Northern Blot Hybridization and Detection

Conditions of hybridization and washing reactions for RNA blots are different from those used for DNA. This is because of the difference in the stability of RNA:DNA hybrids, as compared to DNA:DNA hybrids (see Chapter 10 for detailed discussion of these conditions). Two protocols are described: hybridization using a hybridization oven and hybridization using plastic bags. The hybridization oven procedure rather than plastic bags procedure is recommended because this procedure is less time consuming, requires smaller volume of reagents, and results in much lower hybridization background. The hybridization procedure using plastic bags is essentially the same as the procedure with a hybridization oven. The procedure using bags is given because not all laboratories have hybridization ovens.

Hybridization of RNA blots always uses solutions that permit hybridization at low temperatures. These solutions usually contain formamide, or the commercially available Dig Easy Hyb solution, to lower incubation temperature and prevent destruction of RNA. Because formamide is toxic it is better to use Dig Easy Hyb solution in Northern hybridization experiments. The solution is inexpensive, non-toxic, and gives very low background with both DIG-labeled and radioactive probes. The procedure described uses stringent conditions of hybridization and washing, as calculated from equations 2, 6 and 9 in chapter 10. The %GC of the target is assumed to be 50% and the probe is RNA. To use less stringent conditions, a target with other than 50% GC content or DNA probes, hybridization conditions should be recalculated (see equations 2, 6 and 9 and the discussion of this topic in Chapter 10). A schematic outline of the hybridization and detection procedures is shown in Figure 5.

Materials

- Hybridization oven (e.g., HyBaid Co. # H9320 or equivalent)
  Oven should be capable of rotation at variable speeds.
- Hybridization Bottles 150 x 35 mm (e.g., HyBaid Co. # H98084 or equivalent)
- Plastic bags (e.g., Kapak Co. # 402 or Roche Moleculare Biochemicals # 1666 649)
- Plastic containers with lids

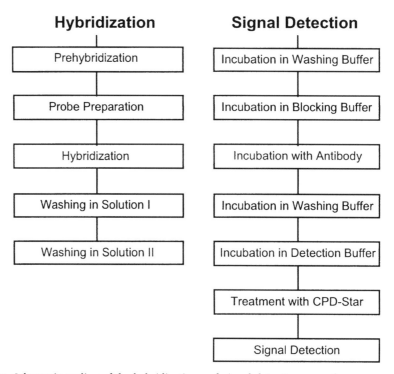

Fig. 5. Schematic outline of the hybridization and signal detection procedures

- Rocking platform shaker
- Rotary platform shaker
- Plastic bag sealer (Fisher Scientific # 01-812-13 or equivalent)
- 15 ml and 50 ml sterilized polypropylene conical centrifuge tubes (for example, Corning # 25319-15 and 25330-50)
- CDP-Star solution (Tropix Co. # MS100R)
- Store solution in the dark at 4 °C. CDP-Star is easily destroyed by ubiquitous alkaline phosphatase. Wear gloves and use sterilized tips to pipette it.
- Dig Easy Hyb solution (Roche Molecular Biochemicals # 1603 558)
- Blocking reagent for nucleic acid hybridization (Roche Molecular Biochemicals # 1096 176)
- Anti-DIG-alkaline phosphatase 750 u/ml (Roche Molecular Biochemicals # 1093 274)
  Stock solution should not be frozen. Store at 4 °C.
- Antifoam C (Sigma Co. # A 8001)
- Maleic acid (Sigma Co. # MD 375)
- Ion exchange resin AG 501-X8 (Bio-Rad # 142-6424)
- Formamide (Fluka # 47671)

- **20 x SSC Solution**
  - 3.0 M NaCl
  - 0.3 M Sodium citrate

  Dissolve 175 g NaCl and 88.2 g of sodium citrate in 900.0 ml of RNase-free water. Adjust pH to 7.5 with 10 N NaOH, and add water to 1000 ml. Store at 4°C.
- **Sarcosyl Stock Solution, 20% (w/v)**

  Dissolve 40 g of Sarcosyl in 100 ml of RNase-free water. Fill up to 200 ml with water. Sterilize by filtration through a 0.45 μm filter. Store at room temperature.
- **SDS Stock Solution, 10% (w/v)**

  Add 10 g of powder to 70 ml of RNase-free water and dissolve by slow stirring. Add water to a final volume of 100 ml and sterilize by filtration through a 0.45 μm filter. Store at room temperature.

  Safety Note: SDS powder is a nasal and lung irritant. Weigh the powder carefully and wear a face mask.
- **Formamide (deionized)**
  - 100% Formamide

  Add 500 ml of formamide to 50 g of ion exchange resin AG 501-X8 and stir slowly for 30 minutes. Remove resin by filtration through Whatman 1MM filter. Store at - 20 °C.
- **Hybridization Solution**
  - 5 x SSC
  - 2.5% blocking reagent
  - 0.001% Antifoam A
  - 0.02% (w/v) SDS
  - 0.1% (w/v) Sarcosyl
  - 50% formamide, deionized

  Prepare 2 x hybridization mixture as follow. Add 5 g of blocking reagent powder to 40 ml of 10 x SSC Add 2 μl of antifoam C, 0.2 ml of 10% SDS and 0.5 ml of 20% Sarcosyl. Dissolve by stirring for 2 hours at 50 °C to 60 °C. Fully dissolved blocking solution will have a uniformly "milky" appearance. Fill up with RNase-free water to 50 ml. Store at -20 °C. Immediately before use, add an equal volume of 100% deionized formamide. Formamide is toxic, wear gloves when using solutions containing it.

Note: Instead of this solution we recommend using Dig Easy Hyb solution.

- **Washing Solution I**
  - 2 x SSC
  - 0.1% SDS

- **Washing Solution II (high stringency)**
  - 0.1 x SSC
  - 0.2% SDS
- **10 x Maleic Acid Solution**
  - 1.0 M Maleic acid
  - 1.5 M NaCl

  Adjust pH to 7.5 with concentrated NaOH. Sterilize by autoclaving for 20 minutes. Store at 4 °C. This solution is available in the Wash and Block Buffer set from Roche Molecular Biochemicals (# 1585 762).
- **10 x Blocking Solution**
  - 10% (w/v) Blocking reagent
  - 1 x Maleic acid solution

  Add 10 ml of maleic acid solution to 90 ml of sterile RNase-free water. Add 10 g of blocking reagent. Dissolve by slowly stirring at 60 °C. Do not boil because it will cause the reagent to coagulate. Autoclave for 20 minutes. Blocking reagent must be completely in solution before autoclaving. Store at 4 °C. This solution is available in the Wash and Block Buffer set from Roche Molecular Biochemicals (# 1585 762)
- **Washing Buffer (Buffer A)**
  - 1 x Maleic acid solution
  - 0.3% (w/v) Tween 20

  Dilute 10 x maleic acid solution in sterilized RNase-free water and add Tween 20. Do not autoclave. Store at 4 °C. This solution is available in the Wash and Block Buffer set from Roche Molecular Biochemicals (# 1585 762)
- **Blocking Buffer (Buffer B)**
  - 1 x Maleic acid solution
  - 1 x Blocking solution

  Dilute 10 x maleic acid solution and 10 x blocking solution in RNase-free water. Prepare only the amount necessary for use. This buffer can be prepared from components of the Wash and Block Buffer set (Roche Molecular Biochemicals # 1585 762).
- **Detection Buffer (Buffer C)**
  - 0.1 M Tris HCl, pH 9.5
  - 0.1 M NaCl
  - 0.05 M $MgCl_2$

  Prepare from stock solutions. Adjust pH before the addition of $MgCl_2$ to avoid precipitation. Sterilize by filtration. Store at 4 °C.
- **Northern Probe-Stripping Solution**
  - 50 mM Tris-HCl, pH 8
  - 60% Formamide

– 1% (w/v) SDS

Prepare solution using deionized formamide and stock solutions of Tris and SDS prepared in RNase-free water. Do not store.

Procedure

The procedure uses Dig Easy Hyb solution but this can be substituted by hybridization solution described in Materials.

Hybridization oven protocol

1. Place the dry membrane into a hybridization bottle. Make sure that the side of the membrane with RNA is facing away from the glass.

Note: Several membranes can be placed in a single bottle. Some overlapping of the filters does not affect hybridization or increase background.

2. Pour 10 ml of Dig Easy Hyb solution into a 15 ml plastic centrifuge tube and add it to the hybridization bottle. Save the centrifuge tube for further use. The minimum amount of liquid per hybridization bottle is 5 to 6 ml.

3. Close the hybridization bottle and place it in the hybridization oven. Incubate for 1-3 hours at 55 °C, rotating at a **slow speed** (3 to 5 rpm). The prehybridization step lowers the background.

4. Ten minutes before the end of prehybridization, begin to prepare the probe. Pour 10 ml of Dig Easy Hyb solution into the 15 ml conical centrifuge tube that you saved. Add RNA probe to a final concentration of 30 to 100 ng/ml.

Note: If a DNA probe is used, denature it by incubation at 80 °C for 10 minutes. The final concentration of a DNA probe cannot exceed 25 ng/ml.

5. Pour off the prehybridization solution. The solution can be stored at 4 °C and used again.

6. Add Dig Easy Hyb, containing the probe, to the hybridization bottle. Return it to the oven and rotate it slowly at 55 °C overnight.

7. Pour the hybridization solution into a 15 ml centrifuge tube. The probe can be stored at -70 °C and reused 3 to 4 times.

8. Add 20 ml of washing solution I to the bottle. Place it into the hybridization oven and rotate it at maximum speed for 15 minutes at room temperature.

9. Pour off solution I and discard it. Drain the liquid well by placing the bottle upright on a paper towel for 1 minute. Repeat the wash one more time and drain the bottle.

10. Add 20 ml of washing solution II that has been heated to 65 °C. Place the bottle into the hybridization oven preheated to 65 °C. Rotate it at a slow speed for 20 minutes.

11. Pour off solution II and discard it. Drain the remaining liquid well by placing the bottle upright on a paper towel for 1 minute. Wash with solution II, preheated to 65 °C, one more time.

12. Discard washing solution II and drain the liquid as described above.

Detection

1. Add 20 ml of buffer A to the hybridization bottle. Cool the oven to room temperature and rotate the bottle at maximum speed for 2 to 5 minutes.

2. Pour off buffer A and add 10 ml of buffer B. Incubate for 1 to 3 hours **rotating slowly** at room temperature.

3. Pour off buffer B and discard it. Invert the bottle over a paper towel and let it drain.

4. Add 10 ml of buffer B to a plastic 15 ml centrifuge tube and add 2 µl of anti-DIG-alkaline phosphatase stock solution. Mix well and add the solution to the bottle with the membrane. Incubate for 30 minutes, rotating slowly at room temperature.

Note: The diluted antibody solution is stable for about 12 hours at 4 °C. Do not prolong incubation with antibody over 30 minutes as this will result in high background.

5. Pour off antibody solution and drain the bottle well. Discard the antibody solution.

6. Add 20 ml of buffer A and wash the membrane for 15 minutes at room temperature rotating at maximum speed.

7. Discard buffer A and drain the bottle well. Add 50 ml of buffer C and equilibrate the membrane in it for 20 minutes rotating at maximum speed. Repeat this procedure one more time.

8. Move the membrane in the bottle towards the opening by shaking the bottle. Open the tube and pour half of buffer C into a plastic bag. Wearing powder-free gloves, transfer the membrane from the tube to the bag. This can be easily accomplished by keeping the wet membrane im-

mersed in a pool of buffer C and moving it with your fingers. Move the membrane to the end of the bag and leave as little space as possible between the membrane edges and the edges of the bag. This will limit the amount of expensive chemiluminescent substrate necessary to fill the bag.

9. Pour buffer C from the bag. Place the bag on a Whatman 3MM paper and remove the remaining liquid from the bag by gently pressing it out with a Kimwipe tissue.

Note: Do not press strongly on the membrane because this will increase background. Most of the liquid should be removed from the bag, leaving the membrane slightly wet. At this point, a very small amount of liquid will be visible at the edge of the membrane.

10. Open the end of the bag slightly, leaving the side of the membrane without RNA sticking to the bag. Add 0.9-1 ml of CDP-Star solution, directing the stream toward side of the bag. Do not add CDP-Star solution directly onto the membrane. Place the bag back on a sheet of Whatman 3MM paper with the RNA-side up and distribute the liquid over the membrane by gently moving the liquid around with a Kimwipe tissue. Make sure that the entire membrane is evenly covered. Do not press on the membrane because this will result in visible "press marks" on the film. Remove excess CDP-Star from the bag by gently pressing it out onto the Whatman paper with a Kimwipe. Make sure that the membrane remains damp. At this point, small liquid droplets are visible on the edge of the membrane, but there is no liquid present on its surface. Seal the bag with a heat sealer.

11. Place the bag in an X-ray film cassette and incubate for 30 minutes at room temperature. In a darkroom, place X-ray film over the bag with the membrane. The initial recommended exposure time is 10 to 15 minutes at room temperature for single copy gene detection. This time can be increased up to 12 hours. See Chapter 10 for a description of membranes and recommended times of exposure. Open the cassette and develop the film using the standard procedure for film development.

Note: Maximum light emission for CDP-Star occurs in 20 to 30 minutes, the light emission remains constant for approximately 24 hours, and the blot can be exposed to film a number of times during this period.

1. Place the bag on a clean sheet of Whatman paper. Place the dry filter on the sheet of Whatman 3MM paper cut to a size slightly larger than the nylon membrane. Insert the nylon membrane and Whatman

Hybridization using plastic bags protocol

paper carefully into the plastic bag. Gently withdraw the Whatman paper from the bag leaving the nylon membrane inside.

2. Add 10 ml of Dig Easy Hyb (or 1 x prehybridization solution) to the bag. Place the bag on the lab bench and carefully eliminate all bubbles.

Note: Use at least 20 ml hybridization solution per 100 cm^2 of membrane.

3. Seal the top of the bag leaving 4 to 8 cm between the top seal and the top of the membrane. It is very important to leave this large space between the filter and the sealed edge.

4. Put a rocking platform shaker into a 55 °C incubator and place the bag on the platform. Incubate for 2-4 hours with constant rocking.

5. Ten minutes before the end of prehybridization, begin to prepare the probe. Pour 10 ml of Dig Easy Hyb solution into the 15 ml conical centrifuge tube. Add RNA probe to a final concentration of 30 to 100 ng/ml.

Note: If a DNA probe is used, denature it by incubation at 80 °C for 10 minutes. The final concentration of a DNA probe cannot exceed 25 ng/ml.

6. Open the bag by cutting it with scissors along the line of the top seal, as far away from the edge of the membrane as possible. Pour off the prehybridization mixture.

Note: Prehybridization mixture can be stored and reused several times.

7. Add Dig Easy Hyb solution, containing probe, to the bag. Remove air bubbles and seal the bag. Return the sealed bag to the 55 °C incubator and leave it rocking overnight.

8. Carefully cut open the top of the bag. Pour the hybridization mixture into a 15 ml plastic centrifuge tube. The probe can be stored at -70 °C and reused several times.

9. Cut open the bag completely, open it like a book and remove the membrane with forceps. Place the membrane in a plastic box fitted with an air tight cover. Work over bench paper and wear gloves.

10. Add 100 - 150 ml of washing solution I to the membrane. Wash at room temperature on a rotary shaker at slow speed for 15 minutes. Drain the solution well.

11. Repeat the washing one more time with washing solution I.

12. Preheat 100 ml of washing solution II to 65 °C, and add it to the plastic dish with the membrane. Close the lid of the plastic container

tightly and place it into a 65 °C water bath. Wash for 20 minutes with occasional shaking. Keep the plastic container immersed in water by placing a weight on the container.

13. Discard the washing solution and repeat the washing procedure one more time.

14. Add 50 ml of buffer A and equilibrate the membrane for 1 minute at room temperature.

15. Remove the membrane from the plastic container and allow excess liquid to drain. Do not allow the membrane to dry.

16. Add 15 ml of buffer B to the plastic bag. Wearing powder-free gloves transfer the membrane to a new plastic bag. This can be easily accomplished by keeping the wet membrane immersed in buffer B in the bag and moving the blot with your fingers. Move the membrane to the end of the bag and leave as little space as possible between the membrane edges and the edges of the bag.

17. Place the bag on the Whatman paper and remove all bubbles. Seal the top of the bag leaving 4 to 8 cm between top of the bag and the top of the filter. Place the bag on the platform shaker and incubate with gentle rocking for 1 to 3 hours at room temperature.

18. Open the bag and remove buffer B from the bag. Place the bag on the sheet of Whatman paper and remove remaining buffer by gently rolling a 10 ml pipette over the bag. Add 10 ml of buffer B to a plastic 15 ml centrifuge tube and add 2 μl of anti-DIG-alkaline phosphatase stock solution. Mix well and add the solution to the bag, remove all bubbles and seal it. Incubate for 30 minutes at room temperature with gentle rocking.

Note: The diluted antibody solution is stable for about 12 hours at 4 °C. Do not prolong incubation with antibody over 30 minutes as this will result in high background.

19. Open the bag and discard the antibody solution. Add 20 ml of buffer A to the bag, close it and incubate for 15 minutes by rocking at room temperature. Repeat the washing procedure one more time.

20. Add 20 ml of buffer C to the bag and incubate for 10 minutes at room temperature. Repeat this procedure one more time.

21. Pour buffer C from the bag. Place the bag on a Whatman 3MM paper and remove the remaining liquid from the bag by gently pressing it out with a Kimwipe tissue.

Note: Do not press strongly on the membrane because this will increase background. Most of the liquid should be removed from the bag, leaving the membrane slightly wet. At this point, a very small amount of liquid will be visible at the edge of the membrane.

22. Open the end of the bag slightly, leaving the side of the membrane without RNA sticking to the bag. Add 0.9-1 ml of CDP-Star solution, directing the stream toward side of the bag. Do not add CDP-Star solution directly onto the membrane. Place the bag back on a sheet of Whatman 3MM paper with the RNA-side up and distribute the liquid over the membrane by gently moving the liquid around with a Kimwipe tissue. Make sure that the entire membrane is evenly covered. Do not press on the membrane because this will result in visible "press marks" on the film. Remove excess of CDP-Star from the bag by gently pressing it out onto the Whatman paper with a Kimwipe. Make sure that the membrane remains damp. At this point, small liquid droplets are visible on the edge of the membrane, but there is no liquid present on its surface. Seal the bag with a heat sealer.

23. Place the bag in an X-ray film cassette and incubate for 30 minutes at room temperature. In a darkroom, place X-ray film over the bag with the membrane. The initial recommended exposure time is 10 to 15 minutes at room temperature for single copy gene detection. This time can be increased up to 12 hours. See Chapter 10 for a description of membranes and recommended times of exposure. Open the cassette and develop the film using the standard procedure for film development.

Note: Maximum light emission for CDP-Star occurs in 20 to 30 minutes, the light emission remains constant for approximately 24 hours, and the blot can be exposed to film a number of times during this period.

Stripping membrane for reprobing

The nylon membrane can be stripped and reprobed at least 4 times. When reprobing of the membrane is required, make sure that the membrane does not become dry at any time. Store the membrane before stripping at 4 °C in 2 x SSC sealed in a plastic bag. The procedure can be carried out in a hybridization oven or a plastic dish.

Stripping protocol

1. Place the membrane into a hybridization bottle and wash in 20 ml of sterile RNase-water for 2 minutes. Discard the water.

2. Prepare 40 ml of Northern probe-stripping solution.

3. Add 20 ml of the solution and incubate for 30 minutes at 68 °C at maximum rotation speed. Discard the stripping solution and repeat the procedure one more time.

4. Add 20 ml of RNase-free water and wash at room temperature at maximum speed for 10 minutes. Discard the water and repeat the washing procedure using 20 ml of 2 x SSC.

5. Continue with probing, starting from the prehybridization step of the hybridization procedure or store the membrane in 2 x SSC in sealed bag at 4 °C.

Troubleshooting

There are two basic problems that occur in hybridization procedures, problems with low sensitivity of detection and problems with high background (Genius System User's Guide, 1995).

- Low Sensitivity.
  Low sensitivity of detection can result from:
  - Incomplete transfer of RNA to the membrane.
    See troubleshooting of this problem in Subprotocol 2.
  - Incorrect estimation of labeling efficiency of the probe.
    Use a new probe or correctly estimate probe labeling efficiency using the protocol for estimation of DIG-dUTP concentration (Chapter 9).
  - Incorrect temperature of hybridization and washing.
    This most frequently results from incorrect estimation of GC content of the target or probe. See equations 2, 6 and 9 in Chapter 10 for the correct calculation of hybridization and washing temperatures.
  - Using inactivated antibody.
    Repeated freezing and thawing can easily destroy antibody. Always store anti-DIG alkaline phosphatase stock solution at 4 °C. Test suspect antibody using the procedure described for testing the efficiency of probe labeling using labeled standard DNA (Chapter 9).
  - Too short exposure time.
    Increase exposure time up to 24 hours.

- High background.
  High background on hybridization blots can result from:
  - Inadequate blocking of the membrane. Increase blocking reagent concentration in the hybridization solution to 5% or use Dig Easy Hyb solutions.
  - Presence of precipitate in the antibody solution, using gloves containing powder or not completely dissolving the blocking reagent in the hybridization solution will cause spotty or uneven background.
  - Too high probe concentration. Perform hybridization with less probe and increase the time or the temperature of the wash. Wash at 68 °C for 30 to 40 minutes.
  - Uneven distribution of chemiluminescent substrate during detection or partial drying of the membrane during any of the hybridization steps. The background will not be uniform and will appear as irregular smears. It is not possible to remove this background without stripping of the probe.
  - Uneven contact of the bag with X-ray film. The background will appear as irregular smears. Make sure that the surface of the bag is smooth and does not contain wrinkles caused by incorrect sealing of the bag.
  - Dirty surface of the bag. This will result in a dark area on the film caused by electrostatic charges outside the bag. Clean the surface of the bag with 70% ethanol and remove all fingerprints.
  - Uneven distribution of the probe due to delivering the probe solution directly onto the membrane. This will result in "clouds" of background on an otherwise clear autoradiograph.
  - Inadequate washing with buffer C. Repeat washing procedure and add fresh CPD-store solution.

References

Alwine JC, Kemp DJ, Stark GR 1977 Method for detection of specific RNAs in agarose gels by transfer to diazobenzyloxymethyl-paper and hybridization with DNA probes. Proc Natl Acad Sci USA 74:5350-5354.

Bailey JM, Davidson N 1976 Methylmercury as reversible denaturating agent for agarose gel electrophoresis. Anal Biochem; 70:75

Liu Y-C, Chou Y-C 1990 Formaldehyde in Formaldehyde/agarose gel may be eliminated without affecting the electrophoretic separation of RNA molecules. BioTechniques 9:558-560.

Davis L, Kuehl M, Battey J 1994 Basic methods in molecular biology. 2nd edition. Appleton & Lange Norwalk, Connecticut.

Gong Z 1992 Improved RNA staining in formaldehyde gels. BioTechniques 12:74-75.

Lehrbach H, Diamond D, Wozney JM, Boedtker H 1979 RNA molecular weight determination by electrophoresis under denaturing conditions critical reexamination. Biochemistry 16:4743-4751.

McMaster GK, Carmicheal GC 1977 Analysis of single- and double-stranded nucleic acids on polyacrylamide and agarose gels and acridin orange. Proc Natl Acad Sci USA 74:4835-4838.

Rave N, Crkvenjakov R, Boedtker H 1979 Identification of procollagen mRNAs transferred to diazobenzyloxymethyl paper from formaldehyde gels. Nucl Acids Res 6:3559-3567.

Southern EM 1975 Detection of specific sequences among DNA fragments separated by gel electrophoresis. J Mol Biol 98:503-517.

Thomas PS 1980 Hybridization of denatured RNA and small DNA fragments transferred to nitrocellulose. Proc Natl Acad Sci USA 77:5201-5205.

DNA Cloning. General Consideration

STEFAN SURZYCKI

1 Introduction

Principle of DNA cloning

DNA cloning is a fundamental technique in recombinant DNA technology. In most general terms, it can be defined as a method of rapid isolation and amplification of DNA fragments that can be used in subsequent experiments. Cloning involves construction of hybrid DNA molecules that are able to self-replicate in a host cell, usually bacteria. This is accomplished by inserting DNA fragments into a plasmid or bacteriophage cloning vector, introducing the vector into bacterial cells and amplifying vector DNA using bacterial DNA replication machinery. The DNA fragment, or insert, can be derived from any organism and obtained from genomic DNA, cDNA, previously cloned DNA, PCR or synthesized in vitro.

The process of cloning relies on performing a set of in vitro enzymatic reactions with well-characterized DNA cleavage enzymes (restriction enzymes) and DNA modifying enzymes to join together discrete DNA molecules. Typically, cloning of any DNA fragment involves the following tasks:

- Preparing vector for cloning.

- Preparing DNA fragment to be cloned.

- Joining the fragment with the vector.

- Introducing hybrid vector into bacteria.

- Selecting for cells with vector.

To accomplish these tasks, the plasmid and DNA fragment are engineered to be linear molecules with termini compatible to be joined by ligation. The fragment and vector are ligated together to form a circular recombinant molecule. Ligated constructs are introduced into E. coli cells by transformation. Finally, transformed E. coli cells are selected from cells without vector. Different procedures to carry out each of these steps have been devised. In general, these procedures can be separated into two groups, the procedures for cloning many different fragments at once, e.g., procedures for construction of DNA libraries, and procedures for cloning one or few DNA fragments at a time.

2 Cloning Vectors

The replication machinery of bacterial cells can be used to clone and amplify specific DNA fragments. Most DNA fragments are incapable of self-replication in bacterial cells. However, any DNA fragment can be easily amplified and replicated when it is a part of an autonomously replicating bacterial element. These elements, or cloning vectors, are manmade replication units derived from extrachromosomal bacterial replicons such as plasmids, bacteriophages or viruses. Most cloning vectors can be categorized according to purpose and type of extrachromosomal element used (Brown, 1991). Vectors used in most cloning experiments are general purpose plasmid vectors that were made from naturally occurring prokaryotic plasmids, primarily *Escherichia coli* plasmids. These plasmid vectors have the following characteristics:

- Cloning vectors are small circular double-stranded DNA molecules. The vector DNA contributes as little as possible to the overall size of recombinant molecules. First, this permits cloning large fragments because there is a size limit of DNA that can be transformed using chemical transformation. Second, this assures that a cloned fragment constitutes a large percentage of amplified and isolated plasmid DNA, making it easier to prepare large quantities of insert DNA.

- Cloning vectors contain a **replicon**, that is a stretch of DNA that permits DNA replication of the plasmid independent of replication of the host chromosome. This element contains the site at which DNA replication begins or origin of replication (ori), and genes encoding RNAs or proteins that are necessary for plasmid replication. The replicon largely determines copy number of the plasmid defined as the number of plasmid molecules maintained per bacterial cell. High-copy-number plasmids are plasmids that accumulate 20 or more copies per bacteria, and low-copy-number plasmids are plasmids that have less than 20 copies per cell (Feinbaum, 1989). High copy number plasmids are the most frequently used plasmids in cloning. Low copy number plasmids are useful for cloning DNA sequences deleterious to host cells or DNA sequences that are prone to rearrangement by host, such as inverted or direct repeats.

Most recombinant plasmids contain either the ColE1 or closely related pMB1 replicon (Bernard and Helinski 1980). Examples of ColE1 replicon containing plasmids are pBluescript or pT series plasmids, and plasmids containing the pMB1 replicon are pBR322, pGEM and pUC series plasmids. For a complete list of plasmids and their replicons see Brown (1991). An antisense RNA transcript and ROP protein regulate copy number for ColE1 and pMB1 replicons. Mutations in either of these elements can result in a higher plasmid copy number. This feature was used in construction of very high copy number, general purpose plasmids such as pUC and pGEM series plasmids that lack the ROP gene.

- Cloning vectors contain selectable markers to distinguish cells transformed with the vector from nontranformed cells. This marker also maintains the presence of the plasmid in cells, especially in the case of low copy number plasmids. The selectable marker is usually a gene conferring resistance to antibiotic on the host cells. This positive selection marker, in most plasmids, is the β-lactamase gene, the product of which inactivates penicillin or its more frequently used derivative ampicillin. When transformed cells are grown in the presence of antibiotic, cells carrying the antibiotic-resistant plasmid survive, while host cells that do not contain the plasmid are eliminated.

- Cloning vectors contain unique cloning sites for the introduction of DNA fragments. The cloning sites, in most general purpose vectors used today, consist of a multiple cloning site (MCS) or a polylinker cloning region where a number of restriction enzyme cleavage sites are immediately adjacent to each other. These sites are chosen to be unique in the vector sequences. Thus, recombinant fragments can be easily introduced into the circular vector by digesting the vector with one or two enzymes present only in the polylinker region and ligating the desired fragment into it. This procedure creates a chimeric molecule without disrupting the critical features of the vector. Examples of this type of vector are pUC, pGEM and pBluescript.

  Polylinker sites are usually flanked by sequences that can be used for priming DNA synthesis with commonly available primers for DNA sequencing or PCR. The general purpose vectors such as pUC, pGEM or pBluescript contain M13 reverse, M13 forward -20, and -40 primers that can be used for amplification or sequencing any DNA fragment inserted into the polylinker.

Other sequences incorporated into flanking regions are promoter sequences for SP6, T7, or T3 RNA polymerases. These promoters can be used for in vitro or in vivo production of large quantities of RNA transcripts from DNA inserted into the polylinker. For example these sequences are present in pBluescript and pGEM plasmids but are absent from pUC plasmids.

• Cloning vectors contain an element for screening for the recombinant clones. The screening procedure, as opposed to selection procedures, permits recognition of colonies transformed with vectors containing insert from those transformed with vector alone (Rodriguez and Tait, 1983).
The most commonly used method is the α-complementation screening procedure for insertional inactivation of β-galactosidase enzyme activity. The α-complementing vectors have a MCS region inserted into a DNA sequence encoding the first 146 amino acids of the lacZ (β-galactosidase) gene α-fragment. Intact α-fragment, with a few additional amino acids, can complement ω fragment of lacZ gene in bacteria restoring a fully active enzyme. The β-galactosidase enzyme metabolizes a chromogenic substrate (X-gal) and causes the formation of blue colonies on indicator plates. Inserting a DNA fragment into the polylinker region disrupts the protein coding region of the α-fragment protein (insertional inactivation) that cannot complement into an active enzyme. Thus, cells carrying plasmid with insert have a gal- phenotype. This allows for rapid identification of bacteria containing plasmids with inserts as white colonies on indicator plates.
Recently several vectors have been developed that allow for direct positive selection of plasmids containing inserts rather then screening for them. These vectors contain a polylinker in a protein, the expression of which kills the cell. Insertional inactivation of the protein by a cloned fragment allows the cell to survive, resulting in positive selection for cells carrying recombinant plasmid. Two plasmids of this type are available. Plasmid CloneSure (CPG, Inc.) contains a region of the eukaryotic transcription factor GATA-1 and, pZErO (Invitrogen, Inc.) has the ccdB (control of cell death) gene. Both genes, if expressed, cause E. coli cell death.

Preparation plasmid for cloning

The first step in plasmid preparation is choosing the vector for cloning. The choice of vector largely depends on the experiments for which the recombinant clone will be used. For example, different types of vectors would be used for generating large quantities of DNA, expressing a fusion protein in bacteria, synthesis of mRNA or preparation of RNA or DNA probes. Once the type of vector is determined, then deciding upon a particular vector is dependent on vector characteristics such as size, copy number, polylinker restriction sites, nature of positive selectable marker and the ability to select and/or screen for inserts. The list of most commonly used cloning vectors is available at http://vectordb.atcg.com.

The second step involves preparation of the plasmid for ligation with the insert. This step is one of the most important in the DNA cloning procedure upon which the success of cloning critically depends. This is because the major difficulty in cloning with plasmids is identifying bacterial colonies carrying hybrid plasmid from those carrying plasmid without insert. Because linearization of the plasmid is an enzymatic reaction, it never goes to completion. As a result, a small number of supercoiled molecules are always present. Since supercoiled DNA molecules are transformed at high frequencies, appropriate care must be taken to remove uncut plasmid from the preparation. In addition, religation of plasmid to itself also will create plasmid molecules lacking insert. Thus, it is important to limit plasmid recircularization during the ligation reaction.

How plasmid and DNA fragments are prepared depends on the choice of ligation procedure to be used to join them. There are two types of ligation procedures: ligation of fragments with compatible cohesive ends and ligation of fragments with blunt termini. Thus, there are two procedures for cloning, cohesive-end cloning and blunt-end cloning.

Cohesive-end cloning requires the generation of ligation-compatible cohesive ends in both plasmid and DNA fragments to be cloned. These ends can be generated by cleaving both vector and fragment DNA with the same restriction enzyme that recognizes a perfect, uninterrupted, palindromic sequence or by cleaving with pairs of restriction enzymes that have different recognition sequences but form identical, or partially identical, staggered ends.

Blunt-end cloning requires creating any blunt termini in both plasmid and DNA fragment since any two blunt ends are compatible for ligation. Cleaving DNA with any two restriction enzymes that create blunt ends, or cleaving DNA with any restriction enzyme and then converting the protruding ends to blunt ends can generate these termini.

Figure 1 presents a schematic representation of various ways to join DNA fragments with T4 DNA ligase. Both types of cloning, cohesive-end cloning and blunt-end cloning have their unique advantages and disadvantages.

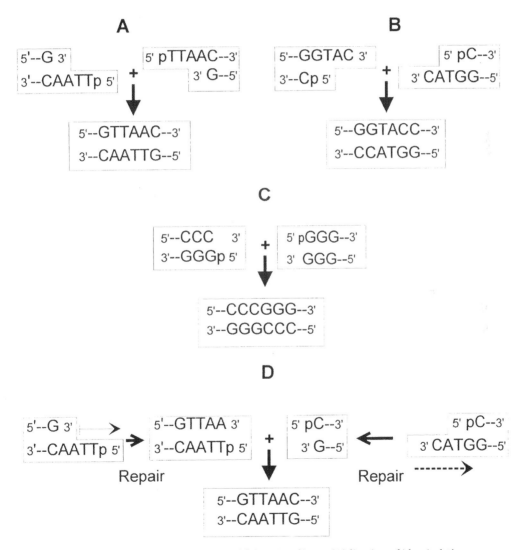

Fig. 1. Ways to join DNA fragment termini with T4 DNA ligase. (A) ligation of identical 3' recessed cohesive ends. (B) ligation of identical 5' recessed cohesive ends. (C) ligation of two non-identical blunt termini. (D) blunt-end ligation of two DNA fragments after conversion of 3' recessed end and 5' recessed end to blunt ends.

There are two advantages to **cohesive-end cloning**. First is its high efficiency due to high efficiency of cohesive-ends ligation. Rates of 10^6 to 10^7 transformants per µg of DNA are routinely obtained. The second is the possibility of directional cloning by creating asymmetric termini for insert and plasmid. This procedure, also called **forced cloning**, is especially useful for cloning a single fragment in a specific orientation. The vector DNA is cleaved with two different restriction endonucleases at two unique sites in the gene (for example, *Eco* RI and *Hind* III in the pUC cloning cassette) to produce cohesive ends (or one cohesive and one blunt-end). The vector cannot recircularize again unless a second (insert) DNA is added with complementary ends compatible with those generated in the vector. This procedure makes it possible to introduce a fragment into a vector in only one orientation. The method is frequently used for cloning in expression vectors.

There are several disadvantages to cohesive-end cloning. One is the necessity to prepare many different plasmids with many different cohesive ends to clone DNA fragments with different cohesive termini. Another is the requirement for the fragment to have specific cohesive ends that are compatible with plasmid ends. It is not always possible to find the appropriate restriction enzyme site in the DNA fragment one wants to clone that matches the available sites in the cloning vector. Finally, ligation of cohesive ends requires several hours incubation at low temperature.

The use of **blunt-end cloning** has several advantages. The most obvious advantage is that any DNA fragment can be cloned regardless of its origin. Only one type of plasmid needs to be prepared for cloning all DNA fragments with blunt termini. The ligation reaction is short (10 to 30 minutes) and is done at room temperatures. A disadvantage of the method is the necessity to prepare most of the fragments for cloning by creating blunt termini. Also, cloning efficiency is low because of the low efficiency of blunt-end ligation and because high concentrations of fragment and plasmid DNA are required for the ligation reaction. However, improvements in blunt-end cloning technology have resulted in increases in blunt-end cloning efficiency to about 10^7 to 10^9 transformants per µg of DNA making this method more attractive for most cloning applications.

During ligation, DNA ligase will catalyze the formation of phosphodiester bonds between adjacent nucleotides only if phosphate is present at the 5' end of DNA molecules. Removing the 5' phosphates from both ends of the linear plasmid DNA and leaving them on the fragment to be cloned can therefore minimize recircularization of the plasmid. The enzyme used for this purpose is either bacterial alkaline phosphatase or calf intestinal phosphatase. The result of this treatment is that neither strand of the vector can form a phosphodiester bond if insert DNA is not present in the reaction mixture. Thus, almost all transformants selected should contain recombinant plasmids.

There are, however, some definite drawbacks inherent in the use of phosphatases. First is the inefficiency of phosphatases in removing phosphate from monoester ends that are either at nicks in duplex DNA, or otherwise covered by an overhanging, complementary strand, precisely the structure created by most restriction enzymes. To aid dephosphorylation of these ends, reaction conditions can be used that promotes dephosphorylation of recessed ends. For instance, one can carry the reactions at an elevated temperature, such as 60 °C for bacterial alkaline phosphatase or use calf intestinal alkaline phosphatase (CIAP) which is somewhat more efficient in these reactions.

Second, phosphatase should be completely removed or inactivated before the ligation reaction. The presence of even residual amounts of phosphatase during ligation will remove 5' phosphates from the insert DNA, preventing the ligation reaction from occuring. Introduction of thermolabile alkaline phosphatases helps in this problem. Heat inactivation of these enzymes makes it possible to remove phosphatase from reactions without phenol extraction. These enzymes are HK thermolabile phosphatase from Epicentre Technologies (# H92025) derived from Antarctic bacterium, and shrimp alkaline phosphatase isolated from North Sea shrimp (Amersham Co. # E70092). However, HK phosphatase cannot be used for dephosphorylation of either blunt ends or protruding 3' ends (Epicentre Technologie. Inc., 1997). Shrimp alkaline phosphatase has properties very similar to CIAP enzyme but has a narrower buffer requirement than CIAP and is more expensive.

3 Preparation of Insert DNA

DNA fragments for cloning are most frequently derived from restriction enzyme digestion of DNA or from PCR. A protocol for preparation of restriction-enzyme-generated fragments for blunt-end cloning is described in Chapter 14.

Preparation of inserts by restriction enzyme digestion involves the following steps:

- Restriction digestion of the desired DNA.

- Gel electrophoresis of the digest to separate DNA fragments.

- Isolation of the DNA fragment from the gel.

- Converting protruding ends into blunt-ends if blunt-end cloning is used.

Restriction endonucleases catalyze sequence-dependent, double-stranded breaks in DNA yielding a homogeneous population of DNA fragments. These enzymes are used in a number of applications in molecular biology, including (a) establishment of an endonuclease map of DNA (b) fragmentation of genomic DNA prior to Southern blotting; (c) generation of fragments that can be subcloned in appropriate vectors; and (d) generation of fragments for labeled probes.

The most frequently used restriction endonucleases belong to type II endonucleases discovered by Smith and coworkers (Smith and Wilcox, 1970). These enzymes are small, monomeric proteins that require only Mg^{++} for activity. Type II enzymes recognize a short nucleotide sequence with dyad symmetry (the 5' to 3' nucleotide sequence of one DNA strand is identical to that of the complementary strand sequences). Most of sites consist 4, 5 or 6 bp (Brooks, 1987), but a few have recognition site of 8 bp or larger or sites smaller than 4 bp (Roberts and Macelis, 1991). In general, there are three possible cleavage positions within a recognition sequence: (a) at the center of the axis of symmetry, yielding "flush" or "blunt" ends (e.g., CCC3' | 5'GGG; (b) to the left of the center giving cohesive termini with a protruding 5' phosphate (e.g., C3' | 5'CGG); or (c) to the right of the center giving cohesive termini with protruding 3' phosphates (e.g., CTGCA3' | 5'G). Other types of cleavage are also possible. For a full description of all types of restriction endonucleases, consult Roberts and Macelis (1997).

An estimate of the number of cleavage sites for a restriction endonuclease within a given piece of DNA, assuming even distribution of bases, is described by the equation: Site Number = $N/4^n$, where N is the number of base pairs in the DNA fragment and n is the number of bases in the recognition site of restriction endonuclease. This should be treated only as an approximation of the number of expected sites (Rodriguez and Tait, 1983). Owing to the complicated nature of the restriction reaction and differences in substrates, it is difficult to define universal units of activity for these enzymes. For these reasons, a convention was adopted to define a unit of enzyme activity as the amount of enzyme required to digest completely 1 μg of bacteriophage lambda DNA in 1 hour.

The choice of restriction enzyme depends on the fragment to be cloned. There are many restriction enzyme suppliers and their recommendation of digestion conditions may differ significantly. Because the manufacturer of a given enzyme optimizes the reaction conditions for their particular enzyme, their recommendations should always be followed.

General rules for working with restriction enzymes and preparing digestion reactions are:

- Store restriction endonuclease at - 20 °C in a non-frost-free freezer at concentrations of 10 u/µl or more.

- The volume of enzyme added should not exceed 10% of the reaction volume. The final volume of 30 µl is convenient for most restriction enzyme digestion reactions.

- The amount of DNA in the reaction should not be larger than 10 µg. Volume of the DNA added should not exceed 1/3 of the reaction volume. Addition of a large volume of DNA dissolved in TE will decrease Mg^{++} ion concentration in the reaction, inhibiting restriction enzyme activity.

- Use 10 units of enzyme per 1 µg of DNA. Although this is far more enzyme than is theoretically required, such excess assures complete digestion in the presence of impurities in the DNA, decreased enzyme activity during storage, pipetting errors during the addition of enzyme, etc. Some enzymes cleave their defined sites with different efficiency, largely due to differences in flanking nucleotides. Cleavage rates for different sites recognized by a given enzyme can also differ by a factor of 10. Using excess enzyme will compensate for these differences. Moreover larger amounts (up to 20-fold more) of some enzymes are necessary to cleave supercoiled plasmid or viral DNA (Fuchs and Blakesley, 1983) as well as large genomic DNA. In addition, a few restriction endonucleases such as NarI, NaeI, SacII, and XmaIII show extreme variability in cleavage rate of different sites. These sites are difficult to cleave using standard enzyme concentrations.

- Enzyme activity is decreased when DNA is stained with ethidium bromide. This can be overcome by increasing enzyme concentration in the reaction by a factor of two. Removing ethidium bromide from stained DNA before digestion with restriction enzyme is not necessary.

- To digest DNA with two enzymes, both reactions can be carried out simultaneously providing that both enzymes work well in the same buffer. If the enzymes have different requirements, DNA should be digested first with the enzyme that requires buffer of low ionic strength. The reaction can be adjusted to higher salt concentration and digestion with the second enzyme carried out. Simultaneous digestion with more than two enzymes, in author's experience, seldom gives satisfactory results.

To prepare DNA fragments for blunt-end ligation DNA polymerases and polynucleotide kinase are used. DNA polymerases "repair" protruding ends generated by restriction enzymes. Endonuclease activity of DNA polymerases remove 3' overhang and polymerization activity of these enzymes "fill up" 5' overhangs. Since polymerization activity leaves 5' ends that are not phosphorylated, polynucleotide kinase is used to add 5' phosphate required by ligase. Commonly used DNA polymerases are Klenow fragment of DNA polymerase I of *E. coli* and T4 DNA polymerase. The polynucleotide kinase most frequently used is T4 polynucleotide kinase. DNA-end repair and kinase reactions are usually carried out simultaneously.

The recommended enzyme for repair of the ends is bacteriophage T4 DNA polymerase rather than the Klenow enzyme customarily used, because it has high exonuclease activity and is active in many different buffers. This enzyme is functionally similar to the Klenow fragment of polymerase I. It possesses 5' to 3' polymerase activity and a potent 3' to 5' exonuclease activity. Thus, it can fill 5' overhangs and digest protruding 3' termini in a single step, efficiently creating blunt termini. Degradation of the double-stranded DNA template in the presence of a high concentration of dNTPs is balanced by polymerization activity of the enzyme, preventing DNA fragment degradation.

4 Ligation Reaction

Ligation, or joining of a foreign DNA fragment to a linearized plasmid, involves formation of new bonds between 5' phosphate and 3'-hydroxyl ends of DNA. There are two enzymes available to catalyze this reaction in vitro: (a) *E. coli* DNA ligase and (b) bacteriophage T4 DNA ligase. For almost all cloning purposes, bacteriophage enzyme is used. The ligation of dephosphorylated plasmid ends with phosphorylated insert results in formation of hybrid molecules with two phosphate bonds and two single-stranded gaps. These gaps are repaired after introduction of plasmid into bacterial cells.

The most significant factor in a ligation reaction is the concentration of DNA ends. Essentially, during ligation, there are two competing reactions: bimolecular concatamerization and unimolecular cyclization. To achieve successful cloning, the first reaction to occur should be bimolecular concatamerization of vector with insert, followed by recirculization reaction to form circular plasmid. Unfortunately these two reactions are mutually exclusive, so some form of compromise must be used.

The ratio of recirculization and concatameric ligation products is dependent on two factors i and j, where i is the total concentration of DNA termini in the reaction mixture, j is the effective concentration of one end of a DNA molecule in the immediate neighborhood of the other end of the same molecule. The value of j is constant for a linear DNA molecule of a given length and is **independent of DNA concentration**. For cohesive-end ligation and blunt-end ligation, the value of i in ends/ml can be calculated from the following equation:

$$i \; = \; 2N_0M \times 10^{-3} \; ends/ml \tag{1}$$

where N_0 is Avogadro number and M the molar concentration of the DNA molecules. The j value can be determined using the following equation (Dugaiczyk et al, 1975):

$$j \; = \; j_\lambda \left(\frac{mw_\lambda}{mw_x}\right)^{3/2} \; end/ml \tag{2}$$

Where j_λ is 3.6×10^{11}, mw_λ is the molecular weight of the λ genome (30.8×10^6 daltons); and mw_x is the molecular weight of unknown molecule. After converting M to DNA concentration in $\mu g/ml$ [DNA], and mw_x to Kbp, the ratio j/i for any given DNA is equal to (rearranged from Rodriguez and Tait, 1983):

$$j/i \; \approx \; \frac{51.1}{0.812 \; (Kbp_x)^{0.5}[DNA]} \tag{3}$$

Equation 3 indicates that to achieve low j/i ratio the concentration of DNA in the reaction should be high. For any ligation reaction, three conditions apply:

- When $i = j$, there is an equal chance of one end of the molecule making contact with the end of a different molecule, and/or, the end of the same molecule.

- When $j > i$, there is a higher chance of one end of the molecule contacting the other end of the same molecule, than of finding the end of another molecule. Therefore, under this condition, the recirculization reaction predominates.

- When $j < i$, there is a greater chance of one end of the molecule contacting the other DNA molecule rather then its own end. Therefore, under this condition, the concatamerization reaction predominates.

Although the theoretical consideration discussed above predicts that conversion from concatamerization to circulization should occur at $j/i = 1$, experimental observation indicates that this conversion actually occurs at $j/i = 2$-3 (Dugaiczyk et al, 1975).

To choose the correct compromise between two types of ligation, one should realize that a ligation reaction is not an instantaneous event. This reaction proceeds through a progressive series of individual ligation events, where each event significantly alters the j/i ratio for the remaining unligated molecules. Thus, as the reaction proceeds, the value of i decreases resulting in increases of j/i because j value is constant. The high j/i ratio favors the formation of circular molecules rather than linear concatamers.

For cloning DNA insert into plasmid, the reaction conditions initially should favor joining the insert to one end of the plasmid vector. The second ligation event, occurring later, should be circularization of the plasmid-insert hybrid molecule to avoid joining the next linear fragment to the first one. To achieve this, the ratio of j/i at the beginning of ligation should be low to allow the concatamerization reaction, and as the reaction proceeds, this ratio should rise to a value greater than 3 to permit circularization of the plasmid. This can be done in three ways:

- Starting the ligation reaction at a **high concentration of vector DNA.** As equation 3 indicates, low j/i value depends on DNA concentration per ml. For the vectors commonly used for cloning (2600 bp to 4000 bp), the concentration of vector DNA is usually adjusted in the range 3-6 µg/ml or 1 to 3 pmoles/ml.

- Adjusting the initial molar ratio of plasmid to linear fragments to be greater than 1. Since during the reaction, the j value is not changed, but the i value will decrease as a result of the ligation of one end of the fragment to the plasmid, the j/i ratio will increase toward the end of the reaction. At a high j/i ratio, the circularization reaction predominates, resulting in the formation of circular hybrid molecules. The insert concentration is usually adjusted to a 2 to 4 fold **molar** excess over plasmid DNA. The j/i ratio for such a reaction can be approximately calculated using equation 3 and assuming a kbp average between vector and insert. For example, using pUC19 plasmid in the ligation reaction (2.68 Kbp) at a concentration of 6 µg/ml and 2.6 times molar excess of 1 Kbp insert (6 µg/ml) the j/i ratio at the beginning of the reaction is approximately 4 ($51.1/[2.68+1.0/2 \text{ bp}]^{0.5}$x 12 µg/ml x 0.812 = 3.8). This ratio will steadily increase as the reaction proceeds resulting in almost complete circularization of the hybrid plasmid molecules.

- Using phosphatase treated vector with phosphorylated insert. Under this condition, vector cannot self-ligate at any *j/i* ratio and therefore the initial concentration of plasmid is not important. The initial concentration of insert should be chosen carefully so as not to allow its cir-

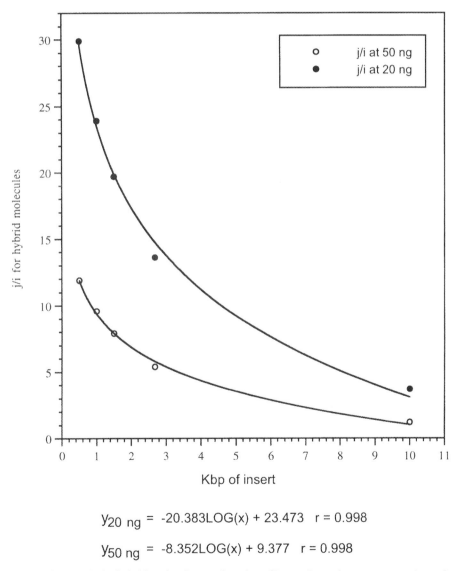

$$y_{20\ ng} = -20.383 LOG(x) + 23.473 \quad r = 0.998$$

$$y_{50\ ng} = -8.352 LOG(x) + 9.377 \quad r = 0.998$$

Fig. 2. The *j/i* ratio for hybrid molecules as a function of insert size and two concentrations of pUC vector.

culization at the beginning of the reaction. Ligation of fragment to vector results in the formation of a new molecule that can be self-ligated into a circle. Cyclization of a hybrid molecule depends on its j value and is independent of insert concentration as long as the molar concentration of insert is equal to or higher than the molar concentration of plasmid. The j/i value for this molecule can be calculated using equation 3 with kbp value equal to the sum of vector and insert size and DNA concentration calculated from the molar ratio of vector to insert. An example of such a calculation for pUC plasmid ligated with different size inserts is given in Table 1. Figure 2 presents a plot of j/i ratio of hybrid molecules as a function of insert size using pUC as the vector at two different concentrations. From this figure it is obvious that, for inserts larger than 4 kb, the initial concentration of plasmid DNA should be less than 20 ng to obtain a j/i ratio favoring circulization of hybrid molecules ($j/i > 5$).

Table 1. Ratio j/i for plasmid and insert for ligation reaction with dephosphorylated 2.68 Kbp vector

Plasmid concentration	Insert size	Insert concentration μg/ml and j/i [a]ratio at molar ratio of vector:insert			Size of hybrid	Hybrid concentration	j/i of hybrid[a]
(μg/ml)	(Kbp)	1:1	1:2	1:3	(kbp)	(μg/ml)	
2.5	0.5	0.46/197	0.9/98	1.8/35	3.18	2.96	11
2.5	1.0	0.93/69	1.8/35	2.7/23	3.68	3.43	9.6
2.5	1.5	1.39/36	2.8/18	4.17/12	4.18	3.89	7.9
2.5	2.68	2.5/15	5/7.6	7.5/5.1	5.36	5	5.4
2.5	10.0	9.3/2.1	18.6/1	27.9/0.7	12.68	11.8	1.2[b]
1.0	0.5	0.18	0.37	0.36	3.18	1.18	30
1.0	1.0	0.37	0.71	1.1	3.68	1.37	24
1.0	1.5	0.56/102	1.12/51	1.68/34	4.18	1.56	20
1.0	2.68	-	-	-	5.36	2	14
1.0	10.0	3.7/9.5	7.4/2.1	11.1/1.7	12.68	4.7	3.7[c]

[a] Calculated from equation 3

[b] No self-ligation of hybrid molecules, circles are not formed

[c] Self-ligation of hybrid molecules, circles are formed

The efficiency of a ligation reaction depends not only on DNA and enzyme concentrations, but also on the purity of the vector and insert, reaction pH and temperature, and the presence of inhibitors of ligase. In general, most deleterious components that frequently contaminate ligation reactions are:

- Monovalent salts (for example, NaCl) or ammonium salts. Concentrations of NaCl above 50 mM and ammonium salt above 10 mM severely inhibit ligation reaction.

- Phosphate. Blunt-end ligation is particularly sensitive to this salt and concentrations of PO_4 greater than 25 mM should be avoided.

- ATP concentration. Blunt-end ligation is inhibited by ATP concentrations above 0.1 mM.

5 Transformation of Bacteria

Three methods for transforming *E. coli* bacteria can be used in cloning procedures. These are: electroporation, chemical transformation and transformation by microwave radiation.

Electroporation is the most efficient method of transformation, with transformation efficiencies as high as 10^{10} transformants/µg plasmid DNA. However, this method requires use of an expensive instrument. Induction of competence for electroporetic insertion of plasmid is not necessary and nearly all bacteria strains can be used (Wai and Chow, 1995).

Chemical transformation is much less efficient than electroporation (10^6 to 10^7 transformants/µg of plasmid) but does not require expensive equipment. Using this method it is only necessary to induce competence for DNA uptake in *E. coli* cells because this bacterium does not possess a natural mechanism for transformation. However, a number of *E. coli* bacterial strains have been developed recently, for which chemical transformation efficiency is very high and can reach 10^9 transformants/µg of DNA.

Transformation using microwave radiation is simple and can be used with all bacterial strains, but transformation efficiency is low, on the order of 10^5 transformants/µg of plasmid (Abbas et al. 1995).

Transformation by electroporation

Electroporation is the most efficient transformation procedure that can be used for bacterial transformation. It involves a brief application of high-voltage electric field to the cells resulting in the formation of transient holes in the cell membrane through which plasmid DNA can enter the cell. The method was originally developed for animal and plant cells (Neumann et al. 1982; Fromm et al. 1985) and later for bacteria (Böttger 1988; Dower et al. 1988; Li et al. 1988; Smith et al. 1990). Transformation efficiencies as high as 10^{10} transformants/µg of plasmid have been achieved for E. coli cells (Dower et al. 1988; Smith et al. 1990).

Maximum efficiency of this process depends on many variables. The most important are: electrical pulse shape, electrical field strength and electrical pulse time. The relationship between these parameters is described by the following equations (Shigekawa and Dower, 1988):

$$V_t = V_0 \left[e^{-(t/\tau)} \right] \tag{4}$$

$$E = \frac{V_0}{d} \tag{5}$$

$$\tau = R \times C \tag{6}$$

Where V_t is voltage at time t, V_0 is initial voltage or peak of voltage, τ is pulse time constant, R is resistance of the circuit (ohms), C is capacitance of the charged capacitor (in Farads), E is electric field strength and, d distance between the electrodes.

In most instruments used today charging a capacitor to a predetermined voltage and its subsequent discharge to electroporation chamber generates electric pulse. Capacitor discharge results in an exponentially decaying pulse shape that is characterized by value of τ. According to equation 4, τ can be defined as the time over which the voltage declines to l/e or $\sim 37\%$ of the initial value (V_0). The electrical field is derived from the peak voltage (V_0) delivered to the chamber at the moment of discharge. At this moment the difference in potential in the chamber is at its maximum, generating membrane depolarization and pore formation. The formation of pores is largely dependent on the field strength parameter determined by equation 5. The introduction of external compounds (e.g., plasmids, DNA, proteins etc.) into the cell critically depends on the voltage drop determined by τ.

The time constant τ is specific for each cell type and, in general, the smaller the cell size, the shorter the τ needed to introduce external elements. The optimal time constant for E. coli was determined to be 5 milliseconds. The field strength for optimal electroporation of different bacterial strains differs and must be experimentally determined.

Transformation results are usually described by the values representing frequency of transformation (f_{tr}) and efficiency of transformation (E_{tr}). Equation 7 describes frequency of transformation and efficiency of transformation is described by equation 8.

$$f_{tr} = \frac{T_{tr}}{S} \qquad (7)$$

$$E_{tr} = \frac{T_{tR}}{[pDNA]} \qquad (8)$$

Where T_{tr} is the number of transformants, S is the number of cells at 80% survival, and [pDNA], the concentration of plasmid DNA used in μg. Frequency of transformation is directly dependent on DNA concentration in the range of 10 pg/ml to 7.5 μg/ml, and at these DNA concentrations, 80% of the cells survive electroporation (Dover et al. 1988). Moreover, because from equation 7, $T_{tr} = f_{tr} \times S$, the efficiency of transformation can be rewritten as:

$$E_{tr} = \frac{f_{tr} \times S}{[pDNA]} \qquad (9)$$

This equation indicates that transformation efficiency (E_{tr}) is high when cell concentration is high (10^9 to 3×10^{10}) and DNA concentration is low (1 – 10 ng/rx). These conditions are commonly used in electroporation of E. coli to achieve high transformation efficiencies and to avoid cotransformation. For example, for a standard electroporation reaction, 30 μl of cells at a concentration of 10^{10} cells/ml are used with 5 ng of plasmid DNA. Transformation efficiency commonly achieved is 10^9 transformants/μg of DNA, giving a transformation frequency of two transformants per 100 surviving cells ($f_{tr.} = E_{tr} \times [pDNA]/ S; f_{tr} = 10^9 \times 0.005 / 2.4\,10^8 = 0.02$). At this transformation frequency, transformation of a single cell with two or more plasmids is very rare.

In addition to the parameters described above, electroporation is affected by a number of other variables such as: temperature, components of electroporation medium and method of cell recovery after electroporation.

Electroporation, at low temperature, is roughly 100 times more efficient than when cells are pulsed at room temperature. Low temperature presumably affects fluidity of the membrane, aiding in pore formation and slowing their closure. Similarly, quick restoration of membrane fluidity and closing pores is crucial for cell survival after the pulse. Thus, cells should be transferred to prewarmed growth medium as fast as possible after pulse application. Delaying this transfer by one minute will cause a threefold drop in transformation efficiency and cell survivability. Moreover, cells should be given some time to recover and rebuild their membrane before plating on solid medium. Recovery time should be short enough to prevent cell division that will result in "cloning" of transformed cells. This is particularly important when electroporation is used for preparation of genomic and sequencing libraries. Recovery time of 45 to 50 minutes at 37 °C appears to be short enough not to allow any cell division.

Chemical transformation

E. coli cells that are used for most cloning applications do not possess a natural mechanism for DNA uptake. This ability or competence has to be induced by chemical means and the method to transform bacterial cells so induced is called **chemical transformation**. Chemical transformation techniques are based on the observation made by Mandel and Higa (1970) that the uptake of λ DNA by bacterial cells can be increased substantially in the presence of calcium chloride. Soon, this method was successfully applied for the introduction of plasmid DNA into bacterial cells (Cohen et al. 1972).

The original procedure involved exposure of growing bacteria to a hypotonic solution of calcium chloride and a brief heat shock (42°C). This method yields 10^4 to 10^6 transformants/µg of plasmid that corresponds to only about one per 10 000 plasmid molecules actually entering the cell.

Many modifications of the original technique to improve transformation efficiencies have been since described. These include prolonged exposure of cells to $CaCl_2$ at high cell density (Dagert and Ehrlich, 1974), substitution of calcium with other cations such as Rb^+ (Kushner, 1978), Mn^{2+}, and K^+ (Inoue et al. 1990), addition of other compounds such as dimethyl sulfoxide, dithiothreitol, cobalt hexamine chloride (Hanahan, 1983) or PEG (Chung et al. 1989).

All of these modifications increased transformation efficiency to about 10^8 transformants/µg of DNA but usually only 3% to 10% of the cells are competent to incorporate plasmid DNA. To improve the degree of competency, many "ultracompetent" *E. coli* strains have been developed by different companies. Transformation frequencies with these cells are about 10^9 transformants/µg but their use requires proprietary solutions and purchasing competent cells from the supplier.

Most of the transformation protocols used today include the following steps:

- **Preincubation step**. In this step the cells are suspended in a solution of cations and incubated at 4 °C. During this time bacterial spheroplasts are formed and cations neutralize the negatively charged phosphates of lipids in the membrane. The low temperature limits fluidity of the cell membrane, stabilizing the distribution of charged phosphates allowing more effective shielding by the cations.

- **Incubation with DNA**. In this step DNA is added to the cells. Cations complexing with negatively charged oxygen shield the negatively charged phosphates on DNA, effectively neutralizing DNA charges. This permits DNA to enter the cells through adhesion zones.

- **Heat shock**. In this step cells are briefly incubated at 42 °C and returned to low temperature. The rapid temperature change creates a thermal imbalance on both sides of the membrane. This creates a flow that accelerates entrance of plasmids into the cell. The use of microwave radiation to achieve quick uptake of plasmid into the cells has been described (Abbas et al. 1995). Some chemical transformation protocols do not use heat shock step (Chung et al. 1989).

- **Recovery step**. Cells are incubated at 37 °C before plating on selective media. During this time cells recover from the treatment, and express the antibiotic-resistance proteins.

In spite of the availability of many protocols and the modification of old protocols, no chemical method used today gives consistent and reproducible results. This is true not only for transformation using different bacterial strains but also when the same strain is used in everyday experiments. Day to day transformation efficiency, using the same strain and identical procedure can vary by as much as factor of 10. Moreover, procedures that work efficiently in some laboratories do not work as efficiently in others (Liu and Rashidbaigi, 1990: Hengen, 1996). The number of modifications and new procedures that continue to be introduced attest to the imperfection of all protocols presently in use.

The method that gives the most consistent results today is a method introduced by Inoue et al. (1990). In this procedure cells are grown at low temperatures and treated with DMSO to induce competency. This frequently can give a transformation efficiency as high as 10^8 transformants per µg of plasmid (Hengen, 1996). However, the method is time consuming and the tradeoff in time for higher efficiency maybe not be worth it, especially for routine cloning. The simplest procedure described today is a single-step procedure described by Chung et al. (1989). This method gives transformation efficiency as high as 10^7 transformants/µg of plasmid (Liu and Rashidbaig, 1990).

In conclusion, it appears that state of the art of chemical transformation is not yet entirely a "closed chapter" and continues to be developed. All procedures described here deliver sufficient transformation efficiency for routine cloning purposes, but are not reliable or efficient enough for preparation of libraries when efficiency and reliability are of the utmost importance. For these applications electroporation still remains the method of choice.

References

Abbas A, Alilat M, Therwath A 1995 Bacterial Transformation using microwave radiation. Method in Molecular and Cellular Biology 5:128-130.

Bercovich JA, Grinstein S, Zorzopulos J 1992 Effect of DNA concentration on recombinant plasmid recovery after blunt-end ligation. BioTechniques 12:190-193.

Bernard HU, Helinski DR 1980 Bacterial plasmid cloning vectors. In: genetic Engineering. Principles and methods. Eds. Setlow JK, Hollaender A Vol. 2 pp. 133-167. Plenum Press. New York and London.

Böttger EC 1988 High-efficiency generation of plasmid cDNA libraries using electro-transformation. BioTechniques 6:878-880.

Brown TA 1991 Cloning vectors. In: Molecular biology Labfax. pp. 193-234. Eds Hames BD and Rickwood D. Bios Scientific Publishers Ltd. and Academic Press Inc. Oxford UK.

Brooks JE 1987 Properties and uses of restriction endonucleases. Methods Enzymol 152:113-129.

Chung CT, Niemel SL, Miller RH 1989 One-step preparation of competent *Escherichia coli*: Transformation and storage of bacterial cells in the same solution. Proc Natl Acad Sci USA 86:2172-2175.

Cohen SN, Chang ACY, Hsu L 1972 Nonchromosomal antibiotic resistance in bacteria: Genetic transformation of *E. coli* by R-factor DNA. Proc Natl Acad Sci USA 69:21110-2114.

Dagert L, Ehrlich CD 1979 Prolonged incubation in calcium chloride improves the competence of *Escherichia coli* cells. Gene 6:23-28.

Dower WJ, Miller JF, Ragsdale CW 1988 High efficiency transformation of *E. coli* by high voltage electroporation. Nucl Acids Res 16:6127-614

Dugaiczyk A, Boyer HW, Goodman HM 1975 Ligation of *Eco*RI endonuclease-generated DNA fragments into linear and circular structures. J Mol Biol 96:171-184

Epicentre Technologies 1997 HK Thermolabile Phosphatase product information.

Feinbaum R 1989 Introduction to plasmid biology. In: Current protocols of molecular biology. John Wiley and Sons, Inc.

Fromm M, Taylor LP, Walbot V 1985 Expression of genes transferred into monocot and dicot plant cells by electroporation. Proc Natl Acad Sci USA 82:5824-5828.

Fuchs R, Blakesley R 1983 Guide to the use of type II restriction endonucleases. Methods Enzymol 100:3-38.

Hanahan D 1983 Studies on transformation of *Escherichia coli* with plasmids. J Mol Biol 166:557-580.

Hengen PN 1996 Method and reagents. Preparation of ultra-competent *Escherichia coli*. Trends in Biochemical Sci. 21:75-76.

Inoue H, Nojima H, Okayama H 1990 High efficiency transformation of *Escherichia coli* with plasmid. Gene 96:23-28.

Kranz R 1989 Convenient and inexpensive system for low temperature preservation of enzymes during storage and use. BioTechniques 7:455-456.

Kushner SR 1978 An improved method for transformation of *Escherichia coli* with ColE1-derived plasmids. In: Genetic Engineering. Eds Boyer HW and Nicoria S pp. 17-23. Elsevier/North Holland. Amsterdam.

Li SJ, Landers TA, Smith MD 1988 Electroporation of plasmids into plasmid-containing *Escherichia coli*. BioTechniques 12:72-74.

Life Technologies Inc 1996 Tech-online. Rapid Ligation protocol for plasmid cloning. www. lifetech.com/cgi-online/techonline?Document=rapidt.txt.

Liu H, Rashidbaig A 1990 Comparison of various competent cell preparation methods for high efficiency DNA transformation. BioTechniques 8:21-25.

Mandel M, Higa A.1970 Calcium dependent bacteriophage DNA infection. J Mol Biol 53:154-330.

Murphy NR, Hellwig RJ 1990 Improved nucleic acid organic extraction through use of a ubique gel barrier material. BioTechniques 21:934-939.

Neumannn E, Schaefer,-Ridder M, Wang Y, Hofschneider PH 1982 Gene transfer into mouse L-cells by electroporation in high electric field. EMBO J 1:841-845.

New England BioLabs. www.neb.com

Roberts RJ, Macelis D 1991 Restriction enzymes and their isoschizomeres. Nucleic Acids Res (suppl) 19:2077-2109.

Roberts RJ, Macelis D 1997 REBASE-restriction enzymes and methylases. Nucleic Acids Res 25:248-262.

Robinson CR, Sligar SG 1995 Heterogeneity in molecular recognition by restriction endonucleases: Osmotic and hydrostatic pressure effects a BamHI, PvuII, and EcoRV specificity. Proc Natl Acad Sci USA 92:344-3448.

Rodriguez RL, Tait RC 1983 Recombinant DNA techniques: An introduction. Addison-Wesley Publishing Co. Reading Mass ISBN 02-201-10870-4.

Shigekawa K, Dower WJ 1988 Electroporation of eukaryotes and prokaryotes: A general approach to the introduction of macromolecules into cells. BioTechniques 6:742-751.

Smith HO, Wilcox K 1970 Restriction enzymes from Hemophilius influenzae. Purification and general properties. J Mol Biol 51:379-391.

Smith M, Jesse J, Landers TA, Jordan J 1990 High efficiency bacterial electroporation: 1 x 10^{10} *E. coli* transformants/µg. Focus 12:38-40.

Wai LT, Chow K-C 1995 A modified medium for efficient electrotransformation of *E. coli*. Trends in Genetics 11:128-129.

DNA Cloning – Experimental Procedures

STEFAN SURZYCKI

Introduction

This chapter describes procedures used in blunt-end and cohesive-end cloning of a single DNA fragment. Chapter 13 describes theoretical considerations of cloning procedures. The required tasks for cloning are divided into separate procedures that can be performed at different times. The procedures described are:

- Preparation of the plasmid.

- Preparation of DNA insert.

- Ligation of plasmid with insert.

- Transformation of bacterial cells and selection of transformants.

Subprotocol 1
Preparation of Plasmid

This protocol describes linearization of the plasmid with restriction enzyme and the removal of 5' phosphate groups with alkaline phosphatase. The plasmid used is a pUC series plasmid, pUC19, but any plasmid with appropriate restriction sites can be used. To create blunt-end termini the restriction endonuclease *Sma* I is used, the recognition sequence for which is located in the middle of the multiple cloning site (MCS) of pUC plasmids. Blunt-end ligation destroys the recognition site for the enzyme, but a fragment cloned into the middle of the MCS can be excised from the plasmid by double-digestion with any pair of restriction endonucleases, with recognition sites in the Multiple Cloning Site. Two other restriction endonucleases, *Hinc* II and *Hind* II, create blunt termini, and can also be used in preparation of plasmids for blunt-end cloning. For cohesive end cloning, use a restriction endonuclease that creates cohesive ends.

Termini of the plasmid are dephosphorylated by calf intestinal alkaline phosphatase (CIAP). CIAP enzyme is used since it is much easier to inactivate than bacterial alkaline phosphatase. Moreover, alkaline phosphatase (CIAP) can be added directly to a restriction enzyme reaction mixture because it is active in most restriction enzyme buffers. A schematic outline of plasmid preparation procedures is shown in Figure 1.

Fig. 1. Schematic outline of the preparation of plasmid procedure

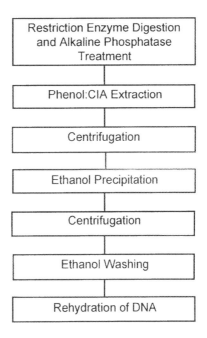

Materials

-   Restriction endonuclease as needed for vector digestion (for example, *Sma* I)
    Restriction endonucleases are supplied with specific reaction buffers.
-   Alkaline phosphatase, calf intestinal (CIAP) (New England BioLabs, Inc. # 290S or equivalent).

Note: Enzymes should always be stored at -20 °C in non-frost-free freezer. All frost-free freezers go through freeze-thaw cycles subjecting enzymes to repeated warming resulting in loss of enzymatic activity.

- Plasmid DNA (for example, pUC, pBluescript, pGM).
  The comprehensive list of most plasmids, their restriction enzyme maps and full DNA sequences can be accessed on http://vectordb.atcg.com/. Plasmids should be purified by any of the procedures described in Chapter 5 or bought ready to use. (for example, pUC plasmid from Pharmacia; pGem plasmid from Clontech Laboratories, Inc.; pBluescript plasmid from Stratagene, Inc.).
- Phenol:CIA mixture (PCI) (Ambion # 9732 # I-737642).
- Phenol water saturated (Ambion # 9712 or equivalent).
- Phase Lock Gel (PLG I) microfuge tubes (Eppendorf 5' # 0032007953). PLG tubes contain a proprietary compound that, when centrifuged, migrates to form a tight barrier between organic and aqueous phases (Murphy and Hellwig, 1996). The interphase material is trapped in and below this barrier allowing the complete and easy collection of the entire aqueous phase free from contamination with organic solvents. The PLG barrier also offers increased protection from exposure to organic solvents used in deproteinization procedures. The tubes are not expensive and are strongly recommend when phenol extraction is carried out in microfuge tubes.

Solutions
- **Chloroform:Isoamyl Alcohol Solution (CIA)**
  Mix 24 volumes of chloroform with 1 volume of isoamyl alcohol. Because the chloroform is light sensitive and very volatile, CIA solution should be stored in a brown glass bottle, preferably in a fume hood.
  Safety Note: Handle chloroform with care. Mixing chloroform with other solvents can involve a serious hazard. Adding chloroform to a solution containing strong base or chlorinated hydrocarbons could result in an explosion. Isoamyl alcohol vapors are poisonous. Handle the CIA solution in a fume hood. Used CIA can be collected in the same bottle as phenol as hazardous waste and discarded together.
- **Phenol:CIA Mixture (PCI)**
  This mixture can be bought directly or made as follows. Equilibrate water saturated phenol with equal volume of 0.1 M of sodium borate. Sodium borate should be used rather than the customary 1 M Tris solution because of its superior buffering capacity at pH 8.5, its low cost, its antioxidant properties and its ability to remove phenol oxidation products during the equilibration procedure. Mix an equal volume of a water-saturated phenol with 0.1 M sodium borate in a separatory funnel. Shake well until the solution turns milky. Wait until the phases separate and collect the bottom, phenol phase. Add the sodium salt of 8-hydroxyquinoline final concentration of 0.1% (w/v). **Do not use**

the hemisulfate salt of 8-hydroxyquinoline. Mix equal volumes of equilibrated phenol and CIA. Phenol:CIA mixture can be stored in a dark bottle at 4 °C for several weeks. For long term storage keep at -70 °C. **Safety Note:** Because of the relatively low vapor pressure of phenol, occupational systemic poisoning usually results from skin contact with phenol rather than from inhaling it. Phenol is rapidly absorbed by the skin and is highly corrosive. It initially produces a white softened area, followed by severe burns. Because of the local anesthetic properties of the phenol, skin burns may not be felt until there has been serious damage. Gloves should be worn at all times when working with this chemical. Because some brands of gloves are permeable to phenol, they should be tested before use. If phenol is spilled on the skin, flush it off immediately with a large amount of water and treat the affected area with a 70% aqueous solution of PEG 300 (Hodge, 1994). **Do not use ethanol.** Used phenol should be collected into tightly closed, glass receptacles and stored in a chemical hood until disposal as hazardous waste.
- **0.5 M EDTA Stock Solution**
  Weigh out an appropriate amount of EDTA (disodium ethylendiaminetetraacetic acid, dihydrate) and stir into distilled water. EDTA will not go into solution completely until the pH is greater than 7.0. Adjust the pH to 8.5 using concentrated NaOH solution or by adding a small amount of pellets. Bring the solution to a final volume with water and sterilize by filtration. EDTA stock solution can be stored indefinitely at room temperature.
- **TE Buffer**
  - 10 mM Tris HCl, pH 7.5 or 8.0
  - 1 mM Na_2EDTA
  Sterilize by autoclaving and store at 4 °C.
- **7.5 M Ammonium Acetate Solution**
  Dissolve 57.8 g of ammonium acetate in 70 ml of double-distilled or deionized water. Stir until salt is fully dissolved, do not heat to facilitate dissolving. Fill up to 100 ml and sterilize by filtration. Store tightly closed at 4 °C. Solution can be stored for one to two months at this condition. Long term storage is possible at -70 °C.
- **70% Ethanol**
  Add 25 ml of double distilled or deionized water to 70 ml of 95% ethanol. Never use 100% ethanol because it contains an additive that can inhibit activities of some enzymes. Store in -20 °C freezer.

- **Loading Dye Solution (Stop Solution)**
  - 15% Ficoll 400
  - 5 M Urea
  - 0.1 M Sodium EDTA, pH 8
  - 0.01% Bromophenol blue
  - 0.01% Xylene cyanol

Prepare at least 10 ml of the solution. Dissolve the appropriate amount of Ficoll powder in water by stirring at 40 - 50 °C. Add stock solution of EDTA, powdered urea and dyes. Aliquot about 100 µl into microfuge tubes and store in a -20 °C freezer.

Procedure

This procedure describes plasmid preparation for blunt-end ligation. Alkaline phosphatase (CIAP) is added to the restriction enzyme reaction mixture together with restriction endonuclease. However, digested plasmid DNA can be first, phenol-extracted, ethanol-precipitated and resuspended in CIAP incubation buffer if desired. Both methods work equally well, but the first one is simpler.

Linerization of plasmid

1. Always observe the following rules when preparing the reaction mixture:
   - Thaw all reagents at room temperature and place them on ice.
   - Calculate the amounts of all reagents needed. Do not include water in these calculations.
   - Calculate the amount of water needed to obtain the desired reaction volume. **Amount of water is always calculated last.**
   - Add water to the reaction tube. **Water is always added first.**
   - Add remaining ingredients in the following order: buffer, cofactors, and substrate.
   - Start the reaction by the addition of enzyme. Do not keep enzymes on ice while assembling the reaction. Take the enzyme from the freezer and place it into a portable - 20 °C laboratory bench "freezer". This can be bought from Stratagene Inc. (StrataCooler # 400015) or made from "liquid ice" packages usually supplied with enzyme shipments (Krantz, 1989). Prepare "portable freezer" as follows: insert a microfuge tube into the package of liquid ice and place it in a -20 °C freezer until it freezes; remove the empty tube from the package and insert

the tube with enzyme. This "portable freezer" will keep enzymes at -20 ˚C on the lab bench for several hours.

2. Total reaction volume is 30 μl. The concentrations of the restriction enzyme and DNA will depend on the concentration of stock solutions. The amount of DNA in a reaction should be between 1 to 5 μg. The amount of the restriction enzyme to be used in a reaction is typically between 10 to 20 units. The amount of alkaline phosphatase depends on the amount of pmoles of DNA ends present in the reaction. Calculate the amount of enzyme, DNA and water to be added using the following chart:

Ingredients	Amount (μl)	Final concentration
10 x Restriction enzyme buffer	3	1x
Plasmid DNA	x	5 μg
Sma I enzyme (10 u/μl)	2	20 units
Alkaline phosphatase (10 u/μl)	1	10 units
Water	to 30 μl	

Note: One unit of CIAP can dephosphorylate 1 pmol of DNA ends. To calculate pmole of DNA ends, use the following equation:

$$\text{pmol DNA ends} = \frac{\mu g \text{ DNA x 2}}{0.662 \text{ x Kbp}} \tag{1}$$

Where Kbp is the size of DNA expressed in kilobase pairs. For example, because the size of pUC18 vector is 2.686 Kbp, 5 μg of this DNA is equal to 5 x 2/0.662 x 2.686 = 5.6 pmoles ends. Therefore, 5 μg of linearized pUC19 plasmid will require a minimum of 6 units of alkaline phosphatase to dephosphorylate its ends. The CIAP enzyme can have slightly lower activity in any reaction buffer containing less than 50 mM salt concentration. Increase the concentration of enzyme above the calculated amounts when using such buffers.

Note: To prepare dephosphorylated plasmid for cohesive end ligation, digest vector DNA with the appropriate restriction endonuclease using the reaction described above and the purification procedure described below.

3. Place a 1.5 ml microfuge tube on ice and add the following: the calculated amount of water, buffer and DNA.

4. Start reactions with the addition of the enzyme. Add 10 units of restriction enzyme to the reaction and mix gently by pipetting up and down several times.

5. Add the appropriate amount of alkaline phosphatase and mix by pipetting up and down.

6. Centrifuge tubes for 10 seconds to remove air bubbles and collect liquid at the bottom of the tube.

7. Place the tube into a 25 °C water bath and incubate the reactions for 1 to 2 hours.

Note: If other than *Sma* I restriction endonuclease is used, incubate the reaction at the temperature required for maximum activity of the enzyme. Do not incubate the reaction at a temperature higher than 37 °C when alkaline phosphatase is present.

8. After 1 to 2 hours of incubation, transfer the reaction to a 37 °C water bath and continue incubating for another hour.

Note: This step is necessary only when using *Sma* I restriction endonuclease for which the optimal reaction temperature is 25 °C. Incubation at 37 °C is required for optimal activity of alkaline phosphatase. Omit this step if the first incubation was carried at 37 °C.

9. Centrifuge for 5 to 10 seconds to collect liquid at the bottom of the tube.

10. Stop the reaction by the addition of 5 µl 0.5 M EDTA and 70 µl TE. Mix well by pipetting up and down.

11. Centrifuge the unopened PLG I tube in a microfuge for 30 seconds at 10 000 rpm to pellet the gel. Orient the tube in the centrifuge rotor with the lid connector pointing away from the center of rotation.

Note: Measure the time of centrifugation from the moment of **starting the microfuge**.

12. Transfer l00 µl of the reaction mixture into the PLG I tube.

13. Add 100 µl of phenol:CIA solution. Mix by inverting several times to form an emulsion.

14. Centrifuge the PLG I tube at 10 000 rpm for exactly 30 seconds. Be sure to orient the tube in the centrifuge rotor the same way as in step 11.

Note: After centrifugation, the organic phase at the bottom of the tube will be separated from the aqueous phase at the top of the tube by the PLG barrier.

15. Add 100 µl of CIA solution. Mix by repeated inversion to form an emulsion. Do not vortex or allow the bottom organic phase to mix with the upper aqueous phase.

16. Centrifuge for 30 seconds to separate phases. Be very careful to place tube into the centrifuge in the same orientation as before.

17. Transfer the aqueous phase, collected from above the PLG barrier, to a fresh tube. Record the volume of the aqueous phase. The volume should be similar to the amount of sample added in step 10.

18. Add 50 µl of 7.5 M ammonium acetate (half the volume). Mix well by inverting the tube 4 to 5 times.

19. Add 300 µl of 95% ethanol (2 x the total volume). Mix well by inverting the tube 4 to 5 times.

20. Place the tube in the centrifuge and orient the attached end of the lid pointing away from the center of rotation. Centrifuge the tube at maximum speed for 10 minutes at room temperature. (See Figure 2, Chapter 2.)

21. Remove the tube from the centrifuge and open the lid. Holding the tube by the lid, touch the tube edge to the edge of an Erlenmeyer flask and gently lift the end to drain ethanol. You do not need to remove all the ethanol from the tube. Place the tubes back into the centrifuge, orienting the tube as before.

Note: When pouring off ethanol do not invert the tube more than once because this can loosen the pellet.

22. Wash the pellet with 700 µl of cold 70% ethanol. Holding a P1000 Pipetman vertically slowly deliver the ethanol to the side of the tube opposite the pellet, that is, the side facing the center of the rotor. Hold the Pipetman as shown in Figure 3, Chapter 2. **Do not start the centrifuge,** in this step the centrifuge rotor is used as a "tube holder" that keeps the tubes at an angle, convenient for ethanol washing. Remove the tube from the centrifuge by holding it by the lid. Pour off the ethanol as described in step 21.

23. Place the tube back into the centrifuge and repeat the 70% ethanol wash one more time.

Note: This procedure makes it possible to quickly wash a large number of pellets without centrifugation and vortexing. Vortexing and centrifuging are time consuming and frequently lead to substantial loss of the material.

24. After the last wash, place the tube into the centrifuge, making sure that orientation is the same as before. Without closing the tube lids, start the centrifuge for 2-3 seconds and collect the remaining ethanol at the bottom of the tube. Remove all ethanol with a P200 Pipetman outfitted with a capillary tip.

Note: Never **dry the DNA pellet** in a vacuum. This will make dissolving the DNA pellet very difficult if not impossible.

25. Resuspend the pelleted DNA in 20 to 30 μl of TE. Store at -20 °C.

26. Determine the concentration of DNA by measuring absorbance at 260 nm. Initially use a 1:100 dilution of the DNA in PBS. The absorbance reading should be in the range of 0.1 to 1.5 OD_{260}. A solution of 50 μg/ml has an absorbance of 1 in a 1-cm path cuvette. Determine the purity of the DNA by measuring the absorbance at 280 nm and 234 nm and calculating 260/280 and 260/234 ratios. See Chapter 1 for a detailed discussion of the validity of this method and for calculation of DNA concentration in the presence of a small amount of protein contaminant.

Note: DNA concentration should never be measured in water or TE buffer. For a discussion about determination of DNA concentration and its dependence on ionic strength and pH see Chapter 1.

27. Analyze the prepared plasmid DNA by agarose gel electrophoresis. Use 0.8% agarose and prepare the gel as described in Chapter 8. In a microfuge tube, mix 1 μl of sample with 10 μl of TE. Add 5 μl loading dye solution and load the whole sample onto a gel. Use the appropriate molecular-weight markers (for example, 1-kb ladder From Life Technologies, Inc.).

Troubleshooting

It is very important to check the efficiency of a plasmid preparation before using it in cloning experiments. Incomplete digestion of the plasmid with restriction enzymes will result in the presence of supercoiled plasmid in the sample. Its presence will severely limit the cloning efficiency because it will decrease the amount of transformants with insert. This is because transformation with supercoiled plasmid is 10 to 100 times more efficient than transformation with a ligation product (circular plasmid) when **chemical transformation procedures** are used. However, when **transformation by**

electroporation is used, the efficiency of transformation depends less on the purity of the plasmid preparation because both circular and supercoiled plasmids can enter bacterial cells with equal probability.

To determine that plasmid was fully digested by restriction enzyme, run agarose gel electrophoresis with at least 1 µg of restricted plasmid. The presence of a small amount of supercoiled (not digested) plasmid will be visible in gel electrophoresis as fast moving supercoiled DNA band. Preparation containing visible supercoiled plasmid cannot be used for chemical transformation experiments. The preparation should be digested one more time with a higher concentration of restriction enzyme or for a longer time.

Removing supercoiled form of plasmid by agarose gel electrophoresis is not recommended because it is time consuming and seldom results in sufficient removal of the supercoiled form. This is because the presence of ethidium bromide in the gel changes the mobility of supercoiled DNA during electrophoresis. When supercoiled DNA enters the gel, progressively more dye is intercalated into DNA molecules, changing the number and direction of superhelical twists. As the amount of intercalated ethidium bromide increases, the negative superhelical turns of supercoiled plasmid are progressively removed and the rate of plasmid migration is slowed. Further intercalation of ethidium bromide into these molecules introduces positive superhelical turns, resulting in an increase in their electrophoretic mobility. Thus, small amounts of supercoiled plasmid will be present throughout the gel. Consequently, for removal of supercoiled contaminates, agarose gel electrophoresis should be run without ethidium bromide present in the gel and the gel must be stained for DNA visualization after electrophoresis. Contamination of prepared plasmid with supercoiled molecules up to 1–2 % does not affect efficiency of transformation by electroporation and therefore complete removal of the supercoiled form of the plasmid is not necessary.

Subprotocol 2
Preparation of DNA Fragments

This procedure describes preparation of DNA fragments for cloning derived from restriction enzyme digestion of the appropriate DNA. In particular, preparation of restriction enzyme-generated fragments for blunt-end cloning is presented. The preparation involves: (a) restriction digestion of desired DNA, (b) gel electrophoresis of the digest to separate DNA fragments, (c) isolation of the fragment from the gel and, (d) converting protruding ends into blunt-ends. Chapter 13 describes theoretical principles of preparation of clonable blunt-end inserts.

T4 DNA polymerase repairs protruding ends generated by restriction endonucleases. This enzyme is used for "end-repairing" instead of Klenow enzyme because: (a) the enzyme has high exonuclease activity that assures rapid removal of 3' protruding ends; (b) the enzyme is active in many different buffers so restriction to blunt-ends and phosphorylation of 5' ends can be done simultaneously with the end-repairing reaction; (c) the enzyme is inexpensive and much more stable than Klenow enzyme.

DNA polymerase repaired ends are devoid of 5' phosphates that are needed to clone the fragment into a dephosphorylated vector. To ensure that all of the 5' ends are phosphorylated, fragments are treated with bacteriophage T4 polynucleotide kinase. A schematic outline of the DNA fragment preparation procedure is shown in Figure 2.

Materials

– Restriction endonuclease as needed for fragment preparation.
  Restriction endonucleases are supplied with specific reaction buffers.
– T4 DNA polymerase, 3 u/µl (New England BioLabs, Inc. # 203S or equivalent)

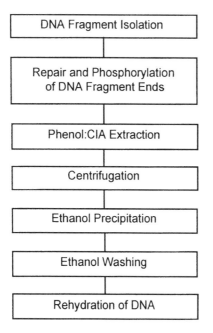

Fig. 2. Schematic outline of the procedure

- T4 polynucleotide kinase, 10 u/µl (New England BioLabs, Inc. # 201S or equivalent)

Note: All enzymes should be stored at -20 °C in a non-frost-free freezer.

- dTTP, dATP, dGTP and dCTP 100 mM stock solutions (New England BioLabs, Inc. # 446 or equivalent).
- ATP 100 mM stock solution (Roche Molecular Biochemicals # 1 140 965 or equivalent)
- Phenol:CIA mixture (PCI) (Ambion # 9732)
- Phenol, water-saturated (Ambion # 9712 or equivalent)
- Phase Lock Gel (PLG I) microfuge tubes (Eppendorf 5′ # 0032007953) PLG tubes contain a proprietary compound that, when centrifuged, migrates to form a tight barrier between organic and aqueous phases (Murphy and Hellwig, 1996). The interphase material is trapped in and below this barrier allowing the complete and easy collection of the entire aqueous phase without contaminating it with organic solvents. PLG barrier also offers increased protection from exposure to organic solvents used in deproteinization procedures.

- **10 x dNTP Solution**   Solutions
  - 2.0 mM each dATP, dCTP, dGTP and dTTP
  Add 2 µl of each dNTP stock solution to 92 µl of sterile 10 mM Tris HCl, pH 7.5 buffer. Store at -20 °C.
- **Loading Dye Solution (Stop Solution)**
  - 15% Ficoll 400
  - 5 M Urea
  - 0.1 M Sodium EDTA, pH 8
  - 0.01% Bromophenol blue
  - 0.01% Xylene cyanol
  Prepare at least 10 ml of the solution. Dissolve an appropriate amount of Ficoll powder in double distilled or deionized water by stirring at 40 - 50 °C. Add stock solution of EDTA, powdered urea and dyes. Aliquot about 100 µl into microfuge tubes and store at -20 °C.
  Other reagents used in the preparation of DNA fragments for cloning are identical to those used for plasmid preparation described in this chapter (Subprotocol 1).

▪▪▪ Procedure

Restriction enzyme digestion

The choice of restriction enzyme used depends on the choice of the fragment to be cloned. There are many restriction enzyme suppliers and recommendation of digestion conditions for the same enzymes may differ significantly between suppliers. Because the manufacturer of a given enzyme has optimized the reaction conditions for their particular enzyme preparation, their recommendations should always be followed. Using buffer of one manufacturer with the enzyme of another is not recommended. The best description of activities of various enzymes with different buffers that can serve as a general guide for running restriction enzyme reactions is available from New England BioLabs, Inc. catalog or web side (www.neb.com).

The general rules for working with restriction enzyme and preparing digestion reactions are:

- Store restriction endonuclease at - 20 °C in non-frost-free freezer at concentration of at least 10 u/μl.

- Remove the enzyme from the freezer and place it immediately on ice. Better still, place the enzyme tube into a portable - 20 °C laboratory bench "freezer". This can be bought from Stratagene Inc. (StrataCooler # 400012) or made from "liquid ice" packages usually supplied with enzyme shipments (Kranz, 1989). To prepare a "portable freezer", insert a microtube into the liquid ice package and place it at -20 °C until it freezes. Remove the tube used to form the hole and insert the enzyme into the frozen package. This "portable freezer" will keep an enzyme at -20 °C on the laboratory bench for several hours.

- The volume of the digestion reaction should be large enough that the restriction enzyme constitutes no more than 10% of the total volume. A 30 μl reaction volume is recommended.

- Use a DNA concentration no greater than 10 μg in a volume not exceeding 1/3 of the reaction volume. If larger amount of DNA is required, concentrate the DNA by ethanol precipitation before adding it to the reaction.

- Use 10 units or more enzyme per 1 μg of DNA. Although this is far more enzyme than is theoretically required, this excess assures complete digestion in the case of impurities in the DNA, decreased enzyme activity from storage, pipetting errors during enzyme addition etc. Some enzymes cleave their defined sites with different efficiency, largely due to differences in flanking nucleotides, and cleavage rates for different sites recognized by a given enzyme can differ by a factor of 10 (New England BioLabs). Using excess enzyme will compensate for these differences.

- Use 20-fold or more excess of enzyme to digest supercoiled plasmid, viral DNA and large genomic DNA (Fuchs and Blakesley, 1983).

- Increase the enzyme concentration by a factor of two when digesting ethidium bromide stained DNA. Removal of ethidium bromide from stained DNA before digestion with restriction enzyme is not necessary.

1. Place a 1.5 ml tube on ice and calculate the amount of ingredients needed using the following table:

Restriction Enzyme Digestion Reaction

Ingredient	Amount (µl)	Final concentration
10 x Restriction enzyme buffer	3	1x
DNA to be digested	x	1 - 10 µg
Restriction enzyme (10 u/µl)	1 - 2	10 - 20 units
Water	to 30 µl	

Always observe the following rules when preparing the reaction:

- Thaw all reagents at room temperature and place them on ice.

- Calculate the amount of all reagents needed using a chart. Do not include water in these calculations.

- Calculate the amount of water needed to obtain the desired reaction volume. **The amount of water is always calculated last.**

- Add water to the reaction tube. **Water is always added first**

- Add the remaining ingredients in the following order: buffer, cofactors, substrate.

- Start the reaction by the addition of enzyme.

2. Add calculated amount of water to the tube. Follow it with 3 µl of 10 x buffer. Mix well by pipetting up and down.

3. Add the appropriate amount of DNA and mix by pipetting up and down or by tapping the tube. Do not vortex.

4. Start reactions with the addition of the enzyme. Mix the enzyme with the reaction mixture by pipetting up and down several times.

5. Centrifuge the tube for 5 to 10 seconds to remove air bubbles and collect liquid at the bottom of the tube.

6. Incubate the reaction for 1-2 hours at the temperature recommended by the manufacturer of the enzyme.

7. Stop reaction by the addition of 5 µl of stop solution. Mix well by pipetting up and down several times. Centrifuge 5–10 seconds to collect liquid at the bottom of the tube. The reaction is now ready to be loaded onto an agarose gel.

Note: Restriction digestion can also be stopped by the addition of 1 µl 0.5 M EDTA. The reaction can be stored at -20 °C for an indefinite time and used for gel electrophoresis when needed.

8. Electrophorese digested DNA to separate DNA fragments. Prepare and run an agarose gel as described in Chapter 8 for large scale agarose gel electrophoresis. Purify the desired DNA fragment using one of the protocols for recovery of DNA fragments from agarose gels described in Chapter 8. Purified fragment can be end-repaired for blunt-end ligation using the protocol described below. Alternatively, the cohesive ends of the fragment can be ligated with a plasmid with compatible and dephosphorylated ends.

End repair of DNA fragments

To prepare DNA fragments for blunt-end ligation, two enzymes, T4 DNA polymerase and T4 polynucleotide kinase are used simultaneously. The buffer used in the reaction should be buffer supplied with polynucleotide kinase. T4 DNA polymerase is fully active in this buffer. When preparing DNA fragments for cohesive end ligation, omit DNA polymerase and 10 x dNTP from the reaction mixture.

1. The total reaction volume is 50 µl. Calculate the amounts of needed ingredients using the following chart:

DNA End-Repair Reaction

Ingredient	Amount (μl)	Final concentration
10 x polynucleotide kinase buffer	5	1x
10 x dNTP solution	5	200 μM
ATP 10 mM	5	1 mM
DNA fragment	1 - 28	0.1 - 5 μg
T4 polynucleotide kinase (10 u/μl)	1	10 - 20 units/μg DNA
T4 DNA polymerase (3 u/μl)	1	1 - 3 units/1 μg DNA
Water	to 50 μl	

2. Add the required amount of water to a 1.5 ml microfuge tube. Next add buffer, ATP, 10 x dNTP solution and DNA. Mix by pipetting up and down.

3. Start the reaction by the addition of both enzymes: Mix by pipetting up and down, being careful not to create air bubbles in the process. Centrifuge for 5 to 10 seconds to collect liquid at the bottom of the tube.

4. Incubate at room temperature for 25 minutes.

Note: If the DNA fragment is shorter than 500 bp, incubation time can be shortened to 10 to 15 minutes.

5. Stop reactions by the addition of 1 μl 0.5 M EDTA and 50 μl of TE. Mix well by pipetting up and down or by tapping the tube.

6. Add 100 μl of PCI solution and mix by inverting several times to form an emulsion.

7. Centrifuge the empty PLG tube in a microfuge for 30 seconds at 10 000 rpm to pellet the gel. Orient the tube in the centrifuge rotor with the lid connector pointing away from the center of rotation.

Note: Measure the time of centrifugation from the moment of **starting the microfuge**.

8. Add the mixture prepared in step 6 to the PLG tube.

9. Centrifuge the PLG tube at 10 000 rpm for exactly 30 seconds. Be sure to orient the tube in the centrifuge rotor as before.

Note: After centrifugation, the organic phase at the bottom of the tube will be separated from the aqueous phase at the top of the tube by the PLG barrier.

10. Add 100 μl of CIA solution. Mix by repeated inversion to form an emulsion. Do not vortex or permit the PLG barrier to be disturbed.

11. Centrifuge for 30 seconds to separate phases. Be very careful to place the tube into the centrifuge at the same orientation as before.

12. Transfer the aqueous phase from above the PLG barrier to a fresh tube. Record the volume of the aqueous phase. The volume should be similar to the volume of aqueous phase originally added to the tube.

13. Add 50 μl of 7.5 M ammonium acetate (half the volume). Mix well by inverting the tube 4 to 5 times.

14. Add 300 μl of 95% ethanol (2 x of the total volume). Mix well by inverting the tube 4 to 5 times.

15. Place the tube in the centrifuge and orient the attached end of the lid pointing away from the center of rotation. Centrifuge at maximum speed for 10 minutes at room temperature. (See Figure 2, Chapter 2.)

16. Remove the tube from the centrifuge and open the lid. Holding the tube by the lid gently lift the end, touching the tube to the edge of an Erlenmeyer flask to drain the ethanol. You do not need to remove all the ethanol. Place the tubes back into the centrifuge orienting them as before.

Note: When pouring off ethanol, do not invert the tube more than once because this can loosen the pellet.

17. Wash the pellet with 700 μl of cold 70% ethanol. Holding a P1000 Pipetman vertically, slowly deliver the ethanol to the side of the tube opposite the pellet. Hold the Pipetman as shown in Figure 3, Chapter 2. **Do not start the centrifuge**, in this step the centrifuge rotor is used as a "tube holder" that keeps the tubes at an angle, convenient for ethanol washing. Remove the tube from the centrifuge by holding it by the lid. Pour off the ethanol as described in step 16.

18. Place the tube back into the centrifuge and repeat the 70% ethanol wash one more time.

Note: This procedure makes it possible to wash a large number of pellets quickly without centrifugation and vortexing. Vortexing and centrifuging are time consuming and frequently lead to substantial loss of the material.

19. After the last wash, place the tube into the centrifuge, making sure that the orientation is the same as before. Without closing the lids, start the centrifuge for 2-3 seconds to collect the remaining ethanol at the bottom of the tube. Remove all ethanol with a P200 Pipetman outfitted with a capillary tip.

20. Resuspend the pelleted DNA in 20 µl of water. Use this DNA for the ligation reaction. The final concentration of DNA solution will be 5 ng to 250 ng per µl.

Note: This concentration can be estimated from the size of DNA insert present in plasmid. For example when 2 µg DNA of 2.6 Kbp is used and the desired insert size is 500 bp, the fragment constitutes 19% of total DNA (0.5/2.6 = 0.19) and should have a mass of 0.38 µg (2 x .19 = 0.38). After elution of the fragment from an agarose gel and repairing the ends, the recovery is about 60% to 70% of initial fragment amount. Thus, there will be about 0.22 to 0.26 µg of fragment in 20 µl giving a concentration of about 10 to 13 ng µl. Such estimation is sufficiently precise for calculating the amount of DNA fragment necessary for a ligation reaction.

Troubleshooting

Restriction enzyme digestion of DNA should be assessed by agarose gel electrophoresis of the digest together with DNA size standards. The intensity of each fragment stained by ethidium bromide is proportional to the mass of the fragment. The fragments should be present in the gel in equimolar quantities. The sum of their molecular sizes should be equal to the size of undigested DNA. A partial digest will generate too many DNA bands, the sum of which will add up to more than the size of original DNA.

Incomplete digestion is likely to be due to one of the following:

- Inactivation of the enzyme during storage or during preparation of reaction mixture. Vigorous vortexing of the reaction can completely inactivate enzymes. Repeat the digestion with other restriction enzyme to confirm that DNA fragment contamination was not the cause of the partial digestion.

- Impurities present in the DNA sample. The presence of residual amounts of phenol, SDS, EDTA or large amounts of salt can cause partial digestion or inhibit restriction enzyme activity altogether. To remove these impurities DNA can be reprecipitated with ethanol.

- Degradation of the DNA sample. Incorporation of a control with undigested DNA will make this cause clearly apparent.

Partial digestion of the sample DNA can be due to one of the problems listed above and usually can be overcome by repeating the digestion with more enzyme or for a longer time rather than repurifying of the DNA.

Digestion of the sample at too many sites can be due to altered cleavage specificity of the enzyme, usually referred to as star activity of the enzyme (Robinson and Sligar, 1995). Star activity of the enzyme can be induced by too high concentration of glycerol, too high pH of the reaction (›8.0), low ionic strength (<25 mM of salt) or a large amount of enzyme relative to DNA (>100 units for 1 µg DNA).

Measurement of the concentration of blunt-ended DNA fragments is necessary for a correct enzyme to DNA ratio in the ligation reaction. The method described in step 20 is not very precise but usually sufficient to make these calculations. It is not possible to measure the concentration of DNA fragments using a standard spectroscopic method of absorption at 260 nm. This is because the spectroscopic method requires a DNA concentration of at least 5 µg / ml ($A_{260} = 0.1$) and a minimum sample volume of 500 µl. DNA fragments prepared for cloning can rarely be obtained at this concentration and volume. For a more precise measurement of concentration, we recommend using DNA concentration measurement kits such as DNA DipStic (Invitrogen Inc. # K5632-01), NUCLEIC dotMETRIC (Geno Technologies, Inc.), FastCheck (Life Technologies, Inc. # 5595UA) or DNA Quick Strip (Eastman Kodak Co. # IB73000). These kits use a small amount of sample DNA (1-2 µl) and are capable of measuring 1 ng to 10 ng of DNA in 1 µl of sample.

Subprotocol 3
Ligation Reaction

In vitro ligation is used to join DNA fragments to a linearized plasmid. For a detailed theoretical description of the ligation reaction see Chapter 13. The protocol describes ligation of dephosphorylated plasmid with phosphorylated insert using T4 DNA ligase. In general, for any ligation of phosphatase-treated insert with dephosphorylated vector, the theoretical calculations for optimal conditions dictate the following practical rules (see Table 1, Chapter 13 for details):

- For all useful molar ratios of vector to insert (1:1 to 1:3) the j/i for insert is always high, favoring circularization of insert rather than concatemerization. Thus, cloning of concatamers of insert does not occur. Using insert concentration higher than 1:3 is not recommended because it may lead to cloning inserts that have been ligated together.

- After formation of the hybrid molecules, conditions favoring cyclization of the hybrid predominates if the insert is not larger than the vector (high *j/i* ratio for hybrid). Practically, lower concentrations of plasmid are preferred because this leads to nearly complete formation of circular hybrid molecules. For example, ligation of plasmid at 2.5 µg/ml with 2.68 Kbp insert gives *j/i* ratio for the construct equal to 5.4, whereas the same ligation at 1 µg/ml plasmid gives value *j/i* equal to 13.6 (see, Table 1, Chapter 13).

- Successful cloning of large fragments requires a **low concentration of plasmid** to assure circularization conditions for linear, vector-insert molecules. For example at 2.5 µg/ml concentration of vector ligated with 10.0 Kbp insert, the j/i for the hybrid molecule is 1.2, the value at which circularization is not favored. The same value for plasmid concentration of 1 µg/ml is 3.7, favoring the circularization reaction for hybrid molecules. A schematic outline of the ligation procedure is shown in Figure 3.

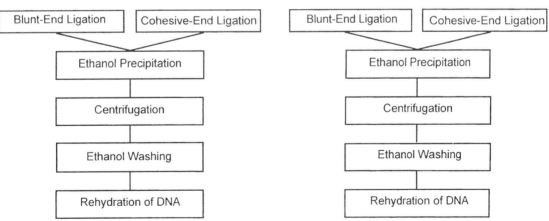

Fig. 3. Schematic outline of the procedure

▓▓▓ Materials

- T4 DNA ligase for blunt-end ligation (5 u/µl) (Roche Molecular Bio-
  chemicals #481 220 or equivalent)
  In author's experience, T4 DNA ligase prepared by this manufacturer
  showed higher conversion of blunt-end fragment into ligated product
  in the course of 10 minutes reaction.
- T4 DNA ligase for cohesive-end ligation (1 u/µl) (Roche Molecular Bio-
  chemicals # 799 009 or equivalent)
- ATP 100 mM (Roche Molecular Biochemicals # 1 140 965 or equivalent)
- PEG 8000 (Sigma Co. # P 4463 or equivalent).
- Dithiothreitol (DTT) (Sigma Co. # D 0632 or equivalent)
- Rapid DNA Ligation Kit (Roche Molecular Biochemicals # 1635 376)

Solutions
- **ATP 5 mM**
  Prepare 100 µl of solution. To 95 µl of sterile 5 mM Tris HCl pH 7.5 add
  5 µl stock ATP solution. Store at -20 °C.
- **ATP 0.5 mM**
  Dilute ATP 5 mM solution 10 times. To 18 µl of sterile water add 2 µl of
  5 mM ATP stock solution. Prepare solution fresh. Do not store.
- **50 mM DTT Solution**
  Prepare 1 ml solution in sterilized water and store at 20 °C.
- **5 x Blunt-End Ligation Buffer**
  - 200 mM Tris HCl pH 7.8
  - 50 mM $MgCl_2$
  - 50 mM DTT
  - 30% (w/v) PEG 8000
  Prepare 10 ml of buffer using stock solutions of Tris, $MgCl_2$ and DTT.
  Add 3 g of PEG 8000 and dissolve completely. Sterilize by filtration. Store
  in 200 µl aliquots at -20 °C.

Note: Some manufacturers of T4 DNA ligase supply enzyme with concentrated stock solution of reaction buffer. These buffers can be used for blunt-
end ligation providing that they do not contain ATP. The ATP concentration necessary for optimal blunt-end reaction is 20 times less than that for
cohesive-end ligation.

- **5 x Cohesive-End Ligation Buffer**
  - 200 mM Tris HCl pH 7.8
  - 50 mM $MgCl_2$
  - 50 mM DTT

Note: Buffers supplied by manufacturers of T4 DNA ligase are usually formulated for cohesive-end ligation. These buffers can be used instead of the buffer described. T4 DNA ligase is active in cohesive-end ligation in all the restriction endonuclease buffers if they are supplemented with 1 mM ATP.

Procedure

The conditions of ligation described in protocols are applicable to most ligations with small plasmids (2.6 to 4.0 Kbp) with inserts that are not larger than the plasmid. Vector concentration used is 1 µg/ml (10 fmol) and molar ratio of vector to insert is between 1:2 to 1:3. The concentration of vector should be lower if insert size is 10 Kbp or larger (see Subprotocol 3 for further explanations). These conditions have been shown to be optimal for blunt-end ligation (Bercovich at al. 1992).

Two ligation protocols are presented. The standard procedure requires 1 hour of reaction time for blunt-end ligation or overnight incubation for cohesive-end ligation. A second rapid protocol requires only 10 minutes reaction for both cohesive-end and blunt-end ligation and uses the rapid DNA Ligation Kit from Roche Molecular Biochemicals. This kit gives the best results for blunt-end and cohesive-end ligations of plasmid with insert. The use of this kit for all ligation reactions is highly recommended.

Both ligation protocols include two control reactions. The first control reaction tests for the presence of undigested vector DNA in plasmid preparations. The second control tests the efficiency of dephosphorylation of vector ends. It is highly recommended to perform these controls when a new vector preparation is used because most failure to ligate results from incorrect preparation of the cloning vehicle.

Standard blunt-end protocol

1. Prepare a master reaction mixture for 4 reactions. Place a 1.5 ml micro-fuge tube on ice and add:

Blunt-End Ligation Master Mixture

Ingredients	Amount (μl)	Final concentration /rx
5 x Blunt-end ligation buffer	16	1x
0.5 mM ATP	8	0.05 mM
Vector linearized, dephosphorylated (10 ng/μl)	8	20 ng (10-30 fmol)
Water	to 40 μl	-

Vortex briefly and centrifuge for 10 - 20 seconds. Place tube on ice. When preparing ligation reactions, always follow these rules:

- Thaw all reagents at room temperature and place them on ice.

- Calculate the amounts of all reagents needed. Do not include water in this calculation.

- Calculate the amount of water needed to obtain the desired reaction volume. **The amount of water is always calculated last.**

- Add water to reaction tube. **Water is always added first.**

- Add remaining ingredients in the following order: buffer, cofactors, substrate.

2. Label the 1.5 ml microfuge tubes 1C, 2C and 3L and place them on ice. Prepare the ligation reaction and two control reactions. First, calculate the amount of insert needed for ligation. Use the following equation for this calculation:

$$\text{Insert (ng)} = \frac{\text{Vector (ng) x Insert (bp)}}{\text{Vector (bp)}} \text{x R} \qquad (2)$$

Where, Vector (ng) is nanogram of vector in reaction and R molar ratio of vector to insert. The R should be between 1 to 3. The optimum ratio may be determined experimentally.

Note: A molar ratio of vector to insert of 1:2.5 works for most blunt-end ligation reactions.

3. Assemble the ligation reaction (3L) and control reactions (1C, 2C) as follows:

Ligation Reactions

Ingredients	Control 1(1C)	Control 2 (2C)	Ligation rx (3L)
Blunt-End Master Mixture	10 μl	10 μl	10 μl
DNA fragments	none	none	as needed
T4 DNA ligase (5 u/μl)	none	1 μl	1 μl
Water	to 20 μl	to 20 μl	to 20 μl

Add water first and follow it with blunt-end master mix. Add insert DNA to the 3L tube only. Mix well pipetting up and down. **Add enzyme last.** Mix enzyme with reaction mixture by gently pipetting up and down.

Note: Most manufacturers use Weiss units of T4 DNA ligase. Cohesive-end ligation units are also used (for example, New England BioLabs, Inc.). One Weiss unit is equal to 67 cohesive-end ligation units. Blunt-end ligation requires 300 to 400 cohesive-end ligation units.

4. Incubate reactions at room temperature for 1 hour.

5. Stop the reactions with the addition of 1 μl of 0.5 M EDTA. Add 60 μl of sterile water to the reaction tubes and mix well by inverting the tubes several times. Use 10 to 20 μl of this reaction for chemical transformation. Reactions can be stored at -20 °C until needed. To prepare the ligation reaction for electroporation, go to step 6.

Note: Do not heat-inactivate ligase. This will result in a large decrease in transformation efficiency.

6. Add 40 μl of 7.5 M ammonium acetate (half the volume) to each tube and mix by inverting several times.

7. Add 200 μl of 95% ethanol (2 x the total volume) to each tube and mix well by inverting 4 - 5 times. Place tubes into the microfuge and centrifuge for 20 minutes at room temperature. **Be sure to orient each tube in the rotor with the attached part of the lid away from the center of rotation.** This will "mark" the position of pelleted DNA since pellets will not be visible. (See Figure 2, Chapter 2.)

8. Remove the tube from the centrifuge and open the lid. Gently invert, touching the tube to the edge of an Erlenmeyer flask to drain the ethanol.

You do not need to remove all the ethanol. Place the tubes back into the centrifuge **in the same orientation as above**.

Note: When pouring off ethanol do not invert the tube more than once because this can loosen the pellet.

9. Wash the pellet with 700 µl of cold 70% ethanol. Holding a P1000 Pipetman vertically, slowly deliver the ethanol to the side of the tube opposite the pellet, that is the side facing the center of the rotor. Hold the Pipetman vertically as shown in the Figure 3, Chapter 2. **Do not start the centrifuge,** in this step the centrifuge rotor is used as a "tube holder" that keeps the tubes at an angle convenient for ethanol washing. Remove the tube from the centrifuge by holding it by the lid. Pour off the ethanol as described in step 8.

10. Place the tube back into the centrifuge and repeat the 70% ethanol wash one more time.

Note: This procedure makes it possible to quickly wash the pellet in a large number of tubes without centrifugation and vortexing. Vortexing and centrifuging the pellet are time consuming and frequently lead to substantial loss of the material and DNA shearing.

11. After the last wash, place the tube into the centrifuge, making sure that the tube side containing the pellet faces away from the center of rotation. Without closing lids, start the centrifuge for 2 – 3 seconds and collect the remaining ethanol at the bottom of the tube. Remove all ethanol with a P200 Pipetman fitted with a capillary tip.

Note: Never **dry the DNA pellet** in a vacuum. This will make dissolving the DNA pellet very difficult, if not impossible.

12. Resuspend the DNA pellet (invisible) in 5 µl of water. This will be successful only if you know the position of the pellet on the side of the tube. It is important to realize that for most microfuges, the pellet will be distributed on the side of the tube. To dissolve DNA, place 5 µl of water in the middle of the tube and move the drop down the side toward the bottom using the end of a yellow tip. Repeat this procedure several times to assure that the invisible pellet at the side of the tube is dissolved. Use all 5 µl for electroporation. Resuspended ligation product can be stored at -20 °C.

Standard cohesive-end protocol

1. Prepare a master reaction mixture for 4 reactions. Place a 1.5 ml micro-fuge tube on ice and add:

Cohesive-End Ligation Master Mixture

Ingredients	Amount (µl)	Final concentration / rx
5 x Cohesive-end ligation buffer	16	1x
5.0 mM ATP	8	0.5 mM
Vector linearized, dephosphorylated (10 ng/µl)	8	20 ng (10-30 fmol)
Water	to 40 µl	

Vortex briefly and centrifuge for 10 - 20 seconds. Place the tube on ice. When preparing reactions, always follow these rules:

- Thaw all reagents at room temperature and place them on ice.

- Calculate the amounts of all reagents needed. Do not include water in this calculation.

- Calculate amount of water needed to obtain the desired reaction volume. **The amount of water is always calculated last.**

- Add water to reaction tube. **Water is always added first.**

- Add remaining ingredients in the following order: buffer, ATP and DNA.

2. Label 1.5 ml microfuge tubes 1C, 2C and 3L and place them on ice. Prepare the ligation reaction and two control reactions. First, calculate the amount of insert needed for ligation. Use equation 2 for this calculation. The molar ratio of vector to insert should be between 1:2 to 1:3.

3. Assemble the ligation reaction (3L) and control reactions (1C, 2C) as follows:

Ligation Reactions

Ingredients	Control 1(1C)	Control 2 (2C)	Ligation rx (3L)
Cohesive-End Master Mixture	10 µl	10 µl	10 µl
DNA fragments	none	none	as needed
T4 DNA ligase (1 u/µl)	none	1 µl	1 µl
Water	to 20 µl	to 20 µl	to 20 µl

Add water first and follow it with blunt-end master mix. Add DNA fragments to 3L only. Mix well by tapping the end of the tube and **add enzyme last to tubes 2C and 3L.** Mix enzyme with reaction mixture by gently pipetting up and down. Note that less T4 DNA ligase is required for cohesive-end ligation (1 Weiss unit/rx).

Note: Most manufacturers use Weiss units for T4 DNA ligase. Cohesive-end ligation units are also used (for example, New England BioLabs, Inc.). One Weiss unit is equal to 67 cohesive-end ligation units.

4. For best results incubate reactions at 16 °C overnight. However, 30 minutes at room temperature can also be used with some decrease in ligation efficiency.

5. Stop the reactions with the addition of 1 µl of 0.5 M EDTA. Add 60 µl of sterile water to the reaction tubes and mix well by inverting the tubes several times. Use 10 to 20 µl of this reaction for chemical transformation. Reactions can be stored at -20 °C until needed. To prepare the ligation reaction for electroporation, go to step 6.

Note: Do not heat-inactivate ligase. This will result in a large decrease in transformation efficiency.

6. Add 40 µl of 7.5 M ammonium acetate (half the volume) to each tube and mix by inverting several times.

7. Add 200 µl of 95% ethanol (2 x the total volume) to each tube and mix well by inverting 4 - 5 times. Place tubes into the microfuge and centrifuge for 20 minutes at room temperature. **Be sure to orient each tube in the rotor with the attached part of the lid away from the center of rotation.** This will "mark" the position of pelleted DNA since pellets will not be visible. (See Figure 2, Chapter 2.)

8. Remove the tube from the centrifuge and open the lid. Gently lift the end, touching the tube to the edge of an Erlenmeyer flask to drain the ethanol. You do not need to remove all the ethanol from the tube. Place the tubes back into the centrifuge **in the same orientation as above.**

Note: When pouring off ethanol do not invert the tube more than once because this can loosen the pellet.

9. Wash the pellet with 700 µl of cold 70% ethanol. Holding a P1000 Pipetman vertically, slowly deliver the ethanol to the side of the tube opposite the pellet, that is the side facing the center of the rotor. Hold the Pipetman vertically as shown in Figure 3, Chapter 2. **Do not start the centrifuge**, in this step the centrifuge rotor is used as a "tube holder" that keeps the tubes at an angle convenient for ethanol washing. Remove the tube from the centrifuge by holding the tube by the lid. Pour off the ethanol as described in step 8.

10. Place the tube back into the centrifuge and repeat the 70% ethanol wash one more time.

Note: This procedure makes it possible to quickly wash the pellet in a large number of tubes without centrifugation and vortexing. Vortexing and centrifuging the pellet are time consuming and frequently lead to substantial loss of the material and DNA shearing.

11. After the last wash, place the tube into the centrifuge, making sure that the tube side containing the pellet faces away from the center of rotation. Without closing the tube lids, start the centrifuge for 2-3 seconds and collect the remaining ethanol at the bottom of the tube. Remove all ethanol with a P200 Pipetman fitted with a capillary tip.

Note: Never **dry the DNA pellet** in a vacuum. This will make dissolving the DNA pellet very difficult, if not impossible.

12. Resuspend the DNA pellet (invisible) in 5 µl of water. This will be successful only if you know the position of the pellet on the side of the tube. It is important to realize that for most microfuges the pellet will be distributed on the side of the tube. To dissolve DNA, place 5 µl of the water

in the middle of the tube and move the drop down the side toward the bottom using the end of a yellow tip. Repeat this procedure several times to assure that the invisible pellet at the side of the tube is dissolved. Use all 5 μl for electroporation. Resuspended ligation product can be stored at -20 °C.

Rapid Protocol

The protocol is identical for blunt-end and cohesive-end ligations.

1. Label one 1.5 ml microfuge tube RX. Also prepare two control reaction tubes labeled 1C, 2C, and one ligation reaction tube labeled 3L. Place the tubes on ice.

2. Prepare the Master Reaction Mixtures for 4 reactions in RX tube as follows:

Master Reaction Mixture

Ingredients	Amount (μl)	Final concentration/rx
2 x Ligation buffer (vial #1)	40	1x
5 x DNA buffer (vial #2)	16	1x
Vector linearized, dephosphorylated (10 ng/μl)	8	20 ng (10-30 fmol)

Vortex tube briefly and centrifuge for 30 seconds in a microfuge. Place tube on ice.

Note: It is absolutely necessary to thoroughly mix the content of vial 1 and 2 just before use.

Note: If possible dissolve DNA to be ligated in 1 x DNA buffer. In this situation **substitute** water for DNA buffer in Master Reaction Mixture.

3. Place tubes 1C, 2C and 3L at room temperature and prepare reactions as indicated below. First, calculate the amount of insert DNA needed using equation 2 and add the DNA fragment to tube 3L. Use a molar ratio of vector to insert 1:3. Second, add water and master reaction mixture to all tubes. **Do not add enzyme at this time.**

Ligation Reactions

Ingredients	Control 1(1C)	Control 2 (2C)	Ligation Rx (3L)
Master Reaction Mixture	16 μl	16 μl	16 μl
DNA fragments	none	none	as needed
Water	to 20 μl	to 20 μl	to 20 μl
T4 DNA ligase (vial #3)	none	1 μl	1 μl

Mix ingredients gently, by pipetting up and down several times. Centrifuge for 5 to 10 seconds to collect liquid at the bottom of the tubes.

4. Start the reaction by the **addition of enzyme to tubes 2C and 3L**. Mix well by gently pipetting up and down several times. Incubate the reactions for 5 to 10 minutes at room temperature.

5. Stop the reactions with the addition of 1 μl of 0.5 M EDTA, pH 8.0 and add 80 μl of water to each tube. Mix well by inverting the tubes several times. Use 20 μl of this reaction for chemical transformation. Reactions can be stored at - 20 °C until needed. To prepare ligation reaction for electroporation go to step 6.

Note: Do not heat-inactivate ligase. This will decrease transformation efficiency by factor ›20.

6. Add 50 μl of 7.5 M ammonium acetate (half the volume) to each tube and mix well by inverting several times.

7. Add 300 μl of 95% ethanol (2 x the total volume) to each tube and mix well by inverting 4 - 5 times. Place tubes into the microfuge and centrifuge for 20 minutes at room temperature. **Be sure to orient each tube in the centrifuge rotor with the lid connector pointing away from the center of rotation**. This will "mark" the position of pelleted DNA since your pellet will not be visible. (See Figure 2, Chapter 2.)

8. Remove the tube from the centrifuge and open the lid. Gently lift the end, touching the tube to the edge of an Erlenmeyer flask to drain the ethanol. You do not need to remove all the ethanol from the tube. Place the tubes back into the centrifuge in the same orientation as above.

Note: When pouring off ethanol do not invert the tube more than once because this can loosen the pellet.

9. Wash the pellet with 700 μl of cold 70% ethanol. Holding a P1000 Pipetman vertically, slowly deliver the ethanol to the side of the tube opposite the pellet, that is, the side facing the center of the rotor. Hold the Pipetman vertically as shown in Figure 2, Chapter 2. **Do not start the centrifuge**, in this step the centrifuge rotor is used as a "tube holder" that keeps the tubes at an angle, convenient for ethanol washing. Remove the tube from the centrifuge by holding the tube by the lid. Pour off the ethanol as described in step 8.

10. Place the tube back into the centrifuge and repeat the 70% ethanol wash one more time.

Note: This procedure makes it possible to quickly wash a large number of pellets without centrifugation and vortexing. Vortexing and centrifuging the pellet are time consuming and frequently lead to substantial loss of material.

11. After the last wash, place the tube into the centrifuge, making sure that the tube position in the rotor is the same as in step 8. Without closing the tube lids, start the centrifuge for 2-3 seconds and collect the remaining ethanol at the bottom of the tube. Remove all ethanol with a P200 Pipetman fitted with a capillary tip.

Note: Never **dry the DNA pellet** in a vacuum. This will make dissolving the DNA pellet very difficult, if not impossible.

12. Resuspend the DNA pellet (invisible) in 5 μl of water. This will be successful only if you know the position of the pellet on the side of the tube. It is important to realize that for most microfuges the pellet will be distributed on the side of the tube. To dissolve DNA, place 5 μl of the water in the middle of the tube and move the drop down the side toward the bottom using the end of a yellow tip. Repeat this procedure several times to assure that the invisible pellet at the side of the tube is dissolved. Use all 5 μl for electroporation. Resuspended ligation product can be stored at -20 °C.

Troubleshooting

Ligation reactions for transformation are best analyzed by transforming bacteria. Analysis of the ligation products by gel electrophoresis is not recommended for two reasons. First, it wastes a large amount of ligation product, in particular valuable insert, quantities of which are frequently difficult to obtain. Second, many ligation products, visible on agarose gels, do not transform cells efficiently. However, if repeated attempts to transform are unsuccessful, this may be due to failure of the ligation reaction. Performing a ligation reaction with molecular size markers (for example, lambda DNA *Hind* III digest) can test this. Ligation products should contain a high molecular weight DNA and very few low molecular weight bands.

Ligation reactions can fail for several reasons (listed in frequency of occurrence):

- Incorrect concentration of insert DNA. Check the concentration of insert using one of the kits for determining DNA concentration described on page 338. Be sure that the vector to insert **molar** ratio is at least 1:1. An insert to vector ratio greater than 1:5 will severely inhibit ligation.

- Too high concentration of vector. Lower concentration of vector to 2 to 5 ng per reaction.

- The presence of a high concentration of monovalent salts (for example, NaCl or ammonium salts). The concentrations of these salts in the final reaction should not exceed 50 mM. Careful washing of DNA pellets with 70% ethanol should remove all salts from plasmid and insert preparations. Neglecting to do so is a frequent cause of failure in ligation reactions.

- The presence of phosphate. Phosphate buffers are used in the storage of some restriction enzymes. Blunt-end ligation is particularly sensitive to this salt since 50 mM PO_4 will inhibit the reaction by 50%. Careful washing of DNA pellets with 70% ethanol should remove this contaminant.

- High concentration of ATP. The blunt-end ligation reaction is inhibited by the presence of ATP above 0.1 mM. Buffers supplied by manufacturers with T4 DNA ligase frequently contain ATP at optimal concentration for cohesive end ligation. The use of these buffers in blunt-end ligation reactions is a very frequent error, causing poor ligation efficiency. Carefully check the content of the ligation buffer supplied when performing blunt end ligation.

- Degradation of ATP in the stock solution or old ligation buffer. Always use freshly prepared 0.5 mM ATP. Aliquot 50 µl portions of a 5 mM stock and store it at -20 °C. Do not thaw it more than 5 to 6 times.

- Degradation of insert or vector by DNases. Use fresh reagents and autoclaved distilled water.

- The presence of restriction endonuclease during cohesive-end ligation causing redigestion of ligated product. After digestion of the insert DNA remove restriction endonuclease by heat inactivation, if possible, or better yet, by extraction with phenol:CIA.

Subprotocol 4
Transformation by Electroporation

Electroporation is the most efficient bacterial transformation technique. Two procedures are described: preparation of bacterial cells for electroporation and electroporation of bacteria. Electroporation is carried out with 30 µl of cells (cell concentration 10^{10} cell/ml), 5 ng of plasmid using an electroporation cuvette with a 0.1 cm electrode gap. This represents standard electroporation conditions for *E. coli* cells. The efficiency of transformation with this method is above 1×10^9 transformants/µg supercoiled plasmid and about 1×10^8 transformants/µg plasmid used in the ligation reaction. The frequency of transformation is 0.02 with both types of plasmids. Low transformation frequency prevents cotransformation of a cell with two or more plasmid molecules. For details about how to calculate and adjust these values, see the description of electroporation in Chapter 13.

The use of electroporation as a transformation technique instead of chemical transformation has the following advantages:

- High transformation efficiency.

- The possibility of using a small volume of cells. A volume of cells as small as 20 µl can be transformed yielding about 10^9 transformants.

- The preparation of cells for transformation is very simple and does not use elaborate and time consuming protocols. Moreover, the cells used for electroporation can be prepared ahead of time and stored indefinitely without losing competency.

- Electroporation transformation frequencies with supercoiled DNA and circular DNA are identical. This makes it unnecessary to use highly purified vector in ligation reactions.

- The molecular transformation efficiency for circular DNA remains very high for plasmids of all sizes up to 50 Kb.

A disadvantage of this method is the requirement for an expensive electroporation instrument.

A special, electrocompetent strain of *E. coli*, ElectroMax DH10B, is used in this protocol, but many other electrocompetent strains can be used. Moreover, a modification of the basic protocol that permits the use of regular strains is presented. This procedure was described by Wai and Chow (1995). A schematic outline of the procedure is shown in Figure 4.

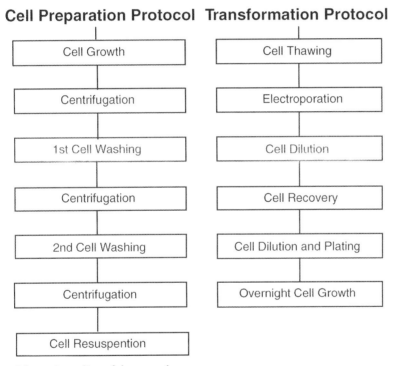

Fig. 4. Schematic outline of the procedure

Materials

- Transformation apparatus (E. coli Pulser, Bio-Rad # 165-2101 or equivalent).
  The electroporation apparatus must be capable of delivering a τ time constant of 5 millisecond and voltage gradient between electrodes (E) of at least 25 kV (for an explanation of these terms see Chapter 13).
- Transformation cuvettes with a 0.1 cm electrode gap (Invitrogen Co. # P410-50 or equivalent).

- Polypropylene culture tubes (Falcon # 2059 or equivalent).
  The "Falcon 2059" tube of Becton Dickson Co. is the standard for transformation experiments. Other equivalent brands are acceptable but batches of tubes are occasionally contaminated with surfactants that inhibit transformation.
- IPTG (isopropylthio-β-D-galactoside) Life Technologies, Inc. # 15529-019 or equivalent.
- X-gal (5-bromo-4-chloro-3-indolyl-β-D galactoside) Life Technologies, Inc # 15520-034 or equivalent.
- Dimethyl-sulfoxide (DMSO) (Sigma Co. # D 8779 or equivalent).
- Ampicillin, sodium salt (Sigma Co. # A 9518 or equivalent).
- Electrocompetent *E. coli* strain (e.g., ElectroMax DH10B from Life Technology, Inc. # 18290-015, or Top10 from Invitrogen Co. # C664-11-55).
- LB agar-ampicillin plates.

Solutions
- **Terrific Broth Medium (TB)**
  - 1.2% Bacto-tryptone
  - 2.4% Yeast Extract
  - 0.4 % Glycerol
  - 10 x Phosphate Stock Solution

  Mix first three ingredients in 900 ml of deionized water and autoclave for 20 minutes, cool to room temperature and add 100 ml of Phosphate Stock Solution. Store at 4 °C.
- **10 x Phosphate Stock Solution**
  - 0.72 M KH_2PO_4
  - 0.17 M K_2HPO_4

  Dissolve in water and autoclave for 20 minutes. Store at 4 °C.
- **10% Glycerol Solution**
  - 10% glycerol

  Add glycerol to sterilized water. Do not sterilize. Prepare fresh and pre-cool to 4 °C before use.
- **Transformation Solution**
  - 10% Glycerol
  - 0.0125% Yeast Extract
  - 0.025% Bacto Tryptone

  Add 2 ml of 100% glycerol and remaining ingredients to 18 ml sterilized water. Stir to dissolve all ingredients. Sterilize by filtration. Store at 4 °C.

- **X-Gal Solution (20 mg/ml)**
  - 2% X-gal
  - DMSO

Dissolve 200 mg X-gal in 10 ml DMSO. Store in the dark at -20 °C.

Note: DMSO is used instead of the commonly used dimethylfomamide (DMF) for X-gal preparation because DMF is very toxic.

- **IPTG Solution (25 mg/ml)**
  - 2.5% IPTG

Dissolve 250 mg of IPTG in sterilized water. Store at -20 °C.

- **1000 x Ampicillin Solution**
  - 500 mg Ampicillin
  - Distilled or deionized water.

Add 500 mg of ampicillin to 5 ml distilled water. Sterilize by filtration and store in small aliquots at -20 °C.

- **LB Agar Amp Plates**
  - 1% Bacto Tryptone
  - 0.5% Yeast Extract
  - 0.5% NaCl
  - 1.5% Difco agar
  - 100 µg/ml Ampicillin

Add the first 3 ingredients to one liter of distilled water in a 2 liter Erlenmeyer flask. Stir to dissolve all ingredients completely. Adjust pH to 7.5 with 1 N NaOH. This will take approximately 4 ml of 1 N NaOH. Add Difco agar and sterilize by autoclaving for 20 minutes. Cool medium to 60-65 °C and add 1 ml of ampicillin stock solution. Mix by swirling the flask and pour the plates. This will make 25 to 30 plates. Plate can be stored for 2 to 3 weeks at 4 °C.

- **PBS Solution, pH 7.4**
  - 137 mM NaCl
  - 2.7 mM KCl
  - 4.3 mM $Na_2HPO_4 \cdot 7H_2O$
  - 1.4 mM KH_2PO_4

Dissolve each salt in double distilled or deionized water. Be sure that one salt has completely dissolved before adding the next. Adjust the pH to 7.4 with 1 N HCl. Autoclave for 20 minutes. Store at 4 °C. If desired, PBS can be made as 10 x concentrated stock solution.

Procedure

TB medium is used for growth of cells. This medium does not contain NaCl making it less laborious to prepare cells for electroporation. The presence of a high concentration of NaCl in other growth medium requires numerous cell washes to remove it completely before electroporation. Small amounts of salt will cause arcing in the electroporation apparatus.

Cell preparation

Both electrocompetent and regular strains can be prepared using this protocol.

1. Inoculate 500 ml of TB medium with 5 ml (1/100 volume) of a fresh, overnight culture of cells. Use a 2 liter Erlenmeyer flask for cell growth.

2. Grow cells at 37 °C with vigorous shaking to an OD_{600} of approximately 0.5 - 0.7. This will take two to three hours. Cool the cells on ice and pour them into three 250 ml centrifuge bottles.

3. Centrifuge the cells at 4000 x g (5000 rpm, HB-4 Sorvall rotor) for 5 minutes at 5° C.

4. Remove as much supernatant as possible. It is better to sacrifice the yield by pouring off some cells than to leave some supernatant behind.

Note: All the following steps should be carried out on ice.

5. Add 10 ml of cold 10% glycerol to each bottle and gently resuspend the pellet by pipetting up and down. Add an additional 60 ml of 10% glycerol to each bottle, mix well and distribute the resuspended cells (220 ml) into two centrifuge bottles. Each bottle should contain about 110 ml of cells. Add 50 ml of 10% glycerol to each bottle for combined total volume of approximately 160 ml.

6. Centrifuge the cells as described in step 3. Discard the supernatant as described in step 4.

7. Add 10 ml of 10% glycerol to each bottle and resuspend the cells by pipetting up and down. After the cells are resuspended, add 80 ml of 10% glycerol to each bottle and combine the cells into one bottle. This bottle should contain about 180 ml of cells.

8. Centrifuge the cells as described in step 3. Remove all of the supernatant. Leave behind about 2 ml of 10% glycerol and gently resuspend the cells in it.

Note: When regular *E. coli* strain is used **remove all** 10% glycerol and re-suspend cells in about 2 ml of transformation solution.

9. Measure absorbance (OD) at 600 nm of a 1/500 dilution (10 µl of cells in 5 ml of water). The OD_{600} should be between 0.25 and 0.4. This corresponds to about 1- 3 x 10^{10} cells/ml.

10. Aliquot the cells in 200 µl portions into microfuge tubes and freeze quickly in a dry ice-ethanol bath. Store at -70 °C. Cells can be stored for at least 1 year.

Electroporation protocol

1. Before starting the electroporation procedure, make the following preparations:
   - Label three electroporation cuvettes 1, 2 and 3 and cool them on ice for at least 5 minutes.

Note: Using cuvettes with an electrode gap 0.2 cm or wider will lower the efficiency of transformation by approximately 40%.

   - Label three Falcon tubes with the numbers corresponding to those of the electroporation cuvettes. Place these tubes in the tube rack at room temperature.
   - Thaw the ligation reaction (3L) and two control reactions (1C and 2C) prepared previously. Set electroporator voltage to 1.8 kV. This setting will depend on the strain used. Use voltage recommended by strain supplier. Warm up 10–50 ml of TB medium to 37 °C.

2. Remove the bacterial cells from -70 °C storage and gently thaw them on ice. Tap the tube gently to mix the cells.

Note: Do not leave cells on ice for an extended time. Use the cells as soon as possible. Cells can be refrozen for later use, but transformation frequency will be significantly lower.

3. Add 30 µl of cells to the tube labeled 1C (control). Pipette the cells up and down two times to mix the DNA with the cells. Be very careful not to create air bubbles during this procedure.

4. Transfer 30 μl of this mixture to a cuvette labeled 1. Holding the cuvette at a 45° angle, deliver the mixture to the bottom of the electroporation chamber of the cuvette. Deliver liquid slowly and do not operate the Pipetman beyond the first stop. This procedure will prevent formation of air bubbles in the electroporation chamber. Tap the cuvette on the laboratory bench several times to distribute the liquid on the bottom of the chamber. Close the cuvette with the cap provided and place it back on ice for 45 seconds.

Note: An electroporation chamber with 0.1 cm gap is very narrow. Frequently the cell sample will stay at the top of the chamber. Vigorous tapping of the cuvette on the laboratory bench will make the sample flow to the bottom. It is important to do this as quickly as possible and not warm up the cuvette and cells.

5. Place the cuvette into the electroporation machine and initiate electroporation.

Note: If a loud "snap" is heard while pulsing, arcing occurred inside the electroporation chamber. Continue with the protocol as usual; some transformation may still have occurred. See troubleshooting section on how to prevent arcing.

6. As quickly as possible, add 1 ml of warm TB medium directly into the cuvette using a P1000 Pipetman. Pipette slowly up and down two times. Gently transfer all the cells from the electroporation cuvette into the appropriately labeled Falcon tube. Treat **the cells very gently, they are very fragile after electroporation**. Leave the tube at **room temperature** until all samples have been electroporated.
   Do not discard electroporation cuvettes. They can be reused after washing for less critical transformations (for example, control transformations). A protocol for washing electroporation cuvettes is given in the troubleshooting section.

Note: Do not use SOC medium for cell recovery as recommended by many protocols. This medium contains NaCl that causes some cell lysis. Use of TB medium increases transformation efficiency by as much as 50%.

7. Electroporate sample 2C following the procedure described in steps 3 to 6. Electroporate 3L sample last, following steps 3 to 6.

8. Transfer all Falcon tubes to an orbital shaker and incubate the cells at 37 °C for 45 minutes to allow cell recovery and expression of the antibiotic resistance. Rotating speed of the shaker should not exceed 240 rpm.

9. Prepare six LB-amp agar plates for plating transformants. Each sample will be plated onto two plates. Add 480 µl of water, 500 µl of X-gal and 20 µl of IPTG to a plastic microfuge tube. You will need 100 µl of the mixture per plate. Add 100 µl of X-gal IPTG mixture onto the surface of each plate. Using sterile technique spread the drop evenly over the surface using a bent glass rod. Solution should soak into the agar for approximately one hour before plating the cells.

Note: X-gal is light sensitive. Use the plates the same day or store plates in the dark at 4 °C.

10. Plate the cells using two dilutions: 1/10 and 1/100. Add 900 µl of PBS to 6 microfuge tubes and label each tube with sample name and dilution factor (for example, 3L 1/10; 3L 1/100, etc.). For each electroporation reaction, prepare two plates labeled the same as the dilution tubes (for example, 1C:1/10; 1C:1/100, etc.).

Note: For optimal cell survival, use only PBS for dilution. Diluting the cells in growth medium instead of PBS will lower cell viability by about 20%.

11. Mix the cells in the tube labeled 1C by gently tapping with fingers. Transfer 100 µl to the first dilution tube (1C 1/10). This will constitute 1/10 dilution of the original cell culture, mix well by pipetting up and down several times. Transfer 100 µl of cells from 1/10 dilution tube into the 1/100 dilution tube using the same yellow tip. Mix by pipetting up and down.

12. Immediately, using the same yellow tip, pipette 100 µl of cells from the 1/100 dilution and then the 1/10 dilution onto the corresponding plates.

Note: Make one set of dilution at a time. Finish plating the bacteria from a single electroporation tube before preparing the next dilution set. This prevents cross-contamination and plating errors.

13. Sterilize the cell spreader by dipping it into a beaker of 95% ethanol, and briefly passing it through a burner flame to ignite the alcohol. Burn off ethanol keeping the spreader **away** from the burner flame.

14. Cool the spreader by touching it to the agar away from the cells. Spread the cells by dragging cell suspension across the agar surface with the spreader back and forth several times. Return spreader to the beaker with ethanol.

15. Replace the plate lids and let them stand until all liquid is absorbed into the agar.

16. Dilute and plate the cells from the remaining two electroporation tubes using the procedure described in steps 11 to 15. Plate the cells from the 3L tube last.

Note: Cells diluted in PBS can be stored for one to two days at 4 °C and used for replating.

17. Place the plates upside down in a 37 °C incubator, and incubate them for 15 to 18 hours.

18. Store the transformed cells for additional plating if necessary. Transfer 850 µl of the cells from 3L tube into microfuge tube and add 60 µl of DMSO. Store the tube at -70 °C. Transformed cells can be stored indefinitely.

19. Count the cells on the 3L plates. Calculate the efficiency of transformation expressed as the number of transformants per microgram of plasmid used. Count only white colonies. Use the following equation for this calculation;

$$E_{tr} = \frac{\text{\# transformed colonies}}{\text{ng plasmid DNA in ligation rx}} \times \text{dilution factor} \qquad (3)$$

20. Inspect two control plates 1C and 2C. They should contain a few blue colonies. The presence of a large number of colonies on both plates indicates failure to linearize the plasmid during vector preparation and/or incomplete dephosphorylation of its 5' ends. If this is the case, the plate with insert ligated to plasmid (3L) will also contain mostly blue colonies.

21. Grow transformants to test for the presence of insert. Prepare several 10 ml glass tubes with 2 ml TB-amp medium (100 µg/ml) and inoculate each with a single white colony from the 3L plate. Using a sterile toothpick, touch the colony and drop the toothpick into the tube. Grow cells overnight, isolate plasmid from 400 µl of cells (Chapter 5, rapid protocol 1) and test for the presence of insert using PCR or restriction enzyme digestion.

22. Store cells with plasmid containing the desired insert. Transfer 930 µl cells from a 2 ml overnight culture (step 21) to 1.5 ml microfuge tube and add 70 µl of DMSO. Mix by inverting 2 to 3 times and store at -70 °C. To inoculate cells from the storage culture, pick up a little "chunk" of frozen cells with a sterile yellow tip and transfer it into medium. **Do not thaw the frozen culture.**

Troubleshooting

The expected efficiency of transformation for a plasmid with an insert no larger than 5 Kbp is about 10^7 to 10^8 transformants/μg plasmid used. This would result in about 50 to 500 colonies on 1/100 dilution plate. If this number is higher, increase the dilution factor (for example, 1/1000 or 1/10 000) using the stored dilution tubes from step 11 and 16 as a starting point. If there are no colonies on 1/10 dilution plate, replate 100 or 200 μl of transformed cells directly from the undiluted transformation that was stored at -70 °C (step 18).

Transformation frequencies of plasmid with large inserts (10 Kbp) should be about 10^5 or 10^6 transformants/μg plasmid. This is because, to clone large fragments, it is necessary to use between 1 to 2 ng of plasmid in the ligation reaction (see Table 1, Chapter 13), the expected number of colonies on a plate from the 1/10 dilution is 10 to 200.

Expect very low transformation frequencies or none, if arcing occurs during pulsing. Arcing can occur for the following reasons:

- Residual salt or buffer in the sample due to inadequate washing of ligation reactions with 70% ethanol.

- Presence of air bubbles in the sample due to incorrect pipetting into electrophoresis chamber.

- Too high concentration of cells used in electroporation. This would result in arcing of all samples. Electroporate cells without plasmid added to test this possibility. If necessary dilute the cells with 10% glycerol.

- Old cell preparations, incorrectly stored cells or thawing the cells too fast. All of these can bring about partial lysis of cells causing arcing.

- Too small volume of cells in the electroporation chamber. A cuvette with 0.1 cm electrode gap requires at least 20 μl of sample.

Lower than expected frequency of transformation can result from:

- Failure of ligation reaction. See troubleshooting of ligation reaction to correct this.

- Incorrect setting of electroporation apparatus. Too high or too low voltage than that recommended for a given bacterial strain. Check the voltage recommendation for the strain and repeat electroporation.

- Incorrect τ constant. This value can be adjusted on some electroporation units. Choose the correct capacitance of capacitor (in Farads) and resistor (in Ohms) to give a τ constant close to 5 milliseconds. Use equation 6 in Chapter 13 to calculate this value. For example using Invitrogen Electroporator II choose 50 μF capacitor and 150 (resistor to get 7.5 ms τ constant (50 x 10^{-6} F x 150 = 0.0075 sec).

- Warm electroporation chamber or sample during pulsing. Warming frequently occurs during loading of the samples. It is better to chill the electroporation cuvettes after loading the sample even if this prolongs incubation of the cells with DNA over the recommended time.

- Excessive volume of the cells in the electroporation chamber. Cell volume should not exceed 40 μl.

- Cloned insert is toxic to the cells. White colonies on the plate will be very small and it will be difficult to grow cells to a high density in 2 ml cultures. Grow cells at 30 °C and/or add glucose in 20 mM to TB medium. If the problem persists reclone the fragment into a low copy number plasmid and use a different bacterial host strain. In general, about 20% of genomic DNA fragments from higher eukaryotes are difficult to clone or are not clonable in *E. coli* cells. Use yeast to clone these fragments.

Other problems frequently met in cloning DNA fragments are:

- Presence of all blue colonies on transformation plates (3L) with concomitant absence of blue colonies on both control plates (1C, 2C). This can happen when the insert was cloned in frame with the α-peptide or when insert DNA is small (<200 bp). In the last instance, color development will be weak and the colonies may appear pale blue. Check pale-blue colonies for the presence of insert as described in step 21 of electroporation protocol.

- Absence of insert or rearranged insert in cells from white colonies. This is frequently a problem when cloning large fragments (›10 Kbp) or fragments containing direct or inverted repeats. To remedy this first, try growing cells with insert on TB medium at 30 °C and collect them for plasmid isolation in the mid to late period of logarithmic growth (OD_{600} = 1-2). If this does not help, retransform plasmid into a host strain specially developed to inhibit rearrangements and elimination of direct and inverted repeats. Several such strains are available: SURE from Stratagene Co., STBL2 from Life Technologies, Inc., or PMC103 (Doherty et al. 1993).
White colonies with recombinant plasmid can be sometimes distinguished from those with plasmid alone by colony morphology. Recombinant colonies appear translucent while non-recombinant colonies are opaque (Austin et al. 1994).

- Appearance of satellite colonies on ampicillin plates. "Feeder" colonies appear near ampicillin resistant colonies because the β-lactamase enzyme, responsible for antibiotic resistance, is secreted from the cell removing antibiotic from the agar in the vicinity of the colony. This can be eliminated by short incubation of the plates (less than 14 hr). To enhance blue color development without bacteria growth, the plates can be incubated at 4 °C for several hours. Alternatively, use carbenicillin (200 µg/ml) or mixture of ampicillin (20 µg/ml) and methicillin (80 µg/ml) instead of ampicillin alone. Both these antibiotics are available from Sigma Co. and Life Technologies, Inc. However, use of these antibiotics substantially increases the cost of plating.

Electroporation cuvettes can be re-used 3 to 5 times. Efficiency of transformation progressively decreases with successive use of the cuvette. This probably is due to some cells being "baked" onto the electrode surface or pitting of the electrode surface by washing procedures.

The washing procedure that works well without apparent contamination with residual plasmid is as follows:

1. Immediately after use immerse the cuvette in 1% Alconox solution. This will prevent drying of bacterial cells onto electrode surface.

2. Rinse the cuvette 6 to 8 times with distilled or deionized water. Do not keep cuvettes in Alconox solution for more than 1 hour.

3. Rinse cuvette 3 times with 70% ethanol. Fill the cuvette with ethanol, cap it and invert several times. This treatment should sterilize the cuvette.

4. Dry the cuvette by filling it with 95% ethanol, inverting several times, pouring off ethanol and drying it upside-down on a paper towel. Replace the cap.

Note: Using 0.25 N HCl for removing DNA is not recommended. This treatment will cause deterioration of the surface of electrodes drastically decreasing transformation efficiency.

Subprotocol 5
Chemical Transformation

A classical procedure to transform of *E. coli* cells with plasmid DNA is described. The cells are incubated in $CaCl_2$ solution that renders them competent to take up DNA. DNA uptake is facilitated by brief heat shock and transformed cells are selected by positive selection on LB plates with the appropriate antibiotic. Each colony on an antibiotic plate represents a single transformation event. The cells carrying plasmid with insert are visually identified on a plate containing chromogenic substrate for β-galactosidase (X-gal) as colorless colonies due to insertional inactivation of enzyme. A simple procedure for preparing and storing competent cells is given.

The efficiency of transformation for this method is between 10^4 to 10^6 transformants/μg of plasmid, depending on the size of the insert and bacterial strain used. This efficiency is generally adequate for most routine cloning procedures. For procedures for which transformation efficiency must be high (i.e., cDNA library preparation, sequencing library preparation, etc.) it is better to use the electroporation procedure. However, if an electroporation apparatus is not available, high efficiency of transformation can be obtained using bacterial strains selected for this. These strains can be bought from several suppliers. The transformation efficiency achieved with these strains is usually 10^9 transformants/μg plasmid. A schematic outline of the procedure is shown in Figure 5.

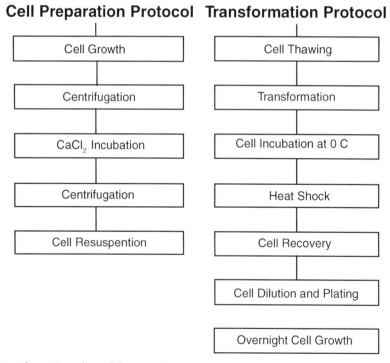

Fig. 5. Schematic outline of the procedure

Materials

- Polypropylene culture tubes (Falcon # 2059 or equivalent).
The "Falcon 2059" tube of Becton Dickson Co. is the standard for transformation experiments. Other equivalent brands are acceptable but batches of tubes are occasionally contaminated with surfactants that inhibit transformation.
- 50 ml Oak Ridge polypropylene centrifuge tubes with caps (for example Nalgene® # 21009).
- IPTG (isopropylthio-β-D-galactoside) Life Technologies, Inc. # 15529-019 or equivalent.
- X-gal (5-bromo-4-chloro-3-indolyl-β-D galactoside) Life Technologies, Inc # 15520-034 or equivalent.
- Dimethyl sulfoxide (DMSO) (Sigma Co. # D 8779 or equivalent).
- Ampicillin, sodium salt (Sigma Co. # A 9518 or equivalent).
- PIPES [piperazine-N,N'-bis(2-hydroxypropanesulfonic acid)] (Sigma Co. # p 6757 or equivalent).
- *E. coli* strain (for example, DH5α) **Do not use an electrocompetent strain of bacteria**. These strains cannot be transformed using chemical methods.
- LB agar-ampicillin plates.
- 100 mg/ml ampicillin stock solution.

- **LB Medium**   Solutions
  - 1% Bacto Tryptone
  - 0.5% Yeast Extract
  - 0.5% NaCl
  Add all ingredients to 1000 ml distilled water and stir until dissolved. Adjust pH to 7.5 by adding 1 N NaOH. This will take about 4 ml of NaOH. Autoclave for 20 minutes. Store at 4 °C.
- **LB Agar Amp Plates**
  - LB Medium
  - 1.5% Difco agar
  Add Difco agar to 1000 ml of LB medium and sterilize by autoclaving for 20 minutes. Cool to 60-65 °C and add 1 ml of ampicillin stock solution. Mix by swirling the flask and pour the plates. This will make 25 to 30 plates. Store at 4 °C. Ampicillin containing plates can be stored for only 2 to 3 weeks.
- **CaCl$_2$ Solution**
  - 50 mM CaCl$_2$
  - 10 mM Tris HCl, pH 8

Solution can be prepared without Tris buffer, but the presence of Tris affords better pH control. Sterilize by autoclaving or by filtration. Store at 4 °C.

Procedure

Preparation of competent cells

The procedure prepares 2.5 ml of competent cells sufficient for 12 transformations. This volume of cells has been chosen so that preparation can be done in a single tube. If larger amounts of cells are required, scale up the procedure appropriately.

1. Grow 5 ml of overnight culture of the bacteria on LB medium.

2. Prepare 40 ml of sterile LB medium in a 250-300 ml Erlenmeyer flask. Inoculate it with 0.4 ml of overnight culture.

3. Grow cells with vigorous shaking at 37 °C to an OD_{600} of 0.4 to 0.6. This will correspond to a cell density about 5×10^7 cells/ml depending on the bacterial strain. Consult Table 1 for the relationship of OD_{600} to viable cell number. This growth usually will take 1.5 to 2 hours.

Table 1. Relation between the number of viable bacterial cells and A_{600}

Strain of E. coli [a]	Cells/ ml at $A_{600} = 1$	A_{600} of 5×10^7 cells/ml
Most of wild type strains	$4 - 8 \ 10^8$	0.1 - 0.2
RecA$^-$ and STBL2 strains [b]	1×10^8	0.5 - 0.6

[a] For example strains C600, DM1, RR1 etc. See for genotypes of these strains in http://www.lifetech.com.

[b] For example strains DHα5, DH10B, STBL2 etc. See genotypes of these strains at http://www.lifetech.com.

Note: For maximum transformation efficiency, it is very important that the bacterial culture be in logarithmic phase of growth, and the cell density be low during calcium treatment.

4. Cool the culture on ice for 10 minutes and transfer it to a sterile 50 ml Oak Ridge centrifuge tube. Cap the tube.

5. Collect cells by centrifugation at 1100 x g (3000 rpm SS-34 rotor, Sorvall) for 5 minutes at 4 °C.

6. Remove the cap from the tube, and briefly flame the mouth. Pour off supernatant. Invert the tube and touch the mouth to a clean Kleenex towel to remove as much as possible of the remaining supernatant.

7. Resuspend the pellet in 20 ml (1/2 the original volume) of a cold $CaCl_2$ solution. Vortex to completely resuspend pelleted cells. It is important that the cell suspension is homogeneous with no visible clumps. Incubate cells on ice tipped at slight angle for 20 to 60 minutes.

Note: For some *E. coli* strains (for example, HB101 or DG75) better transformation efficiencies are obtained when cells are treated with 100 mM $CaCl_2$ solution.

8. Centrifuge cells at 500 to 1000 x g (2000 rpm SS-34 Sorvall rotor) for 5 minutes at room temperature.

Note: A clinical tabletop centrifuge can also be used at 1000 rpm.

9. Pour off supernatant. Be very careful not to disturb the diffuse cell pellet. Remove remaining supernatant with a Kimwipe tissue. Gently resuspend the cell pellet in 2.5 ml (1/15 the original volume) of cold $CaCl_2$ solution. The cells should resuspend very easily by gently flicking the end of the tube with a finger. The suspension should not contain visible clumps of cells. Use 200 µl of cells for a single transformation.

Note: For maximum transformation efficiency, keep cells in $CaCl_2$ solution in 200 µl aliquots at 4 °C for 12 to 24 hours. During this period, the efficiency of transformation increases four to six fold.

10. To store competent cells, add sterile glycerol to final concentration 15% (440 µl glycerol to 2.5 ml cells). Aliquot 200 µl of cells into 1.5 ml microfuge tubes and store at -70 °C. Cells can be stored for several months without losing competency.

Transformation protocol

Three ligation reactions prepared by protocols described in subprotocol 3 will be used. Use recommended amount of ligation reactions for each transformation. Chemical transformation should be carried out with no more than 20 µl of ligation reaction because components of the ligation mixture inhibit transformation.

1. Label three Falcon tubes with numbers corresponding to that of prepared ligation reactions (1C, 2C and 3L). Add 200 µl of competent cells to each tube.

Note: If competent cells were stored at -70 °C, gently thaw them on ice. Tap the tube gently to mix cells.

2. Add the required amount of ligation reaction (between 10 to 20 µl, see description of ligation procedures for details) to the correspondingly numbered tube with cells. Mix by tapping the tubes gently.

3. Place tubes on ice and incubate for 30 minutes.

Note: If bacterial strains JM101, JM103 or JM105 are used, shorten incubation time to 10-15 minutes.

4. Heat shock the cells. Transfer the tubes into a 42 °C water bath for 2 minutes.

Note: The cells should receive a sharp and distinct heat shock. Transfer the tubes directly from the ice into a 42 °C water bath.

5. Add 1 ml of LB medium, prewarmed to 37 °C, to each tube. Incubate the tubes at 37 °C with gentle orbital shaking (about 230 rpm) for 45 minutes. During this time, cells recover and express the antibiotic resistance gene needed for positive selection of transformants.

Cell plating

1. Prepare six LB-amp agar plates for plating transformants. Each sample will be plated onto two plates. Add 480 µl of water, 500 µl of X-gal and 20 µl of IPTG to a plastic microfuge tube. You will need 100 µl of the X-gal IPTG mixture per plate. Add 100 µl of X-gal IPTG mixture onto the surface of each plate. Spread the drop evenly over the surface using a sterile, bent glass rod. The solution should soak into the agar for approximately one hour before plating the cells.

Note: X-gal is light sensitive. Use the plates the same day or store them in the dark at 4 °C.

2. Plate the cells using 1/10 dilution. Add 900 µl of PBS to 3 microfuge tubes and label each tube with the sample name and dilution factor (for example, 3L 1/10, 1C 1/10 etc.). For each transformation reaction prepare 2 plates labeled to correspond to the dilution tubes (for example, 1L, 1L 1/10; 1C, 1C 1/10 etc.).

Note: For optimal cell survival, use only PBS for dilution. Diluting the cells in growth medium instead of PBS will lower cell viability by about 20%.

3. Start plating from the tube labeled 1C. Mix the cells in the tube by gently tapping. Transfer 100 µl to the dilution tube (1C 1/10). This will constitute 1/10 dilution of the original cell culture, mix well by pipetting up and down several times.

4. Immediately, using the same yellow tip, pipette 100 µl of cells from the dilution tube 1C 1/10 onto plate labeled 1C 1/10. Using the same yellow tip withdraw 100 µl from the 1C tube (no dilution) and pipette cells onto plate labeled 1C.

Note: Make one set of dilutions at a time. Finish plating the bacteria from a single tube before preparing the next dilution series. This prevents cross-contamination and plating errors.

5. Sterilize the cell spreader by dipping it into a beaker of 95% ethanol, and briefly passing it through a burner flame to ignite the alcohol. Burn off ethanol keeping the spreader **away** from the burner flame.

6. Cool the spreader by touching it to the agar away from the cells. Spread the cells by dragging the cell suspension across the agar surface with the spreader back and forth several times. Return spreader to the beaker with ethanol.

7. Replace the lid and let the plates stand until all liquid is absorbed into the agar.

8. Dilute and plate the cells from the remaining two tubes using the procedure described in steps 3 to 7. Plate the cells from the **3L tube last**.

Note: Cells diluted in PBS can be stored for one to two days at 4 °C and used for replating if needed.

9. Place the plates upside down in a 37 °C incubator, and incubate them for 15 to 18 hours.

10. Store the transformed cells for additional plating if necessary. Transfer 850 µl of the cells from 3L tube to a microfuge tube and add 60 µl of DMSO. Store tube at -70 °C. Transformed cells can be stored indefinitely.

11. Count the cells on the 3L plate. Calculate the efficiency of transformation expressed as the number of transformants per microgram of plasmid used. Count only white colonies. Use the following equation for this calculation;

$$E_{tr} = \frac{\text{Number of transformed colonies}}{\text{ng plasmid DNA used in transformation}} \text{ x dilution factor} \quad (4)$$

Note: For this calculation, remember that only part of the ligation mixture was used for transformation.

12. Inspect 1C and 2C plates. They should contain a few blue colonies. The presence of a large number of colonies on both plates indicates failure to linearize plasmid and/or incomplete dephosphorylation of its 5' ends. If this is the case, the plate with insert ligated to plasmid (3L) should contain mostly blue colonies.

13. Grow transformants to test for the presence of insert. Prepare several 10 ml glass tubes with 2 ml TB medium supplemented with ampicillin to concentration of 100 µg/ml and inoculate each with a single white colony from a 3L plate. Using a sterile toothpick, touch the colony and drop the toothpick into the medium. Grow cells overnight. Isolate plasmid from 400 µl of cells (Chapter 5, fast protocol 1) and test for the presence of insert using PCR or restriction enzyme digestion.

14. Store cells containing plasmid with the desired insert. Transfer 930 µl cells from a 2 ml overnight culture (step 13) to a 1.5 ml microfuge tube and add 70 µl of DMSO. Mix by inverting 2 to 3 times and store at -70 °C. To inoculate cells from the storage culture, pick up a little "chunk" of frozen cells with a yellow tip and transfer it into medium.

▦ Troubleshooting

Transformation efficiency, using the CaCl$_2$ protocol, primarily depends on using highly competent cells. Preparation of competent cells depends on: (a) harvesting bacterial cultures in logarithmic phase of growth, (b) keeping cells on ice throughout the procedure, and (c) prolonged CaCl$_2$ exposure. Usually only 3% to 10% of cells are competent to incorporate plasmid DNA. The expected efficiency of transformation for plasmid with an insert no larger than 5 Kbp is about 10^4 to 10^6 transformants/µg plasmid. This would result in about 12 to 1500 colonies on a 1/10 dilution plate. If there are no colonies on the 1/10 dilution plate, replate 100 or 200 µl of transformed cells directly using stoned cells prepared in step 10.

Transformation frequencies of plasmid with large insert (10 Kbp) will be at best 10^4 transformants/µg plasmid. This is because uptake of large plasmid by competent cells is very inefficient. Plate 100 µl of cells directly from the transformation reaction after recovery.

Expect a large amount of blue colonies on experimental plate. It is not unusual to get 50% blue colonies on this plate. This does not indicate failure of the ligation reaction, or incorrect plasmid or fragment preparation, providing both control plates contain only a few blue colonies.

Low frequencies in transformation can result from:

- Excess DNA in the transformation reaction. Use at least a 3 x dilution of the ligation reaction for transformation.

- The competent cells were handled improperly. If stored cells are used they should be thawed on ice. Otherwise use competent cells immediately after preparation.

- Tubes other than Falcon were used for heat shock. Heat shock time is calibrated to the size and material of these tubes.

- Viability of competent cells is low. Check cells viability by plating untransformed cells on LB plates without ampicillin. Concentration of competent cells is about 1-5 x 10^8 cells/ml. Plate 100 µl of 10^6 cell dilution. The plate should contain about 50 to 100 colonies.

- The cloned insert is toxic to the cells. White colonies on the plate will be very small and it may be difficult to grow cells to high density in 2 ml cultures. Grow cells at 30 °C and/or add glucose to 20 mM in TB medium. If the problem persists, reclone the fragment into a low copy number plasmid and use a different bacterial host strain. In general, about 20% of genomic DNA fragments from higher eukaryotes are difficult to clone or are not clonable in *E. coli* cells. Use yeast for cloning these fragments.

Other problems frequently met in cloning DNA fragments are:

- The presence of all blue colonies on a transformation plate (3L) with concomitant absence of blue colonies on both control plates (1C, 2C). This can happen when an insert was cloned in frame with the α-peptide or when insert DNA is small (<200 bp). In the last instance, color development will be weak and colonies may appear pale blue. Check pale-blue colonies for the presence of insert as described in step 21 of the electroporation protocol.

- The absence of insert or rearranged insert in cells from white colonies. This is a frequent problem with cloning large fragments (> 10 Kbp) or fragments containing direct or inverted repeats. To remedy this, try growing cells with insert on TB medium at 30 °C and collect them for plasmid isolation in the mid to late period of logarithmic growth ($OD_{600} = 1$-2). If this does not help, transform the plasmid into a host strain specially developed to inhibit rearrangements and eliminate direct and inverted repeats. Several such strains are available: SURE from Stratagene, STBL2 from Life Technologies, Inc., or PMC103 (Doherty et al. 1993).
  Sometimes white colonies with recombinant plasmid can be distinguished from those with plasmid alone by colony morphology. Recombinant colonies appear translucent while non-recombinant colonies are opaque (Austin et al. 1994).

- Appearance of satellite colonies on ampicillin plates. "Feeder" colonies appear near ampicillin resistant colonies because the β-lactamase enzyme, responsible for antibiotic resistance, is secreted from the cell, removing antibiotic from the agar in the vicinity of the colony. When an ampicillin-resistant plasmid is used, dilute cells to plate no more than 500 cells/plate. Incubating the plates for a shorter time (less than 14 hr) can also eliminate appearance of satellite colonies. To enhance blue color development, incubate plates at 4 °C for several hours. Alternatively, use carbenicillin (200 µg/ml) or a mixture of ampicillin (20 µg/ml) and methicillin (80 µg/ml) instead of ampicillin alone. Both of these antibiotics are available from Sigma Company, and Life Technologies Inc. However, use of these antibiotics substantially increases the cost of plating.

References

Austin RC, Singh D, Liaw PCY, Craig HJ 1994 Visual detection method for identifying recombinant bacterial colonies. BioTechniques 18:381-383.
Boehringer Mannheim 1996 Rapid DNA ligation kit manual.
Bercovich JA, Grinstein S, Zorzopulos J 1992 Effect of DNA concentration on recombinant plasmid recovery after blunt-end ligation. BioTechniques 12:190-193.
Böttger EC 1988 High-efficiency generation of plasmid cDNA libraries using electro-transformation. BioTechniques 6:878-
Brooks JE 1987 Properties and uses of restriction endonucleases. Methods Enzymol 152:113-129.
Doherty 1993 *Escherichia coli* host strains SURE and SRB fail to preserve a palindrome cloned in lambda phage: improved alternate host strain. Gene 124:29-35.

Dower WJ, Miller JF, Ragsdale CW 1988 High efficiency transformation of *E. coli* by high voltage electroporation. Nucl Acids Res 16:6127-614

Dugaiczyk A, Boyer HW, Goodman HM 1975 Ligation of *Eco*RI endonuclease-generated DNA fragments into linear and circular structures. J Mol Biol 96:171-184

Epicentre Technologies 1997 K Thermolabile Phosphatase product information.

Fromm M, Taylor LP, Walbot V 1985 Expression of genes transferred into monocot and dicot plant cells by electroporation. Proc Natl Acad Sci USA 82:5824-5828

Fuchs R, Blakesley R 1983 Guide to the use of type II restriction endonucleases. Methods Enzymol. 100:3-38.

Hodge R 1994 In: Protocols for nucleic acids analysis by nonradioactive probes. Method in Molecular Biology Volume 28. Ed Isaac PG Humana Press. Totowa, New Jersey.

Kranz R 1989 Convenient and inexpensive system for low temperature preservation of enzymes during storage and use. BioTechniques 7:455-456.

Li SJ, Landers TA, Smith MD 1980. Electroporation of plasmids into plasmid-containing *Escherichia coli*. BioTechniques 12:72-74.

Life Technologies Inc 1996 Tech-online. Rapid ligation protocol for plasmid cloning. www. lifetech.com/cgi-online/techonline?Document=rapidt.txt

Murphy NR, Hellwig RJ 1996 Improved nucleic acid organic extraction through use of a unique gel barrier material. BioTechniques 21:934-939.

Neumannn E, Schaefer-Ridder M, Wang Y, Hofschneider PH 1982 Gene transfer into mouse myaloma cells by electroporation in high electric field. EMBO J 1:841.

New England BioLabs. www.neb.com

Roberts RJ, Macelis D 1991 Restriction enzymes and their isoschizomeres. Nucleic Acids Res (suppl) 19:2077-2109

Roberts RJ, Macelis D 1997 REBASE-restriction enzymes and methylases. Nucl Acids Res 25:248-262.

Robinson CR, Sligar SG 1995 Heterogeneity in molecular recognition by restriction endonucleases: Osmotic and hydrostatic pressure effects an BamHI, PvuII, and EcoRV specificity. Proc Natl Acad Sci USA 92:344-3448.

Rodriguez RL, Tait RC 1983 Recombinant DNA techniques: An introduction. Addison-Wesley Publishing Co. Reading Mass. ISBN 02-201-10870-4.

Smith HO, Wilcox KW 1970 Restriction enzymes from *Hemophilius influenzae*. Purification and general properties. J Mol Biol 51:379-391.

Smith M, Jesse J, Landers TA, Jordan J 1990 High efficiency bacterial electroporation: 1 x 10^{10} *E. coli* transformants/µg. Focus 12:38-40

Wai LT, Chow K-C 1995 A modified medium for efficient electrotransformation of *E. coli*. Trends in Genetics 11:128-129.

DNA Sequencing

STEFAN SURZYCKI

Introduction

DNA sequencing is the most powerful technique used in molecular biology. Determination of a DNA sequence is the only method in biological science that generates data that are not biased by previous assumption, hypotheses or experimental design. It is therefore not surprising that DNA sequencing has revealed many unexpected facts concerning gene structure, regulation of gene expression, organization of genomes, as well as, discovered new genes never seen before. Advances in large scale sequencing also brought about new scientific disciplines - genomics and functional genomics, devoted to analyzing whole genomes and their function (Hieter and Boguski, 1997). Knowing the complete sequences of a dozen microbial genomes and expected completion of sequencing of many other organisms, including humans, will without doubt change how biology is done in the future (Doolittle, 1998; DeRisi et al. 1997; Tatusow et al., 1997).

DNA sequencing methods

All of the present methods of DNA sequencing are based on the anchored-end principle and most use gel electrophoresis for separation of different size DNA fragments. In these methods, one end of the sequenced DNA molecule remains unchanged (anchored) while the other is generated in base-dependent way. This creates sets of DNA molecules of various lengths having one end common and the other end terminating at a specific base. Separation of these molecules according to length generates the base sequence of the fragment. The separation is usually accomplished by gel electrophoresis using a matrix capable of distinguishing between two DNA molecules differing only by a single nucleotide.

There are two methods that generate DNA molecules of different lengths in a base specific manner. The first method introduced by Maxam and Gilbert (1980), uses base-specific chemical cleavage of the DNA fragment. The second method uses the enzymatic synthesis of DNA fragments (Sanger et al., 1970). In the Maxam and Gilbert method of chemical degradation, one end of DNA is labeled and four separately run base-modification reactions are performed. In each reaction one of the bases (A, C, G, or T) is modified. The DNA backbone is then cleaved at each modified residue resulting in a "nested set" of fragments all labeled at one end and terminating at the location of a specific modified base. Fragments are then separated using gel electrophoresis. This method of DNA sequencing is rarely used at the present time.

The enzymatic method of DNA sequencing utilizes properties of DNA polymerase to implement the anchored-end principle. DNA polymerase can synthesize a complementary strand from a single-stranded template. Initiation of this synthesis is dependent on the presence of a primer with a 3'-hydroxyl group. Also prokaryotic DNA polymerases are able to incorporate a dideoxynucleotide instead of deoxynucleotide into the growing DNA chain. Thus, all synthesized DNA molecules will share an identical 5' end, the 5'end of the primer, and the 3'end will terminate at a specific bases by incorporation of the substrate lacking a 3'-hydroxyl residue.

Four separate primer extension reactions are initiated using the same primer. Each reaction contains all four usual 2'deoxynucleotides (dNTPs), but only one of the four 2', 3' dideoxynucleotides (ddNTPs). By carefully controlling the ratio between dNTP to ddNTP in each of the four reactions, incorporation of the dideoxy nucleotide, and hence chain termination, is random. The end result is the generation of DNA fragments of different lengths, each terminated at the 3'-end at a specific base. These fragments are then separated using gel electrophoresis.

The length of DNA molecule that can be sequenced using the anchored-end principle depends, not on the ability to create different sized "anchored" fragments but on the resolving power of the gel matrix used to separate these fragments. This is because DNA cleavage in chemical sequencing does not depend on the length of the molecule and DNA polymerase is able to synthesize long DNA pieces. The resolving power of matrices presently used permits separation of DNA fragments up to about 1000 bases setting the limit on the length of DNA fragment that can be sequenced in a single reaction.

Sequencing strategies

Sequence determination of a DNA fragment smaller than 1000 bases is relatively simple. It requires cloning the fragment into an appropriate single- or double-stranded DNA sequencing vector, sequencing the fragment in a single sequencing reaction and running gel electrophoresis to separate these fragments. To sequence a large DNA molecule it is necessary to subdivide this fragment into smaller fragments of about 1000 bases long. Each small fragment is sequenced separately in a single sequencing reaction and the sequence of the whole fragment is assembled. The way in which smaller fragments are generated from the large DNA fragment and then assembled into the sequence of the whole fragment is referred to as the **sequencing strategy**.

There are two groups of sequencing strategies presently in use: directed strategies, in which specific starting points are used for sequencing, and random strategies in which starting points for sequencing are random (Hunkapiller et al., 1991). The directed strategies permit the direct sequencing of a large region by generating small fragments, the position of which in the whole molecule is known. Usually sequencing proceeds successively from one end of the large fragment to the other. Random strategies (shotgun strategies) involve: (a) sequencing of fragments generated by random shearing of a large piece of DNA or, (b) random insertions of a universal primer site into target DNA using a transposon.

Primer walking, or primer-directed strategy, is the most frequently used directed strategy and can be used to sequence DNA fragments of 10 000 bases or longer. The entire fragment is cloned into a sequencing vector and the initial sequence data are obtained using a vector-based universal primer. The sequences obtained are used to synthesize a new primer that hybridizes near the 3' end of the newly elucidated sequence. This primer is used to sequence the next DNA fragment (Strauss et al., 1986). The cycle of sequencing and primer synthesis is repeated until the whole fragment is sequenced. The technique is uniquely suited to the dideoxy DNA sequencing method and bypasses the need for subcloning smaller pieces of DNA.

In the "single-barrel" random sequencing strategy, randomly generated fragments are subcloned into a sequencing vector, forming a sequencing library of the fragment. Next, the fragments are randomly chosen for sequencing from this library. Identifying overlaps between small fragments and arranging them into the most probable order assembles the sequence of the original piece of DNA. The number of clones necessary to assemble the whole fragment (**subclone coverage**) and the amount of raw data (**sequence**

coverage) are directly proportional to the size of the target DNA. Equation 1 gives the relationship between these parameters (Deininger, 1983):

$$S = 1 - \left(1 - \frac{i}{L}\right)^n \tag{1}$$

Where S is the fraction of the fragment sequenced; i - average bases read per clone; L - length of the whole fragment (bp); n - number of clones of length i sequenced. Since the average length of sequence read per subclone is usually about 400 bases, to assemble 95% of the sequence requires n = 3L/400 subclones to be sequenced giving a sequence redundancy of 3. For example for a 40 000 bp fragment, sequencing 300 clones of 400 bases gives 95% of the whole sequence [0.95 = 1 - (1- 400/40 000)^n]. For both strand coverage of the same fragment the number of clones should be doubled so n will be equal to n = 6L/400 giving a redundancy of 6 and sequencing of the 99.7% of whole fragment. In sequencing projects redundancy coverage 7 is customary used giving sequence of 99.99% of whole fragment (n = 7L/i).

Random strategy using the chain termination method of DNA sequencing is a method of choice for sequencing large DNA fragments or whole genomes. Directed strategies, such as primer walking and deletion strategy, are used for sequencing shorter DNA fragments or for filling gaps between contigs.

Preparation of a sequencing library

Application of random strategy for sequencing large pieces of DNA requires the preparation of a sequencing library that contains DNA fragments about 1000 to 2000 bp long, randomly generated from the original DNA.

There are four methods used to generate random DNA fragments. The first method employs partial restriction enzyme digestions. The second, involves fragmentation of DNA by DNase I in the presence of Mn^{++}, the third, relies on sonication to physically break the DNA, and the forth uses a nebulization process to shear DNA. There are a number of major disadvantages in using the first three methods. To these belong: (a) these methods frequently do not generate fully random libraries, (b) their application is laborious and requires a large amount of material, (c) methods are difficult to reproduce, and (d) using these methods it is difficult to obtain DNA fragments having the desired size.

The nebulization method avoids all of these difficulties and is frequently used now for preparation of random sequencing libraries (Surzycki 1990).

The method works on the following principle. In the process of nebulization, or reducing liquid to a fine spray, small liquid droplets of uniform size are produced. In the process of droplet formation the liquid being nebulized flows from the liquid surface to the forming bubble. This creates a transient flow between the liquid surface and the droplet through a connecting small capillary. The diameter of this capillary channel is approximately half the diameter of the forming droplet and can be adjusted, as desired, by controlling the size of the droplets. The velocity of flow of liquid in the capillary is not constant across the capillary due to frictional resistance between adjacent layers of flowing liquid. The velocity gradient generated causes liquid in the center to flow faster than liquid in the outer layers, creating flow called **laminar flow**. Because the velocity of flow is not constant across the capillary tube, the DNA molecule that finds itself in two adjacent flow layers will have its ends moving at two different velocities. This results in stretching and rotating of the molecule until it is positioned in a single laminar layer or broken at the point where the stretching force is maximal.

For large, rigid, linear molecules, such as DNA, the stretching force is the greatest at the middle of the molecule, consequently the molecule will have the greatest probability of breaking in half. The nebulization of DNA will result in the breakage of each molecule almost exactly in half in the repeated process of bubble formation. This will continue until the molecule reaches a small enough size that it cannot be positioned across two laminar flow layers. The final size of the broken molecules will depend only on the size of droplet formed, i.e., the size of capillary nebulization channel, but **not on the time of nebulization**.

The nebulization process permits the regulation of the size of DNA fragments generated. The formation of droplets can be considered at two levels of nebulization, the primary nebulization process that results in the formation of primary droplets and, the secondary nebulization process resulting in the formation of secondary droplets by the shattering of the primary droplets on the surface of the nebulization sphere. At both sides droplets are formed by laminar flow of the liquid generating force that is formally described by the equation of liquid capillary flow. Accordingly, this force is directly proportional to the gas pressure applied and the viscosity of the liquid, and inversely proportional to the size of the droplets. Consequently, the smaller the droplets, the higher viscosity and greater gas pressure applied, the larger the force that creates smaller DNA fragments.

The droplet diameter that is necessary to shear DNA molecules is on the order of 0.1 - 2.0 microns. This droplet size is created only at the site of secondary nebulization. Inspection of the formula that describes the droplet size formed during the secondary nebulization process indicates that the size of the droplet depends on: (a) a velocity of the primary droplets, (b) absolute viscosity of the gas used and (c) the diameter of droplets generated in the primary nebulization process.

To decrease the diameter of primary droplets, DNA is nebulized in a solution of 25% glycerol. Moreover, nitrogen or argon is used to achieve droplets with the correct diameter in secondary nebulization. Because

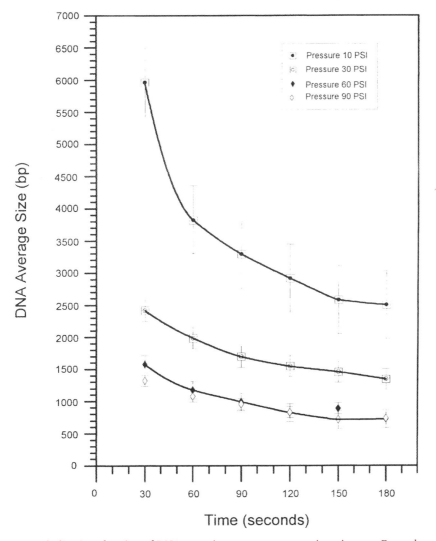

Fig. 1. Nebulization shearing of DNA at various gas pressures using nitrogen. Errors bars represent variation of the average size of DNA fragments obtained in 10 independent experiments. Nebulized sample volume is 2 ml.

the absolute viscosity of these gases is high and the viscosity of nitrogen is very close to the absolute viscosity of argon these gases can be used interchangeably. Use of air, that is a mixture of gases of different absolute viscosity, generates a different size droplet that consequently leads to a broad distribution of fragment sizes. Figure 1 presents the size distribution of DNA fragments as the function of gas pressure and time of nebulization using nitrogen.

Subprotocol 1
Preparation of Clones for Sequencing

Preparation of the DNA fragment for sequencing requires cloning the fragment into a sequencing vector. Vectors used for this task are general-purpose vectors containing universal -forward and -reverse M13 sequencing primers. The pUC series plasmids are usually employed as sequencing plasmids. To avoid recloning DNA fragments for sequencing, it is best to use a pUC plasmid for the original cloning. The techniques used for cloning DNA fragments into plasmid are described in Chapter 14. These are (a) Preparation of plasmid and DNA fragment; (b) Ligation of plasmid with insert; (c) Transformation of bacterial cells with the construct and, (d) selection of transformants.

Similar steps are required for sequencing large DNA fragments using the random sequencing strategy. However, before a large DNA fragment can be sequenced, DNA must be fragmented into a collection of random fragments that are recloned into the sequencing vector. The cloned collection of all random fragments constitutes a **sequencing library** of given DNA. Preparation of a sequencing library involves the following steps: (a) Fragmentation of the DNA using the nebulization procedure. (b) Repairing the ends of the sheared DNA fragments. (c) Blunt-end ligation of repaired fragments into pUC plasmid and, (d) transforming bacteria with chimeric plasmids using electroporation.

This section describes a procedure for DNA shearing using nebulization. Procedures for preparation of the vector, repairing DNA fragments, blunt-end ligation of fragments into a pUC vector and transformation of *E. coli* cells by electroporation are identical to those described in Chapter 14. For preparation of sequencing libraries the best results are obtained when using the rapid ligation protocol described in Chapter 14. A schematic outline of the procedure is shown in Figure 2.

Fig. 2. Schematic outline of the procedure

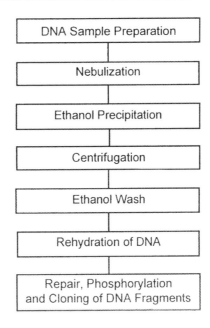

Materials

- AeroMist disposable inhalator (Inhalation Plastic, Inc. # 4207 or Invitrogen Co. #).
- QS-T cap (Isolab, Inc. Akron OH 44303)
  Caps are optional and can be substituted by self-made valves (see protocol).
- Compressed nitrogen gas cylinder with pressure regulator.
  Laboratory in-line compressed air can be substituted for nitrogen gas.

Solutions

- **TE Buffer**
  - 10 mM Tris HCl, pH 7.5 or 8.0
  - 1 mM Na_2EDTA
  Sterilize by autoclaving and store at 4 °C.
- **7.5 M Ammonium Acetate Solution**
  Dissolve 57.8 g of ammonium acetate in 70 ml of double-distilled or deionized water. Stir until salt is fully dissolved, do not heat to facilitate dissolving. Fill up to 100 ml and sterilize by filtration. Store tightly closed at 4 °C. Solution can be stored for one to two months. Long term storage is possible at -70 °C.

▒▒ Procedure

To create a complete sequencing library for a 30 to 100 Kbp fragment, between 0.5 to 2 µg of DNA should be nebulized. Purity and size of DNA is not critical for successful nebulization. A very good library can be prepared using DNA as small as 50 to 30 Kbp. The nebulization volume used here is 2 ml, but procedures can be carried out with only 1 ml of DNA. Because the time of nebulization depends only on the volume of nebulized liquid, when using 1 ml of DNA solution nebulization time should be shortened by half. The breathing hole of the inhalator should be closed to limit DNA lost with escaping mist. A cut end of a 15 ml plastic centrifuge tube can be used (Surzycki, 1990). The opening can also be covered with a QS-T cap (Roe and Crabtree, 1995; Hengen, 1997).

To obtain library of uniform fragment size, nitrogen or argon gas should be used for nebulization. Substituting any other gas or compressed air is possible but will result in a broader size distribution of fragments.

1. Prepare DNA for nebulization shearing. Add the DNA to be nebulized to sterile 15 ml centrifuge tube. Use between 0.5 to 5 µg of DNA. Add 25% glycerol in TE buffer to a final volume of 2 milliliters. Mix by inverting the tube.

2. Transfer the DNA solution into the bottom of the nebulization apparatus. Close the nebulizer top and insert the "closing valve" into breathing hole of the inhalator.

3. Attach the instrument to a nitrogen gas tank pressure regulator using the plastic tubing provided.

Note: The nebulizer can be attached to laboratory compressed air line.

4. Set the gas pressure regulator to the appropriate pressure to obtain the desired size of DNA fragments. See Figure 1 for relation between gas pressure and DNA fragment size. Do not exceed a pressure of 30 psi (2.0 atm). Open the gas pressure regulator valve and nebulize for 180 seconds. You will see a little mist escaping under the "closing valve". This indicates that the nebulizer is operating properly.

5. Tap the nebulization apparatus slightly on the laboratory bench to collect all droplets at the bottom of the reservoir. Alternatively, the whole nebulizer can be placed in Sorvall GSA rotor and centrifuged for 1 minute at 1000 rpm (160 x g).

6. Transfer liquid from the nebulizer chamber into a 15 ml Corex centrifuge tube. Measure the volume of the liquid during this transfer.

7. Add 1 ml of 7.5 M ammonium acetate (half the volume) and mix by inverting tube several times.

8. Add 6 ml of 95% ethanol (2 x combined volume of DNA and ammonium acetate) to precipitate DNA. Mix well by inverting the tube several times. At this step the tube can be stored at - 20 °C overnight.

9. Centrifuge the tube to collect DNA in a Sorvall GS-3 rotor (swinging bucket) at 10 000 rpm (16 000 x g) for 15 minutes at room temperature. Pour off and discard the supernatant.

10. Wash the pellet with 70% ethanol. Slowly add 5 ml of cold 70% ethanol to the tube and keeping the tube at a 45° angle, gently rinse the pellet and sides of the tube. Discard the 70% ethanol and repeat the wash one more time. Be careful not to disturb the pellet at the bottom of the tube during ethanol washing. To collect the ethanol remaining on the side of the tube, centrifuge for 30 seconds. Remove ethanol from the bottom of the tube using a micropipettor fitted with a capillary tip.

11. Dissolve the pellet in 20 μl of TE buffer and transfer the solution to a 1.5 ml microfuge tube. Repair the ends of the entire preparation using the protocol for end-repair of DNA fragments. Ligate repaired fragments using the rapid ligation procedure (see Chapter 14).

12. Clone DNA fragments into pUC plasmid using the electroporation protocol described in Chapter 14.

Results

Figure 3 presents gel electrophoresis of 40 kb DNA nebulized for various times using nitrogen at 60 psi (4 atm). Increasing the time of nebulization to longer than 90 seconds does not generate smaller fragments. This is because the time of nebulization depends on the volume of the liquid to be nebulized. The size of DNA fragments generated depends only on the gas pressure applied and consequently on the droplet size. The minimum amount of the liquid that can be used in an inhalator type nebulizer is 1 ml. A DNA nebulizer especially designed for DNA shearing of smaller volumes can be purchased from Glas-Col Co. USA or Invitrogen Co.

Size fractionation of nebulized DNA by gel electrophoresis is not necessary. More than 85% of DNA fragments are of the desired size because the

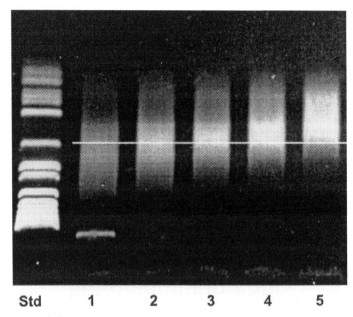

Fig. 3. Agarose gel electrophoresis of 40 kbp DNA nebulized for various times using nitrogen at 60 psi. Lane 1 nebulization for 30 seconds; Lane 2 nebulization for 40 seconds; Lane 3 nebulization for 90 seconds; Lane 4 nebulization for 180 seconds. Std lane is a molecular weight standard of 1 kb ladder from Life Technologies Gibco BRL. White line indicates position of 1 kb DNA fragment.

nebulization process cannot generate DNA fragments smaller than that set by droplet size (see explanation, Introduction of this chapter). Larger fragments can be generated by incomplete nebulization, but even when they are present, their cloning efficiency is low and, these fragments are not present in the sequencing library.

Incomplete nebulization can result from a defective nebulizer, a volume that exceeds 2 ml of nebulized DNA or low gas pressure. It is important to realize **that circular and supercoiled DNA** is not sheared by nebulization.

Subprotocol 2
Sequencing Reaction

Two methods for manual sequencing are described. One uses extended sequencing with the enzyme Sequenase and the other, cycle sequencing using SequiTherm EXCEL II polymerase (Meis and Schanke, 1997). Both methods use internal labeling with ^{35}S dATP and double stranded plasmid templates.

The principal advantage of internal labeling with ^{35}S for manual sequencing is its low-energy β-emission and relatively long half-life. The weaker emission diffuses less during autoradiography permitting more of the gel to be read due to tighter bands in the upper region of the gel. The intensity of the bands is uniform throughout the autoradiogram. Exposure time required is usually 12 - 17 hours (overnight).

Radioactivity is not used in sequencing reactions for automated sequencers. Follow the procedures described by manufacturer of the sequencer when preparing these sequencing reactions.

The method of manual sequencing presented here uses double-stranded plasmid DNA as a sequencing template. It is not necessary to reclone the fragment into a single-stranded vector. To prepare double-stranded template the plasmids are purified using any procedure described in Chapter 5. Plasmid DNA, purified by boiling procedures or organic solvent extraction, often does not serve as a good template for sequencing with Sequenase. A schematic outline of the procedure is shown in Figure 4.

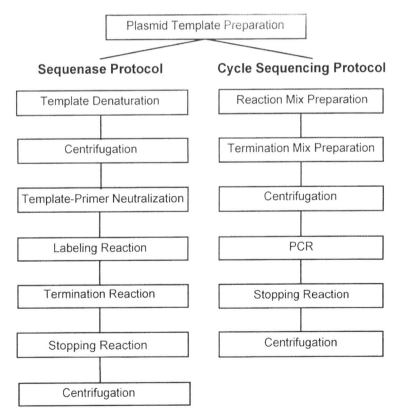

Fig. 4. Schematic outline of the procedure

Materials

- T7 Sequenase version 2 sequencing kit (Amersham Life Science Inc. # US70770)
  This Sequenase sequencing kit contains all reagents necessary for sequencing 100 plasmid templates at a cost lower than the cost of the individual reagents bought separately.
- SequiTherm EXCEL II DNA sequencing Kit. (Epicentre Technologies, Inc. # SEM 79020)
  The kit contains all reagents necessary for sequencing 20 plasmid templates at a cost lower than the combined cost of the individual reagents.
- Universal M13/pUC sequencing primer (-20) (New England BioLabs # 1211, or equivalent).
- Universal M13/pUC reverse sequencing primer (-48) (New England Bio-Labs # 1233, or equivalent).
- ^{35}S-dATP (>1000 Ci/mmol) (Amersham Life Science Inc. # SJ1304, or equivalent)

Solutions
- **10 x HSNA Sequencing Buffer**
  - 0.4 M HEPES pH 7.5
  - 0.2 M $MgCl_2$

  Prepare from stock solutions of 1.0 M HEPES (pH 7.5) and 1.0 M $MgCl_2$. Sterilize by filtration and store at - 20 °C.

Procedure

Sequencing of DNA using Sequenase uses a procedure in which denaturation of the plasmid template and annealing the primer to it is performed in a single-step. This procedure is designed to run many reactions at once by starting them by centrifugation. The protocol of plasmid denaturation does not use the customary ethanol precipitation and washing of denatured plasmid. Plasmid is denatured by NaOH in the presence of primer. This template-primer mixture is incubated at 37 °C to hydrolyze RNA. Thus, removing RNA from plasmid preparation is not necessary when using this protocol. The template-primer mixture is neutralized by quench reagent containing polymerase reaction buffer, $MgCl_2$ and an equimolar amount of HCl. During neutralization the primer anneals to the template and enough NaCl is generated for optimal activity of Sequenase.

The reaction conditions presented here differ from those recommended by the manufacturer of Sequenase. The concentration of the labeling mix is doubled to compensate for the higher Km of the enzyme for substrate with double-stranded template and the concentration of template is increased. To prepare template DNA, use the procedure described in Chapter 5, rapid procedure 2. This procedure isolates plasmid DNA from 1.5 ml of cells with an approximate yield of 10 to 20 µg at a concentration of 1 to 2 µg/µl. It is sufficient for 2 to 3 sequencing reactions. Using the condition described for sequencing reactions it is possible to sequence DNA fragment about 1000 bases long.

Sequencing using the cycle reaction with SequiTherm EXCEL II polymerase follows a protocol recommended by the manufacturer. To prepare template DNA for cycle sequencing, use plasmid prepared by the protocol described in Chapter 5, rapid protocol 2 or rapid protocol 1.

Sequenase protocol

This protocol describes a procedure for sequencing three templates. The amount of ingredients necessary for a single reaction is given for each recipe. If more templates are going to be sequenced scale up the protocol accordingly.

1. Label six 1.5 ml microfuge tubes RX1, RX2, RX3, PD, Q and ML. Place all tubes on ice.

2. Prepare 1 N NaOH from a 10 N stock solution. This solution should be prepared fresh and kept at room temperature.

3. Prepare master **Primer-Denaturation Mix (PD)** for 3 reactions in the tube labeled PD. A small excess of PD mix is prepared to compensate for pipetting losses.

Primer-Denaturation Mix

Ingredient	One reaction	Three reactions
Water	1.5 µl	5.0 µl
Primer (5 pmoles/µl)	1.5 µl	5.0 µl
NaOH 1 N	1.5 µl	5.0 µl

To calculate nonagrams of primer necessary to have 5 pmoles use the following equation:

5 pmoles of primer = (# of nucleotides in primer x 0.33) x 5 ng.

4. Prepare the **Quench Reagent (Q)** in the tube labeled Q. The threefold excess of this mix is prepared to minimize pipetting errors.

Quench Reagent

Ingredient	One reaction	Six reactions
Water	5.5 µl	33.0 µl
HSNA 10 x buffer	1.0 µl	6.0 µl
HCl 1 N	0.5 µl	3.0 µl

5. Add 2.5 µl of the DNA (2 to 3 µg) template to tubes labeled 1, 2 and 3. Add 1.5 µl of the PD reagent, prepared in step 2, to each tube. Mix well by pipetting up and down. Cap the tubes and incubate for 10 minutes at 37 °C.

Note: The amount of template DNA for a single reaction should not be less than 1 µg and should not exceed 4 µg.

6. Remove the tubes from the bath and centrifuge for 30 seconds.

7. Add 7 µl of Quench Reagent to each tube. This mixture constitutes the Template-Primer mix. Mix well by pipetting up and down and return the tubes to the 37°C water bath. Incubate for 5 minutes. Place the tubes on ice.

8. Check that each tube contains 11 µl of solution using a micropipettor set to 11 µl. If the volume of the mix is less than 11 µl, add the required amount of water to obtain a final volume of 11 µl.

9. Prepare enzyme mixture. Dilute Sequenase in glycerol-enzyme dilution buffer and add pyrophosphatase as recommended by the manufacturer. Buffer and pyrophosphatase are provided with the kit. Place the enzyme mixture on ice. This mixture can be stored at -20 °C for several months.

10. Prepare **Master Labeling Solution** in the tube labeled ML as follows. An excess of this mix is prepared to compensate for pipetting losses.

Master Labeling Solution

Ingredient	Single reaction	Three reactions
Water	1.0 µl	4.0 µl
DTT 0.1 M	1.0 µl	4.0 µl
Labeling Mix (5x)	1.0 µl	4.0 µl
^{35}S-dATP (10 mCi/ml)	0.5 µl	2.0 µl

Add reagent in the order indicated and mix by vortexing. Centrifuge for 30 seconds. Place the tube on ice.

11. Label three sets of 1.5 ml microfuge tubes: A1, C1, G1 and T1 for first set, A2, C2, G2 T2 for the second etc. Place all tubes on ice.

12. Add 2.5 µl of the ddATP Termination Mix to three tubes labeled A. Using a fresh tip for each operation, add the ddCTP, ddGTP and ddTTP Termination Mixes to tubes labeled C, G and T, respectively. Cap the tubes to prevent evaporation. Keep them on ice until needed.

13. Place tubes with template-primer mix labeled rx1, rx2 and rx3, prepared in step 7, in a microfuge. Keep the caps of the tubes open.

14. Add 3.5 µl of the Master Labeling Solution, prepared in step 10, to the top of each tube. Make sure that this drop does not "roll" to the bottom of the tube.

15. Mix labeling reactions with DNA template by centrifugation for 5 to 10 seconds.

16. Add 2 µl of enzyme mixture, prepared in step 9, to the top of each tube. Start the reaction by mixing the reagents by centrifugation for 5 to 10 seconds. **Start the timer for 10 minutes immediately after starting centrifugation.** Remove tubes from the centrifuge and place them into tube rack on the bench top.

Note: The following steps of distributing mixtures onto the top of tubes with termination mixtures must be finished within 10 minutes set for the labeling reaction time.

17. Place tubes with termination mixture into the microfuge. Keep the tubes open.

18. Remove 3.5 μl from tube rx1, prepared in step 16, and pipette it to the top of the tube labeled A1 containing ddATP termination mixture. Make sure that drop does not roll down to the bottom of the tube. Remove the next 3.5 μl from the same reaction tube and deliver it to the top of the tube with ddCTP termination mixture labeled C1. Repeat this procedure for the G1 and T1 tubes, always withdrawing 3.5 μl from the reaction tube. The total volume of the reaction mixture in rx1 tube is 16 μl, sufficient to pipette a total of 14 μl to A, C, G, and T termination mixture tubes (4 x 3.5 μl = 14 μl).

Note: You must discard the yellow tips used in the above procedure into the radioactive waste if ^{35}S labeling is used.

19. Follow the procedure described in step 18 for the two remaining tubes with reaction rx2 and rx3. Add 3.5 μl of reaction to corresponding tubes with termination mixture. Use fresh yellow tip for each rx series.

20. When the 10 minutes set in step 16 are up, start termination reaction by centrifuging 5 to 10 seconds. Centrifuge speed should reach only 1000 to 2000 rpm. At this time, start the timer for **another 10 minutes**.

21. Transfer the tubes with termination reaction from the microfuge to a 37 °C water bath. Tubes should be transferred to the water bath as fast as possible. Continue incubation for 10 minutes. Incubation can usually be extended to 30 minutes without problems.

22. Add 4 μl of stop solution, provided in the kit, to each Termination Reaction. Mix thoroughly by pipetting up and down. Use a fresh yellow tip for each addition. Store at - 70 °C until ready to load onto the sequencing gel. Samples labeled with ^{35}S can be stored at this temperature with little or no degradation for at least a week and probably longer.

Cycle sequencing protocol

This protocol describes sequencing a single DNA template. If more templates are going to be sequenced scale up the protocol accordingly.

1. Prepare template DNA. The amount DNA required for single cycle sequencing is 500 fmoles. Calculate amount of DNA using equation:
   500 fmoles DNA = # of Kbp (insert + vector) x 330 ng.
   For example 500 fmoles of pUC plasmid with 1000 bp insert is 1.2 μg (3.7 x 330 ng = 1.2 μg).

2. Label five 0.5 ml microfuge tubes RX, A, C, G and T. Place the tubes on ice.

3. To the tube labeled RX add the following reagents:

Sequencing Reaction Mix

Ingredient	Amount added	Final concentration
Water (deionized)	to 16 µl	
SequiTherm Buffer 10 x	2.5 µl	1x
Primer (40 ng/µl)	2.0 µl	15 pmoles
Plasmid DNA	x	500 fmoles
^{35}S-dATP (10 mCi/ml)	1.0 µl	10 µCi
DNA polymerse (5 u/µl)	1.0 µl	5 u

Add calculated amount of water first. Follow with the remaining reagents in the order indicated. **Do not add enzyme yet.** Mix well by pipetting up and down and centrifuge for 30 seconds. Add 1 µl of SequiTherm EXCEL II polymerase (5 units) and mix again by pipetting up and down several times. Place the tube on ice.

3. Add 2.0 µl of the appropriate termination mix to tubes labeled A (ddA termination), C (ddC termination mix), G (ddG termination mix) and T (ddT termination mix). Keep tubes on ice.

4. Add 4.0 µl of Reaction Mix prepared in step 3 to each of the 4 tubes with termination mix. Mix well by pipetting up and down. Do not remove tubes from ice.

5. Overlay each tube with one drop of mineral oil and centrifuge for 30 seconds at room temperature.

6. Insert tubes into the heating block of the thermal cycler. Start the following temperature cycling program: 95°C for 5 minutes and 30 cycles of: 95 °C for 30 seconds, 50 °C for 15 seconds and 70 °C for 1 minute.

7. Add 3 µl Stop/Loading buffer provided with the kit to each tube. Stop reactions by starting the microfuge for 10-30 seconds. Store tubes at -20 °C or load immediately onto a sequencing gel.

Subprotocol 3
Running Sequencing Gel

The procedure describes the preparation and running of a sequencing gel using Long Ranger® matrix and TTE buffer. This gel material is a high performance alternative to acrylamide for DNA sequencing. Long Ranger® gel has several distinctive advantages over acrylamide gels. These are:

- Long Ranger® gel gives more sequencing information than polyacrylamide gel because it provides more even band spacing,

- Long Ranger® gel is easier to handle and does not require prerunning before gel loading. Long Ranger gels are stronger, therefore gel can be thinner, increasing gel resolution power.

- Long Ranger® gel runs approximately two times faster than equivalent polyacrylamide gels. Thus, one can get more sequencing information in a shorter time.

- Long Ranger® gel material in solution is less toxic than acrylamide polymer (20-30% of toxicity of acrylamide solution). The Long Ranger material after polymerization, unlike acrylamide, is not toxic. This allows for easy disposal.

- Long Ranger® gel does not require special equipment to dry it. Long Ranger gel adheres irreversibly to filter paper and therefore can be air-dried when attached to the Whatman 3MM paper.

- Polymerization of Long Ranger® is not inhibited by oxygen making it unnecessary to degas the gel solution. Also polymerization of Long Ranger requires less time than polymerization of acrylamide.

The reaction mixture described in section 3 contains a high concentration of glycerol introduced with the enzyme. Using a high concentration of glycerol is desirable because it increases polymerase stability during the reaction, contributes to maintaining the single-stranded state of the plasmid template and allows sequencing reactions to be run at a higher temperature. The standard TBE (Tris, Borate, EDTA) buffer cannot be used for sequencing gels when a high concentration of glycerol is used. This is because glycerol can chemically react with borate in the gel buffer creating a number of electrophoresis artifacts. The gel electrophoresis described here uses glycerol tolerant buffer TTE buffer (Tris, Taurine, EDTA) instead of TBE buffer. Taurine, unlike borate, does not react with glycerol.

Two standard sizes of glass plates are used for sequencing. The narrow size gel apparatus uses glass plates about 20 cm wide and 45 to 50 cm long, large enough for about 6 to 10 sequencing samples. The large gel apparatus uses glass plates about 35 cm wide and 45 to 50 cm long. This corresponds to the standard size X-ray film and can accommodate twice as many samples as the small gel. For most small sequencing projects the small gel apparatus is recommended because it is easy to operate.

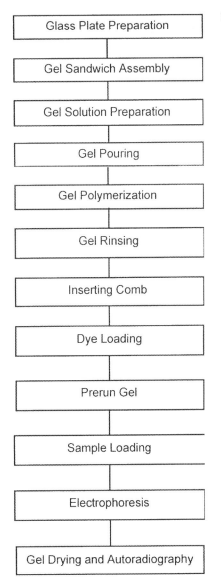

Fig. 5. Schematic outline of the procedure

Glass Plate Preparation

Gel Sandwich Assembly

Gel Solution Preparation

Gel Pouring

Gel Polymerization

Gel Rinsing

Inserting Comb

Dye Loading

Prerun Gel

Sample Loading

Electrophoresis

Gel Drying and Autoradiography

The gel is formed between two glass plates, a long plate and a short plate, clamped together with spacers between them. The long plate is coated with a solution to prevent the gel from sticking to the plate during postelectrophoresis processing and to facilitate pouring the gel. An automobile windshield treatment solution, Rain-X, is used instead of very toxic dimethyldichlorosilane. Rain-X is not toxic, is inexpensive and performs better than dimethyldichlorosilane. A schematic outline of the procedures is shown in Figure 5.

Materials

- Gel sequencing apparatus (Bio-Rad Lab. # 165-3860. Pharmacia Biotech # 80-6301-16, or equivalent).
  In the author's experience Sequi-Gen GT system from Bio-Rad Laboratories is the best manual sequencing apparatus on the market. It is easy to assemble and has an excellent heat dissipation that prevents gel smiling.
- Gel spacers 40 cm long, 0.25 mm thick (Bio-Rad Lab. # 165-3815 or equivalent).
- Sharkstooth comb 0.25 mm (Bio-Rad Lab. # 165-3830, or equivalent).
- High voltage power supply (Bio-Rad Lab. # 165-5056 or equivalent).
  Power supply should be capable of delivering 3000 V and 400 mA.
- Long Ranger® gel solution (FMC BioProducts # 50610).
  Keep solution in a dark bottle at room temperature.
- Rain-X solution (Unelko Corp. USA) or Gel Slick (FMC BioProducts # 50640).
- Ammonium persulfate (Sigma Co. # A 7460).
- N, N, N',N'-tetramethyl-ethylenediamine (TEMED) (Sigma Co. # T 8133).
  Store in the dark at 4 °C.
- Urea, electrophoresis grade (Fisher Scientific # BP169-212, or equivalent).
- Taurine (Sigma Co. # T 0625).
- 60 ml plastic syringe.
- 95% ethanol in a spray bottle.
- 150 ml disposable filtration unit (Nalgen Co. # CN 125-0045)
- Whatman 3MM chromatography paper (Whatman Co. # 3030917 or equivalent)
- Kodak XAR-5 X-ray film (Eastman Kodak Co. # 165 1512, or equivalent).
- Gel temperature monitoring strips (FMC BioProducts # 50642, or equivalent).

– Alconox detergent (VWR Scientific Inc. # 21835-032).

– **10% Ammonium persulfate**
   Add 100 mg to 1 ml water. Store in a dark bottle at 4 °C. This solution can
   be stored for one to two weeks.

– **10 x TTE Electrophoresis Buffer**

Trisma Base	108.0 g
Taurine	36.0 g
Na$_2$EDTA	2.0 g
Water	1000.0 ml

Store buffer at room temperature.

Procedure

The gel running procedure is subdivided into four separate protocols. These
are: (a) Preparation of glass plates; (b) Preparation of Long Ranger®/urea
mixtures and pouring the gel; (c) Loading the gel and electrophoresis of the
samples; (d) Drying the gel and autoradiography.

Preparation of glass plates

Washing the glass plates is critical to obtaining a good sequencing gel. Detergent residue left on the plate or using the incorrect detergent will result in
sequencing artifacts and gel smearing (Li et al. 1993). Dust particles or residues from previous gels will make it extremely difficult to pour a gel that
does not contain bubbles. Bubbles will distort sequencing band pattern and
make the gel difficult to read. Prolonged use of the same glass plates will
result in microscopic pitting of the glass surface due to sodium leakage from
the glass. Plates can be restored by cleaning them in saturated ethanolic
solution of NaOH or treatment with cerium oxide (Millard and de Couet,
1995).

1. Clean the short and long glass plates with Alconox detergent. Wash the
   plates 5 to 6 times with distilled water to completely remove detergent.

Note: Washing plates in other detergents can result in sequencing artifacts and gel smearing.

2. Place each plate on the top of two plastic boxes (e.g., boxes used for holding yellow tips) to lift them above the laboratory bench surface. Spray the surface of the glass plates with 95% ethanol and dry with Kimwipe tissues. Repeat this process until the plates are completely clean. Appearance of Newton's rings when ethanol is drying indicates that glass surface is clean.

3. Treat one side of the short plate with Rain-X Solution. Pour a small amount of Rain-X onto the glass surface and spread it evenly with Kimwipe tissues. Let it dry to a white powdery haze, and then wipe it with a Kimwipe. Repeat this procedure once more, and then clean the glass plates with ethanol as described above.

Note: Use only lint free tissue for cleaning the glass surface.

4. Clean two 0.25 mm spacers with soap and water and then 95% ethanol. Dry thoroughly.

5. Clean a sharkstooth comb with soap and water. Spray with 95% ethanol and wipe it clean several times.

6. Place the long plate, on two plastic boxes. Arrange spacers at both edges of a large glass plate and place the small glass plate on top. Assemble the plate sandwich following the recommendations of the apparatus manufacturer.

Preparation of gel solutions

This protocol describes the preparation of enough solution for the narrow, 5% Long Ranger® gel 0.25 mm thick. If wide gel is used, prepare twice as much of each solution.

1. Prepare the Long Ranger® solution in a 250 ml Erlenmeyer graduated flask. Add ingredients in the order indicated in the recipe.

5% Gel Mix (LR Solution)

Urea	29.4 g
Long Ranger® (50%)	7.0 ml
10 x TTE Buffer	8.4 ml
Water	to 70.0 ml

Note: Fill with water to approximately 70 ml. Use the approximate graduation shown on the flask.

2. Place the flask into a 37 °C water bath and swirl it until urea dissolves (about 5 minutes).

3. Filter with a disposable filtration unit. This procedure removes dust particles from the gel solution.

4. Gather together the following materials next to the location where you intend to pour the Long Ranger® solution into the clamped glass plates
   - 60 ml syringe with 18 g needle.
   - Glass plate sandwich.
   - Washed sharks tooth comb and spacer comb.
   - TEMED solution and Long Ranger mix (LR).
   - 10% ammonium persulfate solution.
   - Two plastic boxes.

Note: Wear gloves throughout the rest of the procedure. Long Ranger® solution is toxic and can easily permeate unprotected skin.

5. Add 350 µl of 10% ammonium persulfate and 30 µl of TEMED to the LR solution. Mix quickly. Withdraw 50 ml into the 60 ml syringe.

6. Hold the clamped plates at a 45 to 50 degree angle to the bench with one bottom corner on the bench. This will result in one top corner of the sandwich being higher then the other. Hold the gel assembly with one hand. Hold the syringe in the second hand and touching the syringe needle to the top of the sandwich at the corner that is lower. Slowly add the LR solution. The solution should flow evenly down the spacer. Allow the solution to flow in one **continuous stream** to avoid the formation of bubbles.

Note: Bubbles will form at the site of the slower moving boundary of the solution. Tapping the glass nearest the slower moving boundary can prevent their formation. This operation may require the help of another person.

7. When the sandwich is about three-quarters full, gently lower the gel to a horizontal position and place it on the top of two plastic boxes. Make sure that LR completely covers the top edge of the short glass plates. A small amount of LR usually dribbles out of the top of the gel at this point.

8. Carefully insert the flat end of the sharkstooth comb to a depth of about three to four millimeters. It is easier to wash and load shallow wells than

deep wells. After gel polymerization, the comb will be removed and the "sharks tooth" comb inserted with the points facing towards the bottom of the gel.

9. Allow the Long Ranger® solution to polymerize for about one hour. The gel can be stored overnight or used immediately. To store the gel overnight, wrap the top and the bottom of the gel with plastic wrap to prevent the gel from drying.

Note: Do not remove the comb when storing the gel overnight.

10. Hold the gel sandwich over a sink and rinse the top of the gel under a stream of water holding the top of the plate sandwich down. Water will dissolve any urea or polymerized gel that may be present around the comb.

Note: Do not remove the comb for this washing and keep the top of the sandwich **down** allowing the gel particles to float away from the polymerized gel.

11. Place the sandwich flat on the plastic boxes and gently withdraw the comb. Repeat washing described in step 10 to remove pieces of polymerized gel. Drain the water very well from the top of the gel sandwich. At this point the long well on the top of the gel should be clearly visible.

12. Place the sandwich back onto the plastic boxes and insert the sharkstooth comb, teeth-side down into the long well on the top of the gel. Hold both sides of the comb vertically using two hands and insert the comb slowly until the tips of the teeth embed themselves about 0.5 mm into the polymerized top of the gel. The gel sandwich is now ready to be inserted into the electrophoresis apparatus. Follow the manufacturer's instructions for this step.

Loading and running the gel

Concentration of TTE buffer in the gel is 1.2 x. The running buffer concentration is 0.6 x. This difference in buffer concentration results in sharp bands and shortens gel electrophoresis time by half.

1. Prepare the required amount of 0.6 x TTE buffer. For most sequencing apparatus, 1600 ml of buffer is sufficient to fill top and bottom chambers. Fill the top reservoir with TTE buffer. Buffer level should be high enough to fill the spaces between the sharkstooth comb entirely. Add the re-

maining buffer to the bottom reservoir to a level of about 1 to 2 cm above the glass plates.

2. Use a 50 ml syringe fitted with an 18-gauge needle to fill buffer between the teeth that form the wells. Remove any trapped air bubbles and gently flush the top of the gel, one well at the time, with TTE buffer. Attach a temperature strip to the middle of the glass plate.

3. Load 2 μl of stop solution into every other well. Place a yellow pipette tip on the edge of the shorter plate **between the sharkstooth** and holding the Pipetman at a slight angle, slowly deliver the dye solution. Do not push the micropipettor further than the first stop. Frequently the solution will flow out of the tip by itself without moving the Pipetman plunger. Mark any wells from which the dye solution leaks. Leakage may occur at the pointed tips of the sharkstooth comb. Mark all good wells in groups of four.

Note: Frequently leakage can be stopped by gently inserting the teeth of the comb a little deeper into the gel. Do not pull out and insert the comb repeatedly or insert the teeth too deeply into the gel. This will damage the surface of the gel causing more sample leakage and uneven bands.

4. Run the gel (without samples) for 15 minutes at about 30 W for narrow gels (35 cm wide) or 50 W for wide gels. For example, for a narrow gel, set the voltage to 2000 volts with a current of about 15 mA (2000 x 0.015 A = 30 W). If possible set the power supply to a constant power (30 W) rather than constant voltage or current for the entire electrophoresis time.

Note: Do not prerun Long Ranger gel longer than the indicated time because this will decrease the resolution of the gel.

5. Just before the gel prerun is over, incubate the reaction mixtures at 80 °C to 85 °C for 2-3 minutes. Chill samples on ice for 1 minute.

6. Turn off current to the gel and disconnect the leads to the power supply. Wash each well with fresh buffer from the reservoir as described in step 2. This removes any urea that may have accumulated. Proceed immediately to load your samples.

7. Draw up 3 μl of cooled sample for the 'A' reaction into a P20 Pipetman fitted with a regular yellow tip. Carefully layer the sample onto the bottom of the well formed by the teeth of the sharkstooth comb using the technique described in step 3. The sample will flow from the top of the glass edge into the well by itself.

Note: Use of commercial loading tips which are long and flat is not recommended because sample loading is slow and air bubbles are frequently delivered into the well.

8. Continue loading the rest of the samples onto the gel in the order A, C, G and T. Do not leave any **unloaded wells** between samples.

Note: If fewer samples are electrophoresed, load adjacent wells in the center of the gel.

9. After all samples are loaded, begin electrophoresis at constant power using the power supply setting described in step 4. Continue electrophoresis until the first blue dye moves close to the bottom of the gel. This will take about two hours. The dye moves with DNA 40 to 50 bases long. This should resolve 300 to 400 bases starting from the insertion site of the fragment (e.g., there are 40 bases from the forward universal primer to the *Sma* I insertion site in pUC plasmid).

Note: To obtain a sequence further away from the insertion point, the gel should be run for a longer time. For example, stopping electrophoresis when the second dye reaches the bottom will place sequences of about 200 bases from the insertion point at the bottom of the gel.

10. Monitor the gel temperature during electrophoresis and adjust the power or voltage accordingly. Temperature of the gel should not exceed 50 °C.

Drying and autoradiography of gel

1. Turn off the power supply and disconnect lead wires. Remove the gel sandwich from the apparatus and cool the glass plates by running tap water over them for at least 5 to 10 minutes. Place the gel sandwich horizontally across two plastic boxes with the longer glass plate down and remove the comb.

Note: Discard the upper and the lower reservoir buffers. Be careful because the lower reservoir buffer may contain unincorporated radioactive nucleotide from the sequencing reactions. Gloves should be worn during this process.

2. While the gel is being cooled, cut a piece of Whatman 3MM paper slightly larger than the gel. Usually for narrow gels, cutting a standard sheet of Whatman paper in half will give the desired size.

3. Remove clamps holding the gel sandwich using the procedure recommended by the gel apparatus manufacturer. Insert a curved spatula between the plates at the bottom of the gel sandwich and slowly pry them apart. Lift the top glass plate with one hand and hold the bottom plate with the other hand until the top glass plate **completely** separates from the gel. Remove top glass and spacer. The Long Ranger gel should stick to the longer plate.

Note: Do not allow the top plate to fall back onto gel during this operation. This may damage the gel.

4. Lay the Whatman 3MM paper over the gel starting from the top of the gel. Firmly attach the Long Ranger gel to the paper by moving a hand in a circular motion, with slight pressure, over the entire surface. The paper should remain dry during this procedure. Lift the Whatman paper from the bottom making sure that the gel sticks to the paper. Place Whatman paper, with the gel, onto a glass plate with the gel side up. Dry the gel in an oven (80 °C) for 30 minutes or leave the gel at room temperature to dry overnight.

5. Place the dry gel in an X-ray film cassette. In the dark room, cover the gel with X-ray film and expose it for 12-17 hours. It is helpful to cut a notch in the film and the membrane before exposure to help orient the film later.

6. Develop the film in a recommended developer (e.g., Kodak D19) or in an automatic developing machine and read the sequencing ladder. From left to right, bands for each template should be in order: A, C, G and T.

Results

Begin reading the DNA sequences at the bottom of the gel where smallest fragments, closest to the primer, are located. A diagram illustrating how to read an autoradiograph is presented in Figure 6. Reading can be done manually or by scanning the autoradiogram and reading the sequence from the screen directly into a computer. The simplest and least expensive program of this type is DNA ProScan from DNA ProScan, Inc. Many different computer programs exist that make it possible to assemble a large sequence from data obtained by random sequencing strategy. The most frequently used programs are programs from DNASTAR Inc. for Intel processors PC and Sequencer from Gene Codes Inc. for Macintosh computers. A description of available programs for computer analysis of sequences by far exceeds the scope of this manual. The reader is directed to an excellent introduction to the subject (Grobskov and Devereux, 1991), and recent description of available programs, methods and databases by Baxevanis et al. (1997).

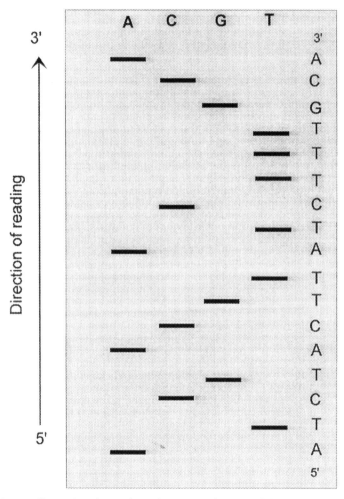

Fig. 6. Diagram illustrating the reading of an autoradiogram of dideoxy DNA sequencing

Troubleshooting

There are many possible problems encountered in DNA sequencing. In general they can be subdivided into two classes:

- problems encountered during gel electrophoresis, and

- problems with the appearance of the sequencing ladder that are only apparent upon analysis of autoradiograms.

The most frequent problem encountered during gel electrophoresis is receding of the gel from the comb and/or distortion of surface of wells. These can result from incorrect gel polymerization or excessive heat during electrophoresis. Temperature of the gel during electrophoresis should never exceed 50 °C and gels that do not polymerize in approximately one hour or polymerize too fast (e.g., 20 minutes) should not be used.

The problems of the second class in order of frequency of occurrence are:

- Anomalous spacing of bands or missing bands. The possible causes are:
  - Sequence compression due to secondary structure formation during electrophoresis. If possible increase temperature of electrophoresis. Use cycle sequencing instead of Sequenase sequencing. Run formamide gels instead of urea gels.
  - Missing bands can result from too long a storage of Sequenase reactions. Use the sequencing reaction as soon as possible and add more pyrophosphatase when storing the reaction for a long time.

- Bands are fuzzy or too wide. The possible causes are:
  - Urea was not washed from the wells prior to loading sample.
  - Too high temperature during sample denaturation prior to loading.
  - Sample loaded too slowly causing some reannealing of sample DNA. All samples should be loaded within 2 to 4 minutes.
  - Incorrect buffer composition in the gel.

- Faint bands or absence of the bands on the gel. The possible causes are:
  - Insufficient amount of template or primer.
  - Incomplete neutralization of sample after denaturation. In most cases samples are too acidic due to use of old NaOH solution. This frequently manifests itself as very light blue or yellow color of the sequencing reaction after addition of the stop solution.
  - Incorrect volume of sequencing reaction due to evaporation. Volume of primer-template mixture should be 11 µl.
  - Lost activity of sequencing enzyme or some components of reaction are missing.

- Bands at the same position in all four lanes. The possible causes are:
  - Contaminated template. Most frequent contaminant is chromosomal DNA from bacteria.
  - Occurrence of nucleotide sequences with strong secondary structure. For reactions with Sequenase, replace dGTP with dITP and add an appropriate amount of pyrophosphatase.
  - Contamination with primers that are less than full-length or hybridization of primer to the secondary site on the template. Use new primer in reaction. For cycling sequencing reaction raise the annealing temperature.

- Faint "shadow bands" present in all or several lines. The possible causes are:
  - Too high concentration of template.
  - Secondary primer binding site on template. In cycling reaction raise the annealing temperature.
  - More than one template present in reaction.
  - Too many cycles in cycling sequencing. Do not use more than 30 cycles. If the problem persists, reduce the cycle number to 15.

- No clear bands except a smear in each lane or high background throughout.
  - Glass plate cleaned with wrong detergent. Use Alconox for detergent cleaning.
  - Residual detergent left on the plates. Wash plates in distilled water to remove all detergent.
  - Polymerization of the gel too fast. Decrease amount of TEMED.
  - Degradation of the gel due to too high temperature during electrophoresis. Long Ranger gel running temperature should not exceed 50 °C.
  - Using old labeling radionucleotide.

- Short sequence reads of less than 150 nucleotides. The possible causes are:
  - Too low concentration of labeling mixture in labeling Sequenase reaction.
  - Breakdown of ddNTP in the termination mix. This frequently results from frequent freezing and thawing of stock solutions. Use fresh termination mix. To prevent breakdown aliquot stock solution into small volumes. Thaw ddNTP termination mixtures on ice.

References

Baxevanis AD, Boguski MS, Ouellette BFF (1997) Computation analysis of DNA and protein sequences. In: Genome analysis. Volume 1. A laboratory manual. Analyzing DNA. Eds. Birren B, Green ED, Klapholz S, Myer RM Roskams J. Cold Spring Harbor Laboratory Press. USA 1997.

Deininger PL (1983) Random subcloning of sonicated DNA: Application to shotgun DNA sequence analysis. Anal Biochem 129:216-223.

DeRisi JL, Vishwanath RI, Brown PO (1997) Exploring the metabolic and genetic control of gene expression on a genomic scale. Science 278:680-686.

Doolittle RF (1998) Microbial genomes opened up. Nature 392:339-342.

Gribskov M, Devereux J (1991) Sequence analysis primer. Stocton Press, 1991. New York. Hengen PN (1997) Method and reagents. Shearing DNA for genomic library construction. TIBS 22:273-274.

Hieter P, Boguski M (1997) Functional genomics: It's all how you read it. Science 278:601-602.

Hunkapiller T, Kaiser RJ, Koop BF, Hood L (1991) Large-scale and automated DNA sequence determination. Science 254:59-67.

Li DK, Rao-Tekamal R, Shin SA (1993) DNA sequencing artifacts: Band smearing and loss. BioTechniques 15:840.

Maxam A, Gilbert W (1977) A new method for sequencing DNA. Proc Natl Acad Sci USA 74:560-564.

Meis R, Schanke JT (1997) The SequiTherm EXCEL II DNA sequencing kit: A new kit for both cycle and non-cycle sequencing. Epicentre Forum 4: # 3. http://www.epicentre.com/f4_3/f4_3xl.htm.

Millard D, de Couet HG (1995) Preparation of glass plates with cerium oxide for DNA sequencing. BioTechniques 19:576.

Roe B, Crabtree JS (1995) Protocols for recombinant DNA isolation, cloning and sequencing. http://www.genome.uo.edu/protocol_partII.html.

Sanger F, Nicklen S, Coulson. (1977) A. DNA sequencing with chain-terminating inhibitors. Proc Natl Acad Sci USA. 74:4298-5467.

Surzycki SJ (1990) A fast method to prepare random fragment sequencing libraries using a new procedure of DNA shearing by nebulization and electroporation. The International conference on the status and future of research on the Human Genome. Human Genome II. San Diego 1990. Abstract 129, p 51.

Tatusov RL, Koonin EV, Lipman DJ (1997) A genomic prospective on protein families. Science 278:631-637.

PCR Analysis

STEFAN SURZYCKI

Introduction

The Polymerase Chain Reaction (PCR) is a powerful method of in vitro DNA synthesis. Large amounts of a specific segment of DNA, of defined length and sequence, can be synthesized from a small amount of template. PCR is a rapid, sensitive and inexpensive procedure for amplifying DNA of specific interest. The technique has revolutionized molecular biology and is used in virtually every area of natural sciences and medicine. The technique has cut across the boundaries separating basic and applied research, commercial technology and medicine in a way few techniques ever have.

The principal of PCR is rather simple and involves enzymatic amplification of a DNA fragment flanked by two oligonucleotides (primers) hybridized to opposite strands of the template with the 3' ends facing each other (Figure 1). DNA polymerase synthesizes new DNA starting from the 3'end of each primer. Repeated cycles of heat denaturation of the template, annealing of the primers and extension of the annealed primers by DNA polymerase results in amplification of the DNA fragment. The extension product of each primer can serve as template for the other primer resulting in essentially doubling the amount of the DNA fragment in each cycle. The result is an exponential increase in the amount of specific DNA fragment defined by the 5' ends of the primers.

The PCR revolution began quietly enough in the mind of Kary Mullis. By his own account, Mullis, then at Cetus Corporation, thought up the technique one Friday night in the summer of 1983 (Mullis, 1990). The following winter, having made some educated guesses about concentrations and reaction times, he carried out his first experiment. He was lucky because the first time he tried the reaction, it worked! The initial procedure used Klenow fragment of *E. coli* DNA polymerase I added fresh during each cycle because the enzyme was inactivated by each denaturation step (Saiki et all 1985; Mullis et all 1986). The introduction of thermostable DNA polymerase from the thermophilic bacterium *Thermus aquaticus* (*Taq*) (Saiki et al.

1988) permitted the development of instruments that automated the cycling portion of the procedure, substantially reducing the work needed to perform PCR. The use of this polymerase also increased the specificity and the yield of the desired product because higher temperatures for annealing and extension could be used.

Since its introduction, numerous modifications and applications of the PCR technique have been developed, the presentation of even a small number of them vastly exceeds the scope of this book. There are books that are

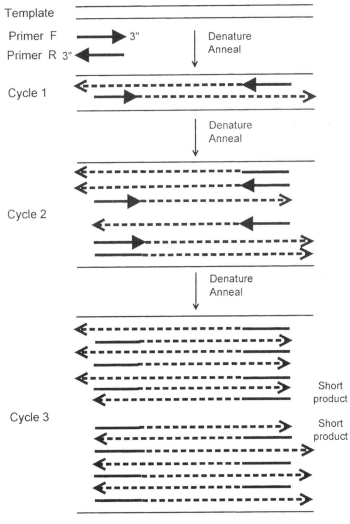

Fig. 1. Schematic outline of DNA amplification resulting from polymerase chain reaction.

dedicated to this subject (Dieffenbach and Dveksler 1995; Innis et all 1995; Griffin and Griffin, 1994; Innis et al, 1990; White et all 1980). In this chapter, the most basic PCR concepts are presented along with general protocols. The aim is to provide an overall introduction to PCR technology that can serve as the basis for understanding more sophisticated applications.

PCR reaction kinetics

During the PCR reaction, products from one cycle serve as template for DNA synthesis in the next cycle. This leads to an exponential accumulation of the product. Theoretically, the amount of product doubles each cycle making the PCR process a true chain reaction that is described by following equation:

$$N = N_0 \times 2^n \tag{1}$$

where: N = number of amplified molecules, N_0 = initial number of molecules and n = number of amplification cycles. This equation holds true if efficiency of amplification (E), defined as the fraction of template molecules that take part in amplification during each cycle, is one. An equation that better describes the amplification process, taking into account the efficiency of the process is:

$$N = N_0 \times (1 + E)^n \tag{2}$$

where: E is amplification efficiency.

Experimentally the accumulation of product during the course of the reaction is far from the case described by equation 2. This is because amplification efficiency varies during the course of the reaction. At the beginning, the efficiency of amplification is close to 1 (0.8 to 0.97) and accumulation of product proceeds exponentially constituting the exponential phase of the reaction. During late PCR cycles, the accumulation of product slows down and eventually stops due to a drastic decrease in amplification efficiency. This effect is usually referred to as the "plateau effect" (Innis et al. 1990; Sardelly 1993). This usually occurs after accumulation of 0.3 to 1.5 pmol of amplification product. The number of cycles it takes to reach plateau depends on the initial number of molecules present in the reaction (N_0).

Table 1 presents the number of cycles, for decreasing amount of initial target DNA, needed to give visible product without reaching plateau. However, plateau can be influenced by a number of conditions, such as the utilization of substrate, temperature of cycling, stability of enzyme and reac-

Table 1. Relation between number of initial target molecules (N_0) and recommended cycle number

Initial Number of Molecule (N_0)	Cycle Number (n)	Number of Molecules Synthesized[1]	Product Amount (pmole)
3.0×10^5	~ 25	7.2×10^{11}	1.2
1.5×10^4	~ 30	7.2×10^{11}	1.2
1.0×10^3	~ 35	8.2×10^{11}	1.4
50.0	~ 45	3.1×10^{11}	0.5

[1] Calculated using equation 2 at 0.8 efficiency of amplification.

tants, inhibition of polymerase by pyrophosphate or the presence of other impurities, reannealing of product at high concentration, incomplete strand separation at high concentrations of product, formation of primer-dimers or the GC content of the template being amplified. Thus, the conditions presented in Table 1 should be treated only as a general guide for optimizing the PCR.

Reaching plateau should always be avoided because it increases the likelihood of obtaining nonspecific amplification products without substantially increasing the concentration of the desired molecules. Thus, choosing the correct number of PCR cycles is one way to avoid false initiation products. This can be achieved by correctly choosing the initial concentration of target DNA molecules. Since most standard amplification protocols use 30 amplification cycles, varying the concentration of target DNA is most frequently employed as a means to avoid reaching plateau. Table 2 presents the recommended amount of template from different organisms needed to amplify a single copy, genomic DNA fragment.

Table 2. Recommended amount of template to use in standard PCR

DNA Source	Amount of DNA/rx	Number of Molecule/rx	Number of Cycles
Plasmids	~1 pg - 10-pg	$3 \times 10^5 - 3 \times 10^6$	25 - 30
Bacteria	~1 ng - 10 ng	$3 \times 10^5 - 3 \times 10^6$	25 - 30
Lower eukaryotes	10 ng -100 ng	$3 \times 10^5 - 3 \times 10^6$	25 - 30
Animals (human)	100 ng - 1000 ng	$3 \times 10^4 - 3 \times 10^5$	30 - 35
Plants (pea)	100 ng - 1000 ng	$3 \times 10^4 - 3 \times 10^5$	30 - 35

Elements of standard PCR reaction

Components of a standard PCR reaction are: thermostable DNA polymerase, DNA template, primers, dNTP substrate, $MgCl_2$ buffer and salt. In addition, PCR reactions frequently include compounds that stabilize the enzyme and reagents that help DNA dissociation or primer annealing.

DNA polymerase   The most commonly used thermostable DNA polymerase is Taq polymerase. Consequently most standard PCR protocols are optimized for maximal activity of this enzyme. However, Taq polymerase exhibits some properties that make it less than ideal for some applications. First, the enzyme has very high error rates due to the lack of 3' to 5' exonuclease activity (Saiki et al. 1998). This may interfere with the preparation of DNA for sequencing, as well as with the inability to obtain long replication products. Second, the enzyme adds nucleotides to 3' ends in a template-independent manner, making the amplification product difficult to clone. Third, the enzyme is quite expensive.

For these reasons a variety of thermostable DNA polymerases from thermophilic or hyperthermophilic bacteria have been isolated, characterized and commercially introduced to PCR. Table 3 presents a list of these polymerases and their major characteristics. Some of these polymerases are used in standard PCR reactions (e.g., Tli, Pvo) while others are used in sequencing (e.g., Pvo, AmpliTherm) or in long PCR reaction mixtures (e.g., Tth).

The concentration of the enzyme in a reaction mixture can affect the fidelity of the product. An excessive amount of polymerase can result in amplification of nonspecific PCR products. The recommended amount of enzyme per reaction for most polymerases is 1 to 5 units. The most frequently used concentration is 2.0 units/100 µl reaction. However, the enzyme requirement may vary with respect to individual template DNA or primer used. It is always prudent to use the enzyme concentration recommended by the supplier since enzyme activity from different manufacturers can vary.

DNA template   One of the most important features of PCR is that it can be performed with very small quantities of relatively impure DNA. Even degraded DNA can be successfully amplified. Therefore a number of simple and rapid protocols to purify DNA for PCR have been developed. In this book some of these methods are described in chapters 2, 3, 4 and 5 about DNA purification. In spite of the fact that DNA need not be absolutely pure, a number of contaminants can decrease the efficiency of amplification. The presence of urea, SDS, sodium acetate and some components eluted from agarose can interfere with

PCR. Most of these impurities can be removed by washing with 70% ethanol or by reprecipitation of DNA in the presence of ammonium acetate. Protocols for removal of these contaminants are described in this book in the chapters on DNA purification.

General guidelines for selection of primers for PCR are: **Primers**

- Optimal primer set should hybridize to the template efficiently with negligible hybridization to other sequences of the sample. To assure this, primers should be at least 20 to 25 bases in length. However, for some applications requiring multiple random initiation (e.g., RAPD analysis), primers are usually 8 to 10 bases in length.

- Primers, if possible, should have a GC content similar to that of the target. Overall GC content of 40% to 60% usually works for most reactions. The GC content of both primers should be similar.

- Primers should not have sequences with significant secondary structure. Therefore, primer sequence should not contain simple repeats or palindromic sequences.

- Primer pairs should not contain complementary sequences to each other. In particular primers with 3' overlaps should be avoided. This will reduce the incidence of "primer-dimer" formation that severely affects efficiency of amplification.

Many commercial computer programs exist to design primer sets. Some excellent freely shared programs are: PRIMER from Whitehead Institute for Biomedical Research (http://www-genome.wi.edu) for PC computers and Amplify and HyperPCR for Macintosh computers (ftp://iubio.bio.indiana.edu.)

The PCR reaction requires a molar excess of primers. Primer concentration is usually 0.2 to 1 μM. Using too much primer may result in false initiations and "primer-dimer" formation.

The distance between primers can vary considerably. However, the efficiency of amplification decreases considerably for distances greater than 3 kbp. In general, short fragments amplify with higher efficiency than long ones. Preferential amplification of short fragments is probably due to more efficient full-length extension by the enzyme in each cycle. Amplification of a very large fragment is possible with modifications of standard reaction conditions and by using a mixture of DNA polymerases (e.g., Taq and Pfu). These conditions are referred to as long PCR. Numerous commercial kits exist that use proprietary reagents to achieve long PCR.

Substrate The concentration of each of dNTP in PCR should not exceed 200 μM. This amount of substrate is sufficient to synthesize 12.5 μg of DNA when half of the nucleotides are incorporated, a quantity far exceeding the amount of DNA synthesized in standard PCR (see Table 1). The four dNTPs should be used at equivalent concentrations, particularly when Taq polymerase is used. This will minimize the error rate of the enzyme. An excess of nucleotides inhibits enzyme activity and can contribute to the appearance of false products. Moreover, when varying dNTP concentration, it should be remembered that dNTPs chelate magnesium ion, decreasing its concentration in the reaction mixture.

MgCl$_2$ Concentration Magnesium ion is a required co-factor for all DNA polymerases. Moreover magnesium ion concentration may affect the following:

- Primer annealing.

- Temperature of strand dissociation for template and product.

- Product specificity.

- Formation of primer-dimer artifacts and enzyme fidelity.

Many templates require optimization of magnesium ion concentration for efficient, correct amplification. The optimal concentrations of Mg^{++} for each thermophilic polymerase are listed in Table 3. They should serve as a starting point for optimizing the magnesium ion concentration in the reaction.

Buffer and salt A standard buffer for PCR is 10 to 50 mM Tris-HCl. The optimum pH is between 8 and 9 for most thermophilic polymerases (Table 3). Since the Δ pK$_a$ for Tris is high (- 0.031/$^\circ$C), the true pH of the reaction mixture during a typical thermal cycle varies considerably (approximately 1 to 1.5 pH units).

   The salt used in most reactions is potassium or sodium. The K$^+$ (Na$^+$) is usually added to facilitate correct primer annealing. For Taq polymerase, the concentration usually used is 50 mM since higher monovalent ion concentrations inhibit polymerase activity (Innis et al. 1988). It is not necessary to use salt when amplifying DNA using a rapid cycle amplification instrument, since fidelity of primer annealing is achieved as a result of the short annealing time. Some other components used in reactions to help stabilize the enzyme are: gelatin, bovine serum albumin or nonionic detergent such as Tween 20 or Triton X 100. However, most protocols work well without the addition of these ingredients.

Table 3. Properties of commercially available thermostable DNA polymerases

| Enzyme | Reaction Conditions | | | Exonuclease Activity | | Error Rate |
	Mg++ (mM)	pH	Tempera-ture (C)	5' - 3'	3' - 5'	(error/bp x 10^{-4})
Taq	2.0 - 4.0	7.8 - 9.4	70 - 80	+	-	0.5 - 2.1
Pfu	1.5 - 8.0	8.9 - 9.0	70 - 80	+	+	0.06
Pwo	2.0	8.0 - 9.0	70 - 80	+	+	0.06
Tli (Vent)	2^1	8.8	72 - 75	+	+	2.0 - 5.0
Tli -exo (Vent -)	2^1	8.8	72 - 75	+	-	2.0
Tth[2]	1.5 - 2.5	8.0 - 9.3	50 - 60	+	-	n/a
Taq (Stoffel frg)	1.5 - 2.5	8.0 - 9.3	70 - 80	-	-	0.5
Ampli-Therm[TM]	1.5-2.5	8.3	70 - 95	-	-	n/a

[1] MgSO$_4$ should be used instead MgCl$_2$

[2] This polymerase has reverse transcriptase activity in the presence of Mn^{++} ion

When using DNA template with a high GC content, the reaction mixture also includes reagents to lower the T_m of the template. Among these are DMSO, acetamide or glycerol. Incorporation of 5% acetamide into the reaction mixture improves results with many complex templates and does not affect activity of Taq polymerase.

Thermal cycling profile

Standard PCR consists of three steps: denaturation, annealing and extension. These steps are repeated or cycled 25 to 30 times. Most protocols also include a single denaturation step (initial denaturation) before cycling begins and single long extension step (final extension) at the end.

An initial denaturation that lasts for a few minutes is added to assure complete denaturation of the template. This preliminary step is important for complex templates or templates with high GC content. The final extension step is usually carried out for 5 to 10 minutes to assure completion of partial extension products by DNA polymerase and to provide time for complete renaturation of single stranded products.

Denaturation step   In this step, complete separation of the template strands must be accomplished. Denaturation is a first order reaction that occurs very fast. A complete description of the denaturation reaction is given in chapter 10. A standard time of 30 to 60 seconds is used to assure uniform heating of the entire reaction volume. When a rapid cycle instrument is used, this time is usually set to 0 for a 10 μl reaction cycled in a glass capillary tube. The temperature of cycle denaturation is usually 94 °C to 95 °C.

Annealing step   In this step, primers anneal to the template. With excess primer the annealing reaction is pseudo first order reaction (see chapter 10 for an explanation) and occurs very fast. The time used for this step, in most protocols, is 30 to 60 seconds, the time required to cool the reaction from the denaturation temperature to the annealing temperature. In a rapid cycle instrument, this time is usually set to 0. The annealing temperature depends on the GC content of the primers and should be carefully adjusted. Low annealing temperatures will result in unspecific primer annealing and initiation. High temperatures will prevent annealing and consequently will prevent DNA synthesis. For most primers, an annealing temperature of 50 °C to 55 °C is used without further optimization.

Elongation step   In this step, DNA polymerase synthesizes a new DNA strand by extending the 3' ends of the primers. Time of the elongation depends on the length of the sequence to be amplified. Since Taq polymerase can add 60-100 bases per second under optimal conditions, synthesis of a 1 Kbp fragment should require a little less than 20 seconds. However, most protocols recommend 60 seconds per 1 Kbp DNA to account for time needed to reach the correct temperature and to compensate for other unknown factors that can affect reaction rate. The shortest possible time should be used to preserve polymerase activity. In air rapid cycle instruments, the elongation time is usually set to that calculated for the theoretical elongation rate of polymerase (15 to 20 seconds per 1 Kbp of desired fragment).

An extension temperature of 72 °C is used in standard amplification protocols. This temperature is close to the optimal temperature for Taq polymerase (75 °C) but low enough to prevent dissociation of the primers from the template. However, it should be remembered that Taq polymerase has substantial elongation activity at 55 °C, the temperature used in most annealing steps. This residual synthesis stabilizes the interaction of primer with template and indeed permits the use of an elongation temperature higher than 72 °C. Higher temperatures of elongation are used when the template can form secondary structures that interfere with elongation.

Instrumentation

Cycle times should be as short as possible to reduce overall reaction time, prevent inactivation of enzyme and increase fidelity of the reactions. These times usually depend on performance characteristics of the instrument used. Performance differences between instruments generally reflect the rate at which heating and cooling of the sample occurs. In the best instruments, use of small volumes of sample and thin-walled tubes can reduce cycle times to a few seconds.

Most of the commercial instruments utilize metal blocks or water for thermal equilibration and samples are contained in plastic microfuge tubes. In these instruments, very fast temperature changes are not possible. A significant fraction of the cycle time in these machines is spent on heating and cooling blocks and tubes and the liquid contained in them. This extends amplification times to several hours as opposed to the time that is really necessary to carry out amplification cycles.

Moreover, extended amplification times and long transition times, make it difficult to determine optimal temperature and times for each stage of the PCR reaction. This leads to many false initiations by polymerase and consequently to poor product specificity. Rapid temperature transitions and small sample volumes improve both specificity and the time necessary to carry out PCR reactions.

Three engineering approaches have been used to decrease cycling time.

- The use of very small reaction volumes in specially developed thin-walled tubes or single-piece 96-well plates. This solution increases cost and does not address the problem of temperature transition due to the heat transfer inertia of metal blocks or water.

- The use of several stationary blocks, each maintained at different temperatures, instead of a single heating block. A mechanical arm moves the tubes from one block to another. This approach solves the problem of heat inertia but is expensive and necessitates use of expensive thin-walled tubes.

- The use of an air thermal cycler (Wittwer and Garling, 1991; Wittwer et al., 1994). This rapid thermal cycler instrument uses heat transfer by hot air to samples contained in thin, glass capillary tubes. Because of the low heat capacity of air, the thin walls and increased surface area of capillary tubes, samples can be cycled very fast. Total amplification time in these instruments, for 30 PCR cycles, is 15 to 20 minutes. Moreover, the specificity of amplification is also dramatically increased. The cost of the instrument is low compared to standard thermal cyclers. The use of these inexpensive instruments is highly recommended.

Outline

This chapter describes a standard protocol to amplify DNA by PCR in a standard thermal cycler. Reaction volume is 50 μl and the protocol incorporates a negative control that contains all reaction components except template DNA. This control is necessary to detect DNA contamination. Incorporation of a positive control, when amplifying DNA fragments from genomic DNA, is also encouraged to test for degradation of reaction components. A schematic outline of the procedure is described in Figure 2.

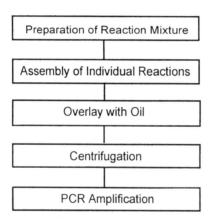

Fig. 2. Schematic outline of DNA amplification resulting from polymerase chain reaction.

Materials

- Taq DNA polymerase (1 to 5 u/μl)
- DNA template (see Table 2 for recommendations)
- Primers (1-10 μM; 1 - 10 pmoles/μl)
- dATP, dCTP, dGTP and dTTP 100 mM stock solution set (Roche Molecular Biochemicals # 1277 049 or equivalent).
- Acetamide (Sigma Co. # A 6082)
- Triton X-100 or Nonidet P40 (Sigma Co. # T 8787)
- Mineral oil (Sigma Co. # M 8662 or equivalent)

Solutions   - **10 x PCR Buffer**
    - 100 mM Tris-HCl; pH 8.3

- 500 mM KCl
- 15 mM MgCl$_2$

The buffer should be prepared using a 1 M stock solution of Tris-HCl adjusted to pH 8.3 at room temperature, sterilized by autoclaving for 20 minutes and stored at -20 °C. This buffer can be bought from different companies (e.g., Roche Molecular Biochemicals # 1271 318; Sigma Co. # P2192). 10 x PCR buffer is frequently supplied with Taq polymerase.

- **10 x dNTP Solution**
  - 2 mM dATP
  - 2 mM dCTP
  - 2 mM dGTP
  - 2 mM dTTP

  Use 100 mM stocks of dATP, dCTP, dGTP and dTTP to prepare 10 x dNTP solution. Add 2 µl of each dNTP to 92 µl of sterilized water. Store at -20 °C in small aliquots.
- **10 x Acetamide Solution (50% v/w)**

  Add 5 g of acetamide powder to 10 ml sterile distilled or deionized water. Store at -20 °C.
- **Triton X 100 or Nonidet NP40 Solution (1% v/v)**

  Add 1 ml of detergent to 100 ml of sterile deionized water. Store at -20 °C in small aliquots.
- **Loading Dye Solution**
  - 15% Ficoll 400
  - 5 M Urea
  - 0.1 M Sodium EDTA, pH 8
  - 0.01% Bromophenol blue
  - 0.01% Xylene cyanol

Procedure

1. Label three 0.5 ml microfuge tubes RSM, 1 and 2. Place tubes on ice.

2. Prepare Reaction Stock Mixture for 3 PCR reactions to account for pipetting loses. First, calculate the amount of water necessary for the desired volume and add it to the tube. Then add buffer and the remaining components. Add enzyme last and mix by pipetting up and down several times. Never mix by vortexing. Taq polymerase is very sensitive to vortexing.

Reaction Stock Mixture (RSM)

Reagent	One Reaction (50 μl)	Three Reactions	Final Concentration
10 x PCR Buffer	5 μl	15 μl	1x
10 x dNTP (2 mM)	0.5 μl	1.5 μl	200 μM each
Upstream primer	0.5-2.0 μl	1.5-6.0 μl	0.1- 1.0 μM
Downstream primer	0.5-2.0 μl	1.5-6.0 μl	0.1- 1.0 μM
Template DNA	1-3 μl	none	
Taq polymerase (1 u/μl)	1.0 μl	3.0 μl	2 u/100 μl
Water	to 50 μl	to 150 μl	-

3. Prepare reaction mixtures to run PCR reactions as follows:

PCR Reaction Mixtures

Ingredient	Tube # 1	Tube # 2 (Control)
RSM	47.0 μl	47.0 μl
DNA template	3.0 μl	none
Water	none	3.0 μl

Add DNA and water and mix gently, pipetting up and down.

4. Add 25 μl of mineral oil to each tube and centrifuge the tube for 5 seconds in a microfuge. Addition of oil is easier if you cut off the end of the yellow tip.

5. Place tubes into the thermal cycler and amplify DNA using standard conditions. Typically these are: initial denaturation at 95 °C for 3 minutes followed by 30 to 35 cycles consisting of denaturation (D) at 94 °C for 30 seconds, annealing (A) at 60 °C for 30 seconds, elongation (E) at 72 °C for 2 minutes. The reaction is finished with an extension step at 72 °C for 5 minutes.

6. Remove tubes from the thermal cycler and add 5 μl of dye solution to each tube. Centrifuge tubes for 5 to 10 seconds in a microfuge to bring dye solution through the oil. Reactions are ready for analysis by gel electrophoresis. It is not necessary to remove oil from the tubes.

Products of PCR are usually analyzed using agarose gel electrophoresis. Because PCR products for most reactions are small, it is necessary to use high resolution agarose electrophoresis to analyze them. The procedure for this electrophoresis is described in Chapter 8.

Troubleshooting

Many artifacts of PCR can be detected during gel electrophoresis of the amplified sample. These are:

- Failure to amplify product or very weak amplification for a reaction that "worked" before. This usually results from a failure of some component of the reaction. Most frequently this is due to primer degradation. Dilute new primer and repeat PCR.

- Using too high concentration of DNA template is the second most frequent cause of reaction failure. Check DNA concentration of the sample or run a series of reactions with a decreasing concentration of template.

- Next check the integrity of the template and freshness of the substrate. Working solution of dNTPs deteriorate rapidly after they have been frozen and thawed a few times. Dilute fresh template DNA and/or working dNTP solution.

- Failure to amplify product or very weak amplification for a "new" amplification reaction. In this case running a positive control reaction can determine if this is a problem with reaction mixture components or incorrect PCR parameters. If the latter is the case, this can result from incorrect primers, wrong annealing conditions for the primers chosen or wrong magnesium concentration. Insufficient denaturation of the template or presence of inhibitors in the DNA sample can also cause this failure. Optimization of the PCR reaction conditions could remedy this problem.

- Presence of faint, diffused bands in the 100 to 200 bp region of the gel. This indicates the presence of specific single-stranded products. This can result from incorrect primer concentration or degradation of one primer. Repeat the reaction with fresh primers.

- Smears that end at the correct band size positions. This indicates priming of the reaction by specific products of amplification and usually results from too low primer concentration. Repeat PCR with higher concentration of primers.

- Absence of a specific band and the presence of smears throughout entire gel. This indicates nonspecific multiple priming and usually occurs when annealing temperature is too low. Increase annealing temperature to increase specificity of the product.

- Presence of many distinct bands having lengths different from the expected. Sometimes these bands are the major products of amplification and the correct size band is very weak. This results from incorrect non-random priming. Increase specificity of priming by changing to a higher annealing temperature and/or shorten annealing step.

- Generation of primer-dimers and little or no amplification of the desired band. Primer-dimers will appear slightly above the electrophoresis front in the region of 50 bp. The intensity of these bands will be inversely proportional to the intensity of the desired band. The less correct synthesis, the stronger the primer-dimer intensity will be. To remedy this problem, first decrease the amount of polymerase and/or primer concentration in the reaction. If the problem persists, design new primers.

Less frequent causes of failure are: destruction of enzyme or contamination of the reagents with EDTA. Prepare fresh buffer or use new enzyme.

References

Dieffenbach C, Dveksler G, eds. (1995) PCR primer: A laboratory manual. Cold Spring Harbor Laboratory Press, Cold Spring Harbor, New York.

Griffin H, Griffin A, eds. (1994) PCR technology: Current innovations. CRC Press, Boca Raton, Florida.

Innis MA, Gelfand DH, Sninsky JJ, eds. (1995) PCR strategies. Academic Press, San Diego, California.

Innis MA, Gelfand DH, Sninsky JJ, White TJ, eds. (1990) PCR protocols: A guide to methods and applications. Academic Press, San Diego, California.

Innis MA, Myambo KB, Gelfand DH, Brow AD (1988) DNA sequencing with *Thermus aquaticus* DNA polymerase and direct sequencing of polymerase chain reaction-amplified DNA. Proc Natl Acad Sci USA 85:9436-9440.

Mullis KB (1990) The unusual origin of the polymerase chain reaction. Scientific American April 91:56-65.

Mullis KF, Faloona F, Scharf S, Saiki R, Horn G, Erlich H (1986) Specific enzymatic amplification of DNA in vitro: the polymerase chain reaction. Cold Spring Harbor Symp Quant Biol 51:263-273.

Saiki RK, Scharf S, Faloona F, Mullis KB, Horn GT, Erlich HA, Arnheim N (1985) Enzymatic amplification of β-globin genomic sequences and restriction site analysis for diagnosis of sickle cell anemia. Science 230:1350-1354.

Saiki RK, Gelfand DH, Stoffel S, Scharf SJ, Higuchi R, Horn GT, Mullis KB, Erlich HA (1988) Primer-directed enzymatic amplification of DNA with a thermostable DNA polymerase. Science 239:487-491.

Sardelli AD (1993) Plateau effect - Understanding PCR limitations. Amplification 9:1-5.

White TJ, Arnheim N, Erlich HA (1989) The polymerase chain reaction. Trends in genetics 5:185-189.

Wittwer CT, Garling DJ (1991) Rapid cycle DNA amplification: Time and temperature optimization. BioTechniques 10:76-83

Wittwer Ct, Reed GB, Ririe KM (1994) Rapid cycle DNA amplification. In: The polymerase chain reaction. Ed. Mullis KB. pp 174-181. Birkhauser, New York.

Subject Index